Bacillus thuringiensis: A Broader View of Its Biocidal Activity

Bacillus thuringiensis: A Broader View of Its Biocidal Activity

Editors

Leopoldo Palma
Diego Herman Sauka
Jorge E. Ibarra

MDPI

Basel • Beijing • Wuhan • Barcelona • Belgrade • Novi Sad • Cluj • Manchester

Editors
Leopoldo Palma
Universitat de València
Burjassot
Spain

Diego Herman Sauka
Instituto Nacional de
Tecnología Agropecuaria
(INTA)
Buenos Aires
Argentina

Jorge E. Ibarra
CINVESTAV
Unidad Irapuato
Irapuato
Mexico

Editorial Office
MDPI
St. Alban-Anlage 66
4052 Basel, Switzerland

This is a reprint of articles from the Special Issue published online in the open access journal *Toxins* (ISSN 2072-6651) (available at: https://www.mdpi.com/journal/toxins/special_issues/NPM8FL40W7).

For citation purposes, cite each article independently as indicated on the article page online and as indicated below:

Lastname, A.A.; Lastname, B.B. Article Title. *Journal Name* **Year**, *Volume Number*, Page Range.

ISBN 978-3-7258-1341-4 (Hbk)
ISBN 978-3-7258-1342-1 (PDF)
doi.org/10.3390/books978-3-7258-1342-1

Cover image courtesy of Leopoldo Palma and Diego Herman Sauka

Contents

About the Editors

Leopoldo Palma

Leopoldo Palma studied Microbiology and completed his Ph.D. in Biotechnology at the Public University of Navarra, Spain. His doctoral research focused on studying novel strains and insecticidal proteins of *Bacillus thuringiensis* (Berliner). Following his doctoral studies, Dr. Palma pursued a postdoctoral contract position at UPNA, where he expanded his expertise. Afterward, Dr. Palma obtained Associated Researcher and Associated Professor positions at CONICET and Universidad Nacional de Villa María, respectively, in Argentina. At CONICET, he supervised a research group dedicated to the discovery of native strains of *B. thuringiensis* with potential pesticidal activities, in close collaboration with international colleagues and the agribusiness sector. In 2022, Dr. Palma returned to Spain to join the University of Valencia, where he currently holds the position of Distinguished Researcher María Zambrano at the Institute BIOTECMED. His research focuses on elucidating cellular receptors for insecticidal proteins from *B. thuringiensis*. With a background in the topic and extensive research experience, Dr. Palma has been honored to be a member of the Bacterial Pesticidal Protein Resource Center since 2021, alongside other top researchers in the field.

Diego Herman Sauka

Diego Herman Sauka studied Biochemistry and completed his Ph.D. in Microbiology at the University of Buenos Aires, Argentina. His doctoral research focused on studying pesticidal proteins and genes from native *Bacillus thuringiensis* isolates in Argentina. His work has contributed to the understanding of the distribution and toxicity of these proteins in agricultural pests. Following his doctoral studies, Dr. Sauka secured a research position at the National Agricultural Technology Institute and CONICET (Argentina). One of his notable achievements is curating the largest bacterial entomopathogen collection in Argentina, which includes selected strains sourced from the environment. Currently, Dr. Sauka's interests lie in Microbial Control and Plant Protection against insect and nematode pests. His laboratory prioritizes the practical use of entomopathogenic and nematicidal bacteria in agriculture and public health. In addition to his research, Dr. Sauka has been involved in part-time teaching activities on Microbiology at the Faculty of Pharmacy and Biochemistry of the University of Buenos Aires since his student days until February 2024.

Jorge E. Ibarra

Jorge E. Ibarra's passion for the study of insects began in childhood. His calling to study Biology was defined during high school, and he later graduated as a Biologist in 1976. He pursued a master's degree in Entomology and then became a professor at the Autonomous University of Nuevo León (México), leading studies on apple tree pests. In 1982, he began his Ph.D. in Entomology at the University of California, Riverside (California, United States), alongside studies in Genetic Engineering, which he completed in 1986, receiving the Harry S. Smith Award in Biological Control of Insects. That same year, he joined the Irapuato Unit of Cinvestav, where he has worked for 37 years, conducting studies on Entomology, Molecular Biology, and Microbiology. He has developed the largest entomopathogen collection in Mexico of selected strains from the field, developed coffee plants that are resistant to the coffee borer beetle, and pioneered novel techniques for the genetic transformation of mosquitoes. In 1999, he spent a sabbatical year at the Pasteur Institute in Paris. His studies have been disseminated in 390 publications of various types, and he has supervised 81 undergraduate, master's, and doctoral theses. In 2023, he was granted the status of Emeritus by the National System of Researchers in Mexico.

Preface

Bacillus thuringiensis (Bt) is the most extensively studied Gram-positive entomopathogenic bacterium, with numerous strains carrying plasmids containing a diverse array of pesticidal genes. This characteristic has established Bt-based products as the predominant commercially available microbial insecticides today. Among these products are insecticidal proteins, both crystalline and vegetative, which are well-known for their high toxicity against a wide range of invertebrates. Many of these proteins have been incorporated into crops, offering protection against some of the most devastating insect pest species worldwide.

However, insects targeted by Bt crops have been continuously exposed to selective pressures, leading to the development of resistance to some of the most commonly used insecticidal proteins. Consequently, ongoing global screening programs seeking strains with new insecticidal proteins to address resistance and expand the variety of susceptible hosts are continuously developed.

This reprint aims to compile information that is relevant to this issue, encompassing descriptions of novel isolated Bt strains and invertebrate active proteins exhibiting previously uncharacterized biocidal activities, which will benefit numerous stakeholders in agribusiness, the economy, and wider society.

Leopoldo Palma, Diego Herman Sauka, and Jorge E. Ibarra
Editors

toxins MDPI

Editorial

Bacillus thuringiensis: A Broader View of Its Biocidal Activity

Leopoldo Palma [1,2,*], Diego Herman Sauka [2,3,*] and Jorge E. Ibarra [4,*]

1 Instituto de Biotecnología y Biomedicina (BIOTECMED), Universitat de València, 46100 Burjassot, Spain
2 Consejo Nacional de Investigaciones Científicas y Técnicas (CONICET), Ciudad Autónoma de Buenos Aires 1425, Argentina
3 Instituto de Microbiología y Zoología Agrícola (IMYZA), Instituto Nacional de Tecnología Agropecuaria (INTA), Hurlingham, Ciudad Autónoma de Buenos Aires 1686, Argentina
4 Departamentos de Biotecnología y Bioquímica, Centro de Investigación y de Estudios Avanzados del IPN-Unidad Irapuato, Apartado Postal 629, Irapuato 36500, Guanajuato, Mexico
* Correspondence: leopoldo.palma@uv.es (L.P.); sauka.diego@inta.gob.ar (D.H.S.); jorge.ibarra@cinvestav.mx (J.E.I.)

Bacillus thuringiensis (Bt) is a Gram-positive bacterium that forms spores and produces parasporal crystalline inclusions containing Cry and Cyt proteins [1]. These proteins exhibit toxicity against various insect orders, nematodes, and human cancer cells [2,3]. Widely utilized as bioinsecticides, Bt strains and their insecticidal proteins effectively control caterpillars, beetles, flies, mosquitoes, and blackflies. During vegetative growth, Bt can also secrete insecticidal proteins targeting lepidopterans (Vip3) and coleopterans (Vpab1/Vpab2). Another less-explored secretory protein, Mpp5Aa1 (formerly Sip1A), has also been described to exhibit activity against coleopteran pests [4]. These features have bestowed Bt as the most specific and effective tool for the control of insect pests for several years, either through insecticidal formulations (a mix of spore and parasporal crystals) or by the production of insecticidal proteins in transgenic plants (Bt plants) [5]. However, some species, such as *Plutella xylostella* (Lepidoptera), have developed field resistance to both formulated products and insecticidal proteins expressed in transgenic plants [2], making screenings for novel strains and pesticidal proteins highly essential in order to provide novel tools for the control of pests and the management of insect resistance.

The aim of this Special Issue, "*Bacillus thuringiensis*: A Broader View of Its Biocidal Activity", was to gather information on novel Bt strains and proteins showing novel pesticidal properties to provide biotechnological tools with useful resources for pest control in agriculture and to incentivize researchers to perform such necessary research. This subject has been of great interest, allowing the publication of 12 research papers from top researchers working in the field worldwide, which have shed light on the diverse and multifunctional properties of novel (unreported) Bt strains and proteins. Beyond the conventional focus, these studies delve into various aspects, including structural insights, insecticidal proteins, toxin interactions, and the evaluation of novel strains.

Additionally, this editorial aims to provide an overview of key findings from these papers; for example, Unzue et al. showcased the broad spectrum of Bt applications and their potential implications, such as the multifunctional properties of Bt strain BST-122, encompassing the biocidal properties beyond those of the well-known pesticidal parasporal crystals (contribution 1). Li et al. presented a deeper understanding of the structure of Cry5B, unraveling the active form of Cry5B that could contribute to the development of more effective and targeted nematicidal products (contribution 2). The study by Best et al. showed the crystal structure of Bt Tpp80Aa1 protein (formerly Cry80Aa1) and its interaction with galactose-containing glycolipids, unveiling molecular details that may have implications for understanding the specificity and selectivity of Bt toxins (contribution 3). Two papers about binary Mpp23Aa/Xpp37Aa proteins and the novel Bt strain Bt_UNVM-84 were presented by de Oliveira et al. and Sauka et al., respectively, highlighting the potential application of these resources in pest control, particularly against *Anthonomus*

Citation: Palma, L.; Sauka, D.H.; Ibarra, J.E. *Bacillus thuringiensis*: A Broader View of Its Biocidal Activity. *Toxins* **2024**, *16*, 162. https://doi.org/10.3390/toxins16030162

Received: 13 March 2024
Accepted: 18 March 2024
Published: 20 March 2024

grandis (Coleoptera), a harmful pest causing high economic losses in the cotton industry in the Americas (contributions 4 and 10). Covering proteins with activity against mosquitoes, Lai et al.'s paper examines the role of Cyt proteins in enhancing the activity of other Bt toxins against *Aedes albopictus*, emphasizing the cooperative interactions between different toxin classes. Understanding these synergies could contribute to the development of more effective mosquito control strategies (contribution 5). Yang's study described a Cry protein with activity against the rice leaffolder *Cnaphalocrocis medinalis*. This paper explored the processing properties and potency of Cry toxins in the context of rice leaffolder control. Insights into the interaction between a Cry protein and this target pest are crucial for optimizing their efficacy in agricultural settings (contribution 6). The study by Xue et al. covered novel synergistic interactions beyond those known among different insecticidal proteins. In this paper, a new synergistic interaction between the extracellular polysaccharide from Bt subsp. *kurstaki* HD-270 and the insecticidal protein Cry1Ac is described, examining this particular synergistic activity and providing valuable information on the interplay between different components in Bt formulations (contribution 7). Trisyono et al. presented a paper covering the baseline susceptibility of field populations of *Ostrinia furnacalis* in Indonesia to Cry1A.105 events and Cry2Ab2 proteins were also covered, assessing the susceptibility of this pest to specific Cry proteins. This research lays the groundwork for understanding the dynamics of resistance and informs us about the sustainable use of Bt technologies in agriculture (contribution 8). Interesting outputs about the field evaluation of transgenic cotton expressing Mpp51Aa2 (formerly Cry51Aa2) as a management tool for the cotton fleahopper *Pseudatomoscelis seriatus* were also published in this Special Issue, contributing valuable data for its implementation. Finally, two papers presented by Hou et al. and Shao et al. describe interesting aspects related to the yet poorly understood mode of action of Vip3A proteins, focusing on molecular details of Vip3A proteins and highlighting specific domains involved in receptor binding and liposomal membrane disruption, respectively (contributions 11 and 12).

In conclusion, the collection of papers featured in this editorial underscores the versatility and complexity of Bt as a bioinsecticide. From structural insights to practical field applications, researchers continue to unravel the potential of Bt strains in addressing agricultural and public health challenges. These findings collectively contribute to the ongoing efforts to harness the full potential of Bt-based technologies for sustainable pest management.

Funding: This research received no external funding.

Acknowledgments: The editors express their gratitude to the authors who made significant contributions to the Special Issue entitled "*Bacillus thuringiensis*: A Broader View of Its Biocidal Activity." A special acknowledgement goes also to peer reviewers for their expertise contributions in their evaluations, enhancing the overall quality of the research works and reviews featured in this Special Issue. Leopoldo Palma would like to express gratitude to the Spanish Government-Universities Ministry, the Next Generation EU and Recovery, Transformation, and Resilience plans for funding his awarded María Zambrano contract (ref. ZA21-003). Lastly, we extend our thanks to the MDPI management team and staff for their invaluable contributions, organization, and editorial support.

Conflicts of Interest: The authors declare no conflicts of interest.

List of Contributions:

1. Unzue, A.; Caballero, C.J.; Villanueva, M.; Fernández, A.B.; Caballero, P. Multifunctional Properties of a *Bacillus thuringiensis* Strain (BST-122): Beyond the Parasporal Crystal. *Toxins* **2022**, *14*, 768. https://doi.org/10.3390/toxins14110768.
2. Li, J.; Wang, L.; Kotaka, M.; Lee, M.M.; Chan, M.K. Insights from the Structure of an Active Form of *Bacillus thuringiensis* Cry5B. *Toxins* **2022**, *14*, 823. https://doi.org/10.3390/toxins14120823.
3. Best, H.L.; Williamson, L.J.; Lipka-Lloyd, M.; Waller-Evans, H.; Lloyd-Evans, E.; Rizkallah, P.J.; Berry, C. The Crystal Structure of *Bacillus thuringiensis* Tpp80Aa1 and Its Interaction with Galactose-Containing Glycolipids. *Toxins* **2022**, *14*, 863. https://doi.org/10.3390/toxins14120863.

4. de Oliveira, J.A.; Negri, B.F.; Hernández-Martínez, P.; Basso, M.F.; Escriche, B. Mpp23Aa/Xpp37Aa Insecticidal Proteins from *Bacillus thuringiensis* (Bacillales: Bacillaceae) Are Highly Toxic to *Anthonomus grandis* (Coleoptera: Curculionidae) Larvae. *Toxins* **2023**, *15*, 55. https://doi.org/10.3390/toxins15010055.
5. Lai, L.; Villanueva, M.; Muruzabal-Galarza, A.; Fernández, A.B.; Unzue, A.; Toledo-Arana, A.; Caballero, P.; Caballero, C.J. *Bacillus thuringiensis* Cyt Proteins as Enablers of Activity of Cry and Tpp Toxins against *Aedes albopictus*. *Toxins* **2023**, *15*, 211. https://doi.org/10.3390/toxins15030211.
6. Yang, Y.; Wu, Z.; He, X.; Xu, H.; Lu, Z. Processing Properties and Potency of *Bacillus thuringiensis* Cry Toxins in the Rice Leaffolder *Cnaphalocrocis medinalis* (Guenée). *Toxins* **2023**, *15*, 275. https://doi.org/10.3390/toxins15040275.
7. Xue, B.; Wang, M.; Wang, Z.; Shu, C.; Geng, L.; Zhang, J. Analysis of Synergism between Extracellular Polysaccharide from *Bacillus thuringensis* subsp. *kurstaki* HD270 and Insecticidal Proteins. *Toxins* **2023**, *15*, 590. https://doi.org/10.3390/toxins15100590.
8. Trisyono, Y.A.; Aryuwandari, V.E.F.; Rahayu, T.; Martinelli, S.; Head, G.P.; Parimi, S.; Camacho, L.R. Baseline Susceptibility of the Field Populations of *Ostrinia furnacalis* in Indonesia to the Proteins Cry1A.105 and Cry2Ab2 of *Bacillus thuringiensis*. *Toxins* **2023**, *15*, 602. https://doi.org/10.3390/toxins15100602.
9. Arthur, B.P.; Suh, C.P.; McKnight, B.M.; Parajulee, M.N.; Yang, F.; Kerns, D.L. Field Evaluation of Cotton Expressing Mpp51Aa2 as a Management Tool for Cotton Fleahoppers, *Pseudatomoscelis seriatus* (Reuter). *Toxins* **2023**, *15*, 644. https://doi.org/10.3390/toxins15110644.
10. Sauka, D.H.; Peralta, C.; Pérez, M.P.; Molla, A.; Fernandez-Göbel, T.; Ocampo, F.; Palma, L. *Bacillus thuringiensis* Bt_UNVM-84, a Novel Strain Showing Insecticidal Activity against *Anthonomus grandis* Boheman (Coleoptera: Curculionidae). *Toxins* **2024**, *16*, 4. https://doi.org/10.3390/toxins16010004.
11. Hou, X.; Li, M.; Mao, C.; Jiang, L.; Zhang, W.; Li, M.; Geng, X.; Li, X.; Liu, S.; Yang, G.; et al. Domain III β4–β5 Loop and β14–β15 Loop of *Bacillus thuringiensis* Vip3Aa Are Involved in Receptor Binding and Toxicity. *Toxins* **2024**, *16*, 23. https://doi.org/10.3390/toxins16010023.
12. Shao, E.; Huang, H.; Yuan, J.; Yan, Y.; Ou, L.; Chen, X.; Pan, X.; Guan, X.; Sha, L. N-Terminal α-Helices in Domain I of *Bacillus thuringiensis* Vip3Aa Play Crucial Roles in Disruption of Liposomal Membrane. *Toxins* **2024**, *16*, 88. https://doi.org/10.3390/toxins16020088.

References

1. Schnepf, E.; Crickmore, N.; Van Rie, J.; Lereclus, D.; Baum, J.; Feitelson, J.; Zeigler, D.R.; Dean, D.H. *Bacillus thuringiensis* and its pesticidal crystal proteins. *Microbiol. Mol. Biol. Rev.* **1998**, *62*, 775–806. [CrossRef]
2. Jouzani, G.S.; Valijanian, E.; Sharafi, R. *Bacillus thuringiensis*: A successful insecticide with new environmental features and tidings. *Appl. Microbiol. Biotechnol.* **2017**, *101*, 2691–2711. [CrossRef] [PubMed]
3. Ohba, M.; Mizuki, E.; Uemori, A. Parasporin, a new anticancer protein group from *Bacillus thuringiensis*. *Anticancer Res.* **2009**, *29*, 427–433.
4. Chakroun, M.; Banyuls, N.; Bel, Y.; Escriche, B.; Ferré, J. Bacterial Vegetative Insecticidal Proteins (Vip) from Entomopathogenic Bacteria. *Microbiol. Mol. Biol. Rev.* **2016**, *80*, 329–350. [CrossRef]
5. Bravo, A.; Likitvivatanavong, S.; Gill, S.S.; Soberon, M. *Bacillus thuringiensis*: A story of a successful bioinsecticide. *Insect Biochem. Mol. Biol.* **2011**, *41*, 423–431. [CrossRef] [PubMed]

toxins

MDPI

Article

N-Terminal α-Helices in Domain I of *Bacillus thuringiensis* Vip3Aa Play Crucial Roles in Disruption of Liposomal Membrane

Ensi Shao [1,2], Hanye Huang [2], Jin Yuan [1], Yaqi Yan [1], Luru Ou [1], Xiankun Chen [1], Xiaohong Pan [2], Xiong Guan [2] and Li Sha [1,2,*]

1 China National Engineering Research Center of JUNCAO Technology, College of Life Science, Fujian Agriculture and Forestry University, Fuzhou 350002, China; es776@fafu.edu.cn (E.S.); yuanjin8956@126.com (J.Y.); yyq961202@163.com (Y.Y.); olr13395087102@163.com (L.O.); cxiankun@126.com (X.C.)

2 State Key Laboratory of Ecological Pest Control for Fujian and Taiwan Crops & Key Laboratory of Biopesticide and Chemical Biology of Ministry of Education, Fujian Agriculture and Forestry University, Fuzhou 350002, China; hhy2460838406@163.com (H.H.); panxiaohong@163.com (X.P.); guanxfafu@126.com (X.G.)

* Correspondence: shal@fafu.edu.cn

Abstract: *Bacillus thuringiensis* Vip3 toxins form a tetrameric structure crucial for their insecticidal activity. Each Vip3Aa monomer comprises five domains. Interaction of the first four α-helices in domain I with the target cellular membrane was proposed to be a key step before pore formation. In this study, four N-terminal α-helix-deleted truncations of Vip3Aa were produced and, it was found that they lost both liposome permeability and insecticidal activity against *Spodoptera litura*. To further probe the role of domain I in membrane permeation, the full-length domain I and the fragments of N-terminal α-helix-truncated domain I were fused to green fluorescent protein (GFP), respectively. Only the fusion carrying the full-length domain I exhibited permeability against artificial liposomes. In addition, seven Vip3Aa-Cry1Ac fusions were also constructed by combination of α-helices from Vip3Aa domains I and II with the domains II and III of Cry1Ac. Five of the seven combinations were determined to show membrane permeability in artificial liposomes. However, none of the Vip3Aa-Cry1Ac combinations exhibited insecticidal activity due to the significant reduction in proteolytic stability. These results indicated that the N-terminal helix α1 in the Vip3Aa domain I is essential for both insecticidal activity and liposome permeability and that domain I of Vip3Aa preserved a high liposome permeability independently from domains II–V.

Keywords: Vip3Aa; Cry1Ac; domain exchange; insecticidal activity; membrane permeability

Key Contribution: Our results verified the crucial role of helix α1 of Vip3Aa in insecticidal activity, proteolytic stability, and membrane permeability. Domain I of Vip3Aa preserved a high liposome permeability independently from domains II–V.

Citation: Shao, E.; Huang, H.; Yuan, J.; Yan, Y.; Ou, L.; Chen, X.; Pan, X.; Guan, X.; Sha, L. N-Terminal α-Helices in Domain I of *Bacillus thuringiensis* Vip3Aa Play Crucial Roles in Disruption of Liposomal Membrane. *Toxins* **2024**, *16*, 88. https://doi.org/10.3390/toxins16020088

Received: 7 January 2024
Revised: 29 January 2024
Accepted: 4 February 2024
Published: 6 February 2024

1. Introduction

Bacillus thuringiensis (Bt) is an entomopathogen that has been used as a microbial biopesticide since 1930s [1,2]. Insecticidal Bt toxin genes have been engineered into plants to confer insect resistance for the management of insect pests. The vegetative insecticidal proteins (VIPs) from Bt have high insecticidal activities in a range of lepidopteran pests, and Bt Vip3 genes have been expressed in plants to pyramid with Bt Cry proteins to improve the protection of plants from insect damage and delay the development of resistance to Bt toxins in target pests [3–5].

In general, Vip3A proteins are proteolytically activated in the midgut of insect larvae, and they then bind to the midgut receptor to form pores in midgut epithelial cells [6–8].

Vip3A proteins contain from 786 to 803 amino acid residues with a molecular weight of around 89 kDa [9]. Previous studies have indicated that the 89 kDa Vip3A protoxin may be proteolytically cleaved by gut proteases to become active fragments before exerting insecticidal activity [10–12]. The formation of a Vip3Aa protein complex, composed of ~19 kDa and ~66 kDa fragments, has been identified to be required for the insecticidal activity [12,13]. As the molecular size of both the Vip3A protoxin and the activated toxin have been determined to be ~360 kDa by gel filtration, the native form of Vip3 has been believed to be a tetramer [11,13]. Recently, the molecular architectures of Vip3A and Vip3B have been revealed by Cryo-EM and X-ray crystallography to provide the tetrameric organization of both the protoxin and the trypsin-activated toxin [14–16]. The 3D structure of the Vip3A and Vip3B proteins also confirmed that Vip3 is composed of five distinct domains [11,14,15]. The N-terminal domain I contains four α-helices and is proposed to play a functional role in insertion of the toxin to the epithelial cell membrane [15]. Proteolytic digestion analysis also confirmed that domain I is a 19 kDa fragment of Vip3 toxins [11]. The crucial role of helix α1 in domain I has been shown for the insecticidal activity of the Vip3Aa toxin [17]. Domain II is composed of five α-helices, being the core of the tetramer that stabilizes the oligomeric structure [14,15,18]. Domain III contains three anti-parallel β-sheets, forming a β-prism fold similar to that found in Cry toxins [7,14,19]. The two C-terminal domains (domain IV and V) are highly variable in the Vip3 family and have been proposed to play roles in protecting Vip3 from digestion by proteases in the insect gut [7,15,18]. The conformational change in proteolytically activated Vip3Aa by releasing domain I α-helices from the tip of the tetrameric protoxin to form a new helical bundle at the bottom of a needle-like structure is believed to be the key step before it interacts with the membrane [14]. The stalk formed by four α-helices in domain I of trypsin-activated Vip3Ba toxins was proposed to play crucial role in the pore formation of artificial liposomes [15].

Proteolytically activated Cry1A toxin is composed of three domains [20]. Domain I, containing seven α-helices, plays a crucial role in both oligomerization and interactions with insect gut membranes. Domains II and III, rich in β-sheets, participate in receptor binding and target insect selection [21,22]. Similar to Vip3 toxins, Cry toxins were proposed to form 150- to 250-kDa oligomers after binding with receptor proteins located on the midgut epithelial membrane of target insects, although the precise mechanism of pore formation in this process remains elusive [23].

This study sheds light on the crucial role of the N-terminal helix α1 in the insecticidal activity and membrane permeation capabilities of Vip3Aa. Four N-terminal truncations of Vip3Aa were created and analyzed. Subsequently, the full-length domain I and its truncated variants were fused to green fluorescent protein (GFP) to assess their membrane permeation activity. Additionally, domains I and extensive α-helices in domain II of Vip3Aa were fused to the N-terminus of Cry1Ac domains II and III (generating Vip3Aa-Cry1Ac recombinant proteins). Although none of these recombinants exhibited insecticidal activity against *Plutella xylostella* and *Spodoptera litura* larvae, most of them displayed a high liposome permeability.

2. Results

2.1. Proteolytic Processing and Bioactivity of N-Terminus Truncated Vip3Aa Proteins

To determine the proteolytic stability of N-terminus truncated Vip3Aa proteins, four N-terminal truncations of Vip3Aa were prepared by the truncation of helix α1, helix α1 and α2, helix α1–α3, and helix α1–α4 (Figure 1a). SDS-PAGE analysis of the trypsin processed samples showed that a ~66 kDa fragment was observed in lanes containing the full-length Vip3Aa, Δα1 Vip3Aa, and Δα1–α2 Vip3Aa, respectively (Figure 1b). Numerous protein fragments, mostly ranging from 15 kDa to 66 kDa, were also observed in these three lanes showing a similar pattern, while a ~19 kDa fragment was present only in the full-length Vip3Aa samples but not in any of the four Vip3Aa truncations. Most of the protein fragments in the samples of Δα1–α3 Vip3Aa and Δα1–α4 Vip3Aa were degraded after treatment by trypsin. Only fragments at ~37 kDa, ~23 kDa, and~17 kDa were observed in the lanes containing Δα1–α3 Vip3Aa, and fragments at ~23 kDa and ~17 kDa were observed in the lanes containing Δα1–α4 Vip3Aa.

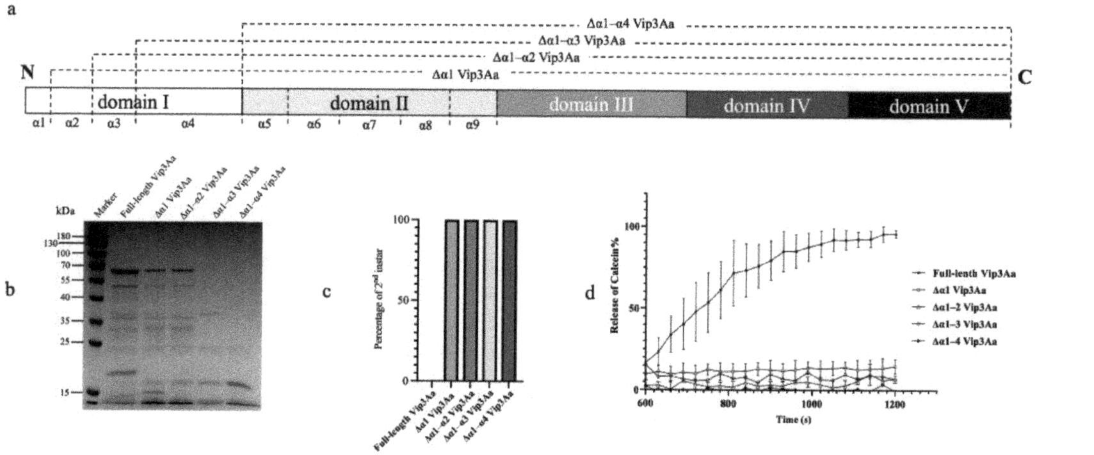

Figure 1. Proteolytic stability and activity assays of four Vip3Aa truncations. Panel (**a**): Schematic presentation of four N-terminal truncations of Vip3Aa proteins. Four N-terminal truncations of Vip3Aa were prepared by truncation of helix α1, helix α1 and α2, helix α1–α3, and helix α1–α4. Panel (**b**): SDS-PAGE analysis of four Vip3Aa truncations after tryptic processing at a ratio of 1.2:1 (trypsin–Vip3Aa, *w*/*w*) at 37 °C for 1 h. Panel (**c**): Neonates of *S. Litura* were fed with 100 μg/mL of N-terminal truncations of Vip3Aa and the full-length Vip3Aa, respectively, by surface overlay bioassays. Numbers of larvae developed to the second instar were recorded after 72 h. Panel (**d**): Calcein release assays by treating liposomes with 16 nM of four N-terminal truncations of Vip3Aa after tryptic processing. Fluorescence signals of released calcein were recorded every 30 s. The error bars represent the standard error of mean of the measurements from triplicate experiments.

In the neonatal larval bioassays, none of the four Vip3Aa truncations exhibited toxicity in *S. litura* larvae. All the neonatal larvae that fed on a truncated Vip3Aa at 100 μg/mL became second-instar larvae after 72 h of feeding, while no neonates survived after ingesting full-length Vip3Aa (Figure 1c). The results of liposome calcein release assays indicated that none of the N-terminus truncated Vip3Aa proteins, which were processed with trypsin, had the activity to permeate liposomes, while a significant release of calcein was determined in the samples processed by the full-length Vip3Aa toxin (Figure 1d).

2.2. Liposome Permeability of Vip3Aa Domain I and GFP Fusion Proteins

To investigate whether the Vip3Aa domain I could independently permeabilize liposome membranes, the Vip3Aa domain I and its N-terminal truncations were fused with a GFP, respectively, creating four Vip3Aa-GFP combinations (Figure 2a). SDS-PAGE analysis indicated that the molecular weights of the four combinations were consistent with the expected sizes from 65 kDa to 49 kDa (Figure 2b). Calcein release assays were then used to assess their liposome permeation activity. Only the Trx-His-Vα1–4-GFP (containing the full-length domain I) displayed significant liposome permeabilization, in which calcein leakage began at 20 s and reached over 90% by 270 s (Figure 2c). None of the other fusion proteins carrying N-terminal truncations of the Vip3Aa domain I exhibited any detectable calcein release.

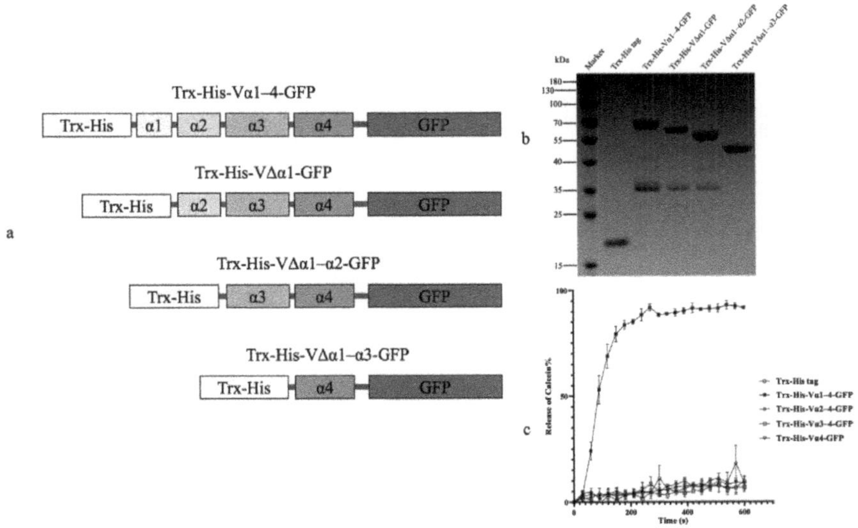

Figure 2. Calcein release assays by treating liposomes with Vip3Aa-GFP fusion proteins. Domain I of Vip3Aa and its N-terminal truncations were fused with a GFP and a Trx-His tag to produce four fusion proteins. Panel (**a**): schematic presentation of four Vip3Aa-GFP recombinant proteins. Panel (**b**): each fusion protein was purified and loaded on an SDS-PAGE gel to determine their molecular weight. Panel (**c**): four Vip3Aa-GFP fusion proteins and Trx-His tag protein were added to the liposome solution (DOPC–DOPE, 1:1) at a final concentration of 10 μM/mL, respectively. Fluorescence signals of released calcein were recorded every 30 s. The error bars represent the standard error of mean of the measurements from triplicate experiments.

2.3. Bioactivity and Proteolytic Stability of Vip3Aa-Cry1Ac Recombinant Proteins

Seven recombinant proteins were generated by fusing N-terminal fragments of Vip3Aa (domain I and II) with domains II and III of Cry1Ac. These recombinants contained the Vip3Aa domain I and additional α-helices from domain II, as shown in Figure 3a. The calcein release assays revealed that five of the seven recombinants exhibited liposome permeability, while VdIα5–7-CdIIdIII and VdIdII-CdIIdIII did not (Figure 3b,c). The liposomes treated with VdI-CdIIdIII began leaking calcein within 400 s, reaching 100% leakage by 680 s. Interestingly, the calcein release curve for the VdI-CdIIdIII-treated liposomes closely resembled that of the liposomes treated with trypsin-activated Vip3Aa. While VdI-CdIIdIII and trypsin-activated Vip3Aa exhibited similar potencies, the other permeable fusion proteins (VdIα5-CdIIdIII, VdIα5–6-CdIIdIII, VdIα5–8-CdIIdIII, and VdIα5–9-CdIIdIII) displayed lower calcein release rates. Despite their membrane permeation activity, none of the seven Vip3Aa-Cry1Ac fusions exhibited insecticidal activity against *S. litura* or *P. xylostella* larvae. Feeding larvae with these fusion proteins resulted in nearly no mortality, while Vip3Aa and Cry1Ac individually caused 100% and 75% mortality in *S. litura* (Figure 4a) and 97% and 100% mortality in *P. xylostella* (Figure 4b), respectively.

To determine proteolytic stability of these Vip3Aa-Cry1Ac recombinants, each recombinant protein was incubated with trypsin in vitro and analyzed by SDS-PAGE. Although part of the recombinant proteins was degraded into fragments at a molecular weight between 25 kDa and 70 kDa, the majority of them remained integrated at a molecular weight from 80.8 kDa to 95.1 kDa after the treatment without trypsin (Figure 5a). Processing the Vip3Aa protoxin with trypsin resulted in a ~66 kDa fragment, a ~20 kDa fragment, and multiple protein fragments between 25 kDa and 40 kDa (Figure 5b). A significant band at ~60 kDa was observed in the lane containing gut proteases processed Cry1Ac protein. In comparison, all the Vip3Aa-Cry1Ac recombinants were degraded, and no protein bands could be observed in the SDS-PAGE analysis.

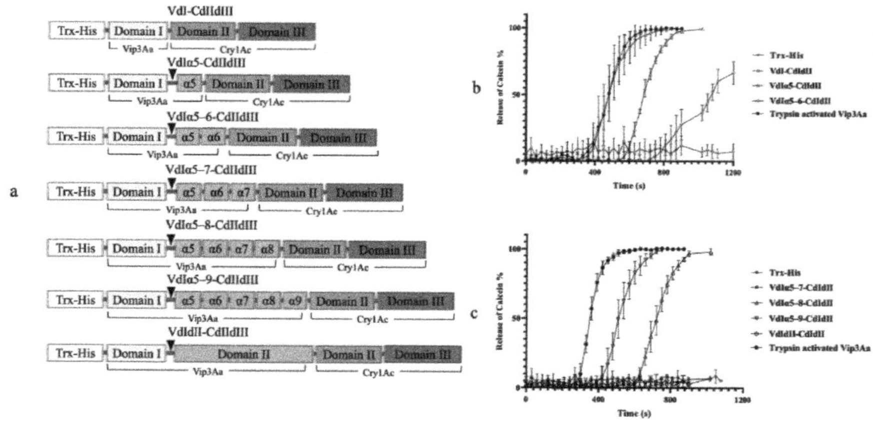

Figure 3. Calcein release assays by treating liposomes with Vip3Aa-Cry1Ac fusion proteins. Domain I and domain II extensions of Vip3Aa were fused with a Trx-His tag and domains II and III of Cry1Ac to produce seven Vip3Aa-Cry1Ac recombinant proteins. The trypsin cleavage site between domain I and II of Vip3Aa is presented by a triangle. Liposome (DOPC–DOPE, 1:1) permeability of seven Vip3Aa-Cry1Ac fusion proteins was determined through calcein release assays. The final concentration of each assayed protein was 10 μM/mL. Fluorescence signals of released calcein were recorded every 30 s. The error bars represent the standard error of mean of the measurements from triplicate experiments. Trx-His tag and trypsin-activated Vip3Aa were used as the negative control and the positive control, respectively. Panel (**a**): schematic presentation of seven Vip3Aa-Cry1Ac recombinant proteins. Panel (**b**): calcein release rate of recombinant proteins, including VdI-CdIIdIII, VdIα5-CdIIdIII, and VdIα5–6-CdIIdIII. Panel (**c**): calcein release rate of recombinant proteins, including VdIα5–7-CdIIdIII, VdIα5–8-CdIIdIII, VdIα5–9-CdIIdIII, and VdIdII-CdIIdIII.

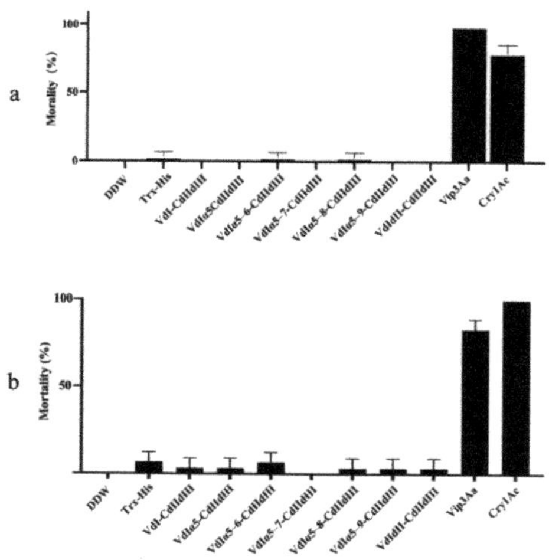

Figure 4. Mortality of *Spodoptera litura* and *Plutella xylostella* larvae by feeding with seven Vip3Aa-Cry1Ac recombinant proteins. Panel (**a**): Neonates of *S. litura* were fed with 100 μg/mL of seven Vip3Aa-Cry1Ac recombinant proteins, respectively, by surface overlay bioassays. Mortality was recorded after 72 h. Panel (**b**): The second-instar larvae of *P. xylostella* were fed with 100 μg/mL of seven Vip3Aa-Cry1Ac recombinant proteins, respectively, by surface overlay bioassays. Mortality was recorded after 48 h. Mortality of larvae fed with the full-length Vip3Aa and Cry1Ac toxin, respectively, was recorded as the positive controls. The error bars represent the standard error of mean of the mortality from triplicate-assayed groups. DDW: double-distilled water.

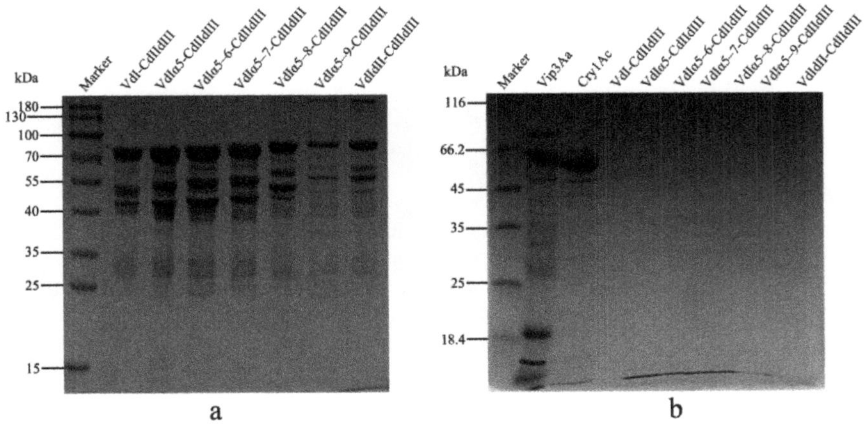

Figure 5. Proteolytic stability of seven Vip3Aa-Cry1Ac recombinant proteins. Panel (**a**): Seven Vip3Aa-Cry1Ac recombinant proteins were incubated at 30 °C for 6 h without any proteases and loaded to a 15% SDS-PAGE gel, respectively. Panel (**b**): seven Vip3Aa-Cry1Ac recombinant proteins, the full-length Vip3Aa, and Cry1Ac protoxins were incubated with bovine pancreas trypsin at ratios of 1.2:1 (trypsin–Vip3Aa, *w*/*w*) for 1 h at 37 °C, respectively, and loaded to a 15% SDS-PAGE gel to separate the digested fragments.

3. Discussion

Cryo-EM studies have revealed the tetrameric structure of Vip3Aa protoxin and trypsin-activated toxin, with each monomer being composed of five domains [14]. Domain I, containing four α-helices, plays a crucial role in both oligomerization and insecticidal activity [17]. This domain undergoes a conformational change during proteolytic activation, where helices α1 to α3 unfold and form a new N-terminal coiled-coil interacting with membranes [15]. To investigate the functional roles of these helices, we generated truncated Vip3Aa proteins lacking α1, α1–α2, α1–α3, or α1–α4. The trypsin treatment revealed the significant impact of these α-helices on proteolytic activation. Helix α1–α3 and α1–α4 deletions led to complete degradation, highlighting their essential role in maintaining protein stability during proteolysis (Figure 1b). The deletion of helix α1 or α1–α2 resulted in fragmentation, similar to the protoxin, but it lacked the characteristic ~19 kDa fragment, which has been known as the fragment of released domain I and plays significant role in insecticidal activity [7,11]. Functionally, truncating either helix α1 or a larger N-terminal segment abolished both the liposome permeation and insecticidal activity against *S. litura* (Figure 1c,d). These observations agree with previous findings on N-terminal deletions abolishing insecticidal activity [24,25] and the crucial role of helix α1 in Vip3Aa's structure and function [17]. Structural analyses suggest that domain I–III maintains the spontaneous conformation of both the protoxin and activated toxin [11]. In this study, deleting the first three α-helices (α1–α3) likely destabilized the protein, leading to complete degradation by trypsin or midgut proteases. The deletion of the first one or two α-helices, however, preserved the proteolytic stability but prevented r processing into active fragments. These findings indicate that the first three N-terminal α-helices of Vip3Aa domain I are not only functional in pore formation but also crucial for maintaining structural stability during proteolytic processing.

Previous studies have estimated the crucial role of Vip3Aa domain I in membrane permeability [11,17]. We further verified this by fusing domain I with GFP, demonstrating its independent activity in membrane perturbation. Only the liposomes treated with the full-length domain I fusion showed significant calcein release (Figure 2c). This confirms that domain I can independently perforate artificial liposome membranes, while the lack of activity in N-terminal truncations underscores the critical role of helix α1 in membrane permeability. Despite the Trx-His tag promoting solubility and proper folding [26], the

Vip3Aa-GFP fusions still aggregated and formed inclusion bodies. However, the high activity of renatured Trx-His-Vα1–4-GFP in permeating liposomes highlights the highly hydrophobic nature of Vip3Aa domain I.

The striking structural homology between the Vip3 and 3D-Cry toxins suggests similar functional roles for their respective domains [15]. Similar to Vip3A domain I, Cry1Ac domain I is responsible for pore formation [27,28]. Additionally, domains II and III of Cry1Ac are involved in receptor binding, target selection, and structural maintenance, analogous to the proposed roles of domains III–V in Vip3A [7,29–31]. According to the individual insecticidal activity of Vip3Aa and Cry1Ac against *S. litura* and *P. xylostella* larvae [12,32–34], we fused domain I and additional α-helices in domain II of Vip3Aa with domains II and III of Cry1Ac to generate multiple chimeric proteins (Figure 3a). The calcein release assays showed that apart from VdIα5–7-CdIIdIII and VdIdII-CdIIdIII, the other recombinant proteins exhibited membrane permeability against artificial liposomes (Figure 3b,c). We hypothesize that the highly hydrophobic helix α1 of Vip3Aa might be folded and become unexposed due to structural changes in VdIα5–7-CdIIdIII and VdIdII-CdIIdIII, leading to the loss of liposome permeation activity. The fusions of Vip3Aa domain I with either GFP or Cry1Ac domains were found to be largely insoluble in native conditions. Urea denaturation and the subsequent renaturation may not lead to the formation of the tetrameric complex as the full-length Vip3Aa toxin. The activity of these fusions to rupture artificial liposome membranes indicated that the monomeric form of Vip3Aa domain I, the core component of the coiled-coil structure in the Vip3Aa tetramers, possesses sufficient membrane-destabilizing activity even in the absence of pore formation. The α-helical architecture of Vip3Aa domain I might facilitate direct interaction with the phospholipid bilayer, similar to the observed mechanism of colicins interacting with phospholipid vesicles [35,36]. Previous studies have investigated the permeability of both Cry1A and Vip3A toxins against artificial liposomes [37–39]. The highly hydrophobic α-helices in domain I have been associated to the pore-forming activity of Cry1A and Vip3A toxins [15,27,40]. Cry1A's activity was shown to be enhanced when liposomes contained soluble BBMV (brush border membrane vesicle) proteins, representing potential Cry1Ac toxin receptors [38]. In this study, we focused solely on assessing the membrane disruption activity of fusion proteins carrying Vip3Aa domain I against artificial membranes. However, the epithelial membrane in the midgut of lepidopteran insects is undoubtedly more complex than a simple liposome model. Therefore, future research is necessary to determine the permeability of these recombinant proteins in eukaryotic cells expressing specific receptors.

Chimeric proteins composed of the full-length Vip3Aa and toxic core of Cry1Ac resulted in a 150 kDa chimera showing increased toxicity against *Ephestia kuehniella* [41]. Fusing Vip3Aa7 to the N-terminus of Cry9Ca also improved insecticidal activity against *P. xylostella* [33]. In this study, although the liposome permeability of these Vip3Aa-Cry1Ac recombinant proteins were determined, none of them displayed insecticidal activity against either *S. litura* or *P. xylostella* larvae (Figure 4a,b). The proteolytic processing results showed a significant degradation of most of the recombinant proteins after treatment by trypsin (Figure 5b). In addition to VdI-CdIIdIII and the other Vip3Aa-Cry1Ac recombinant proteins containing domain I and extension α-helices in domain II of Vip3Aa, the trypsin cleavage site between domain I and domain II was consequently involved. Exposure of this trypsin cleavage site in these fusion proteins may be responsible for the degradations. However, the recombinant protein of VdI-CdIIdIII without this trypsin cleavage site was also fully degraded. These results indicated that fusing the N-terminal domains of Vip3Aa with the C-terminal domains of Cry1Ac may have led to the loss of protection of the active fragments under the proteolytic processing. This drastic reduction in proteolytic stability could be a potential reason for the loss of insecticidal activity.

4. Conclusions

In conclusion, this study highlights the crucial roles of N-terminal α-helices in Vip3Aa domain I for their proteolytic stability and membrane permeation. Domain I is a highly hydrophobic fragment. Renatured domain I fragments could show activity in membrane permeating against liposomes without the help of other domains in Vip3Aa. The combination of α-helices in domains I and II with domains II and III of Cry1Ac eliminated insecticidal activity, while the liposome permeability of most of the recombinant proteins remained.

5. Materials and Methods

5.1. Construction of Expression Vectors

The full-length *vip3Aa* gene (NCBI accession no. AF500478.2) derived from the Bt WB5 strain producing Vip3Aa12 protein was exercised from plasmid pGEX-Vip3Aa and cloned into the vector pET32a at the downstream of a 3C PreScission protease cleavage site [42]. To construct pET32a expression vectors expressing truncated Vip3Aa, fragments of *vip3Aa* with different truncations in the domain-I-coding region were amplified from pGEX-Vip3Aa by PCR using the iProof^{TM} High-Fidelity Master Mix DNA polymerase (Bio-RAD, Hercules, CA, USA) with the primer sets listed in Table 1. The PCR-amplified fragments were cloned to pET32a at the downstream of a 3C PreScission protease cleavage site, as described above. Four expression vectors were constructed to express Vip3Aa with a truncation of the α1 to α4 helices in domain I of Vip3Aa (Figure 1a).

Table 1. Primers for the cloning of vectors expressing Vip3Aa N-terminal truncations and Vip3Aa-GFP fusions.

Primers	Sequences (5′ to 3′)
Vip3Aa-F	acaaggccatggctgatatcGGATCC*CTGGAGGTGCTGTTCCAGGGGCCC*ATGAACAAGAATAATACTAAATTA
Δα1 Vip3Aa-F	acaaggccatggctgatatcGGATCC*CTGGAAGTTCTGTTCCAGGGGCCC*ACGGATACAGGTGGTGATCTAA
Δα1–α2 Vip3Aa-F	acaaggccatggctgatatcGGATCC*CTGGAAGTTCTGTTCCAGGGGCCC*GGAAACTTAAATACAGAATTAT
Δα1–α3 Vip3Aa-F	acaaggccatggctgatatcGGATCC*CTGGAAGTTCTGTTCCAGGGGCCC*AAGTTGGATATTATTAATGTAA
Δα4 Vip3Aa-F	acaaggccatggctgatatcGGATCC*CTGGAAGTTCTGTTCCAGGGGCCC*GAAACTAGTTCAAAAGTA
Vip3Aa-R	agtggtggtggtggtg CTCGAGTTACTTAATAGAGACATCGT
Vip3A-DI-GFP-R	*agtgaaaagttcttctcctttactcat*AGCAAAAGTTAATTCCTCAAATT
GFP-F	TTGAGGAATTAACTTTTGCT*atgagtaaaggagaagaacttttcact*
GFP-R	gttagcagccggatctcagtggtggtggtggtggtgCTCGAG*ttatttgtatagttcatccatgccatgt*

Sequences overlap to pET32a are lowercased; cleavage sites of *Bam*H I, *Xho* I are underlined; cleavage site of 3C PreScission protease is in italics; sequences overlap to *gfp* are lowercased and in italics.

Four fragments of Vip3Aa domain I, including the full-length domain and three N-terminal truncations, were cloned, and fused to the 5′ end of the *gfp* gene using overlap PCR [43]. Specific primer sets listed in Table 1 facilitated the construction of these fusions. The resulting PCR products containing the truncated Vip3Aa domains and *gfp* were then inserted into the pET32a expression vector at the downstream of a Trx-His tag. Four expression vectors were constructed to prepare fusion proteins: Trx-His-Vα1–4-GFP (full-length domain I), Trx-His-VΔα1-GFP (α1-truncated), Trx-His-VΔα1–α2-GFP (α1 and α2-truncated), and Trx-His-VΔα1–α3-GFP (α1 to α3-truncated). Schematic representations of these fusion proteins are presented in Figure 2a.

To generate Vip3Aa-Cry1Ac combinations, nucleotide fragments encompassing Vip3Aa domains I and II (VdIdII) and its C-terminal truncations were amplified using primers listed in Table 2. Plasmid DNA from Bt HD73 strain, expressing the Cry1Ac1 protein (NCBI Accession No. AAA22331.1) was extracted as described previously [44]. Fragments containing domains II and III of Cry1Ac (CdIIdIII) were also amplified using specific primers (Table 2). Overlap PCR, as described above, facilitated the ligation of Vip3Aa and Cry1Ac fragments. The resulting PCR products were then inserted into pET32a at the downstream of the Trx-His tag, generating seven recombinant proteins: VdI-CdIIdIII (Vip3Aa domain I, without the K198 trypsin cleavage site, fused with Cry1Ac domains II and III), VdIα5-

CdIIdIII (α5-extended at the C-terminus of Vip3Aa domain I), VdIα5–6-CdIIdIII (α5–6 extended at the C-terminus of Vip3Aa domain I), VdIα5–7-CdIIdIII (α5–7 extended at the C-terminus of Vip3Aa domain I), VdIα5–8-CdIIdIII (α5–8 extended at the C-terminus of Vip3Aa domain I), VdIα5–9-CdIIdIII (α5–9 extended at the C-terminus of Vip3Aa domain I), and VdIdII-CdIIdIII (Vip3Aa domains I and II fused with Cry1Ac domains II and III). The expression vectors for these Vip3Aa-Cry1Ac combinations are schematically illustrated in Figure 3a.

Table 2. Primers for the cloning of vectors expressing Vip3Aa-Cry1Ac chimeric proteins.

Primers	Sequences (5′ to 3′)
Vip3Aa-F	acaaggccatggctgatatcGGATCC*CTGGAGGTGCTGTTCCAGGGGCCC*ATGAACAAGAATAATACTAAATTA
VdI-Cry-R	aattgggaaactgttcgaattggTTCTGTAGCAAAAGTTAAT
VdI-Cry-F	TTCTTGATGAGTTAACTGAGccaattcgaacagtttcccaatt
VdIα5-Cry-R	aattgggaaactgttcgaattggTACACTTTTCGCTAGTTCAGTT
VdIα5-Cry-F	TTAACTGAACTAGCGAAAAGTGTAccaattcgaacagtttcccaatt
VdIα5–6-Cry-R	aattgggaaactgttcgaattggCATTACATCGTGGAATGTATTAA
VdIα5–6-Cry-F	TTAATACATTCCACGATGTAATGccaattcgaacagtttcccaatt
VdIα5–7-Cry-R	aattgggaaactgttcgaattggATTTTCTTTAGTAATTAATTCC
VdIα5–7-Cry-F	GGAATTAATTACTAAAGAAAATccaattcgaacagtttcccaatt
VdIα5–8-Cry-R	aattgggaaactgttcgaattggTAATAATTTTCGGCATGTTGTTAA
VdIα5–8-Cry-F	TTAACAACATGCCGAAAATTATTAccaattcgaacagtttcccaatt
VdIα5–9-Cry-R	aattgggaaactgttcgaattggGTTTACTCTAAATTCCTCTTTTTC
VdIα5–9-Cry-F	GAAAAAGAGGAATTTAGAGTAAACccaattcgaacagtttcccaatt
VdIdII-Cry-R	aattgggaaactgttcgaattggAGAAAGTGTAGGGAGGATGTTTACTC
VdIdII-Cry-F	GAGTAAACATCCTCCCTACACTTTCTccaattcgaacagtttcccaatt
Cry1Ac-R	gtggtggtggtggtggtgctcgagttatgcagtaactggaataaattcaaatc

Sequences overlapping pET32a are lower-case; cleavage sites of *Bam*H I, *Xho* I are underlined; cleavage site of 3C PreScission protease is in italics; sequence overlapping *cry1Ac* are lower-case and underlined.

5.2. Expression and Purification of Recombinant Proteins

To prepare Vip3Aa proteins, a 250 μL overnight culture of *Escherichia coli* BL21 (DE) cells carrying a plasmid of pET32a-Vip3Aa was inoculated to 400 mL of LB in a 2 L flask. The cultures were incubated at 37 °C and 150 rpm in a shaking incubator until the OD_{600} reached 0.5. Protein expression was then induced by adding 0.8 mM IPTG (Isopropyl-D-thiogalactoside), followed by incubation at 16 °C for 24 h. Cells were harvested by centrifugation at 14,000× *g* for 1 min and resuspended in Tris–NaCl buffer (50 mM Tris, 0.5 M NaCl, pH 8.6) for washing. The cell suspension, supplemented with 1 mM PMSF (phenylmethanesulfonyl fluoride), was disrupted by sonication using a VC-50 sonicator (Sonics & Materials Inc., Newtown, CT, USA). Cell debris was pelleted by centrifugation at 21,000× *g* for 10 min. The supernatant containing soluble Trx-His-3C-Vip3Aa fusion protein was purified using a Ni-Sepharose 6 Fast Flow column (Cytiva, Shanghai, China) according to the manufacturer's instructions. The fusion protein was eluted with 500 mM imidazole in 50 mM Tris–NaCl buffer (pH 8.6). The recombinant proteins after dialysis were collected and reloaded to the Ni-Sepharose column to remove cleaved Trx-His tag and residual fusion proteins. The GST-tagged 3C PreScission protease used for cleavage of the fusion proteins was removed from the recovered recombinant proteins using a GSTrap column (Cytiva, Shanghai, China).

To prepare the Vip3Aa-GFP and Vip3Aa-Cry1Ac fusion proteins, *E. coli* BL21 (DE3) cells carrying the corresponding plasmids were cultured and induced with 0.8 mM IPTG as described above. After harvesting, the cells were washed and resuspended in Tris–NaCl buffer (pH 8.6) containing 1 mM PMSF. Cell suspensions were sonicated, and the inclusion bodies containing insoluble target proteins were collected by centrifugation. These inclusion bodies were dissolved in 6 M urea solution (50 mM Tris, pH 7.5), and the resulting solution was loaded onto a Ni-Sepharose 6 Fast Flow column for purification of the Trx-His-tagged recombinant proteins. Elution was achieved with 3.6 M urea solution containing 500 mM imidazole in 50 mM Tris–NaCl buffer (pH 8.6). Renaturation of the target proteins was carried out through a stepwise urea gradient dialysis using buffers containing 3 M, 2 M, 0.5 M, and 0 M urea in Tris–NaCl buffer (pH 8.6). After dialysis, the supernatant containing soluble recombinant proteins was collected by centrifugation at 12,000× *g* for 2 min at 4 °C and stored at −20 °C for further use.

5.3. Insect Rearing and Bioassays

An inbred lab colony of *S. litura* maintained in the laboratory for over 4 years (~40 generations) [12] was used in this study. The *S. litura* colony was reared on a soybean-based artificial diet at 27 °C with 50% humidity and a photoperiod 16 h of light and 8 h of darkness. The larvae of Bt-susceptible *P. xylostella* was kindly provided by the Institute of Applied Ecology (IAE), Fujian Agriculture and Forestry University [45].

For bioassays using trypsin-processed proteins, assayed proteins were processed by bovine pancreas trypsin (Sigma, St. Louis, MO, USA) at a mass ratio of 1.2:1 (trypsin/Vip3Aa, *w*/*w*) at 37 °C for 6 h. Bioassays were conducted using a diet overlay method [46]. Briefly, assayed proteins were diluted in distilled water to 100 μg/mL, respectively. A 200 μL aliquot of diluted proteins was overlaid on the surface (~7 cm^2) of the diet in each cup (30 mL plastic cup containing ~5 mL diet). Ten neonates of *S. litura* or 2nd-instar larvae of *P. xylostella* were placed into each cup, and each concentration of Vip3Aa included 5 replications. Cups containing diet overlaid by distilled water were used as negative controls. Diet cups containing assayed larvae were covered with lids and kept in a rearing room at 27 °C and at 50% humidity and with a photoperiod of 16:8 (light–dark). The mortality of the assayed *S. litura* and *P. xylostella* was recorded in 72 h and 48 h, respectively.

5.4. In Vitro Proteolytic Processing of Recombinant Proteins

In vitro proteolytic processing was conducted as described previously [12,47]. To investigate the proteolytic stability of Vip3Aa truncations and Vip3Aa-Cry1Ac recombinant proteins, 17.5 μg of Vip3Aa truncations was incubated with bovine pancreas trypsin at ratios of 1.2:1 (trypsin/Vip3Aa, *w*/*w*) in 100 μL of 150 mM NaCl in 50 mM Tris–HCl buffer (pH 8.6) for 1 h at 37 °C. Target proteins incubated in buffer only were used as controls. Proteolytically processed proteins were separated by electrophoretic analysis on 15% SDS-PAGE gels.

5.5. Liposome Fluorescence Permeability Assay of Vip3Aa Complex Activities

The activity of the target proteins to permeate the lipid membrane was determined using liposomes for a calcein release assay [37,38,48]. To prepare liposomes, 1,2-dioleoyl-sn-glycero-3-phosphocholine (DOPC) and 1,2-dioleoyl-sn-glycero-3-phosphoethanolamine (DOPE) were mixed at a ratio of 1:1 (*w*/*w*), and the lipid mixture was dissolved in chloroform at a final concentration of 10 μM/mL. The solvent (chloroform) in the lipid solution was evaporated with a flow of nitrogen gas and then under vacuum in a lyophilizer for 2.5 h. Approximately 15.2 mg of dried lipids was suspended in 200 μL of 30 mM calcein in 50 mM Tris–HCl at pH 8.0. After 5 cycles of freezing and thawing, the solution was passed through a polycarbonate membrane (0.1 μM pore size) for 35 passes using a two-syringe extruder (Avanti Polar Lipid, Alabaster, AL, USA). The lipid vesicle solution was loaded to a HiTrap desalting column with Sephadex G-25 resin (Cytiva) to remove free calcein, and the liposome preparation was recovered and stored until use.

To examine the membrane permeation activity of Vip3Aa truncations, the calcein-encapsulated liposome solution was carefully pipetted to a 1.0 cm light-path quartz cuvette and then mixed with $MnCl_2$ at a final concentration of 10 μM. Twenty microliters of trypsin-processed Vip3Aa truncations at a concentration of 0.5 μM were added to the cuvette containing 250 μL of calcein-encapsulated liposome solution. After an incubation of the mixture for 300 s for to allow the liposomes to stabilize, the release of calcein from the liposomes was monitored as decreasing in fluorescence for 10 min using a fluorescence spectrophotometer (F-4600, HITACHI, Tokyo, Japan) at excitation and emission wavelengths of 485 and 520 nm, respectively, with a slit width of 5 nm. Triton X-100 (0.1%, *v/v*) was added to the cuvette to solubilize liposomes at the end of the assay to completely release the calcein. THe liposomal permeability of Vip3Aa-GFP fusions or Vip3Aa-Cry1Ac recombinant proteins without being processed with trypsin was assayed through the same procedure described above. Trx-His tag was used in the assays as a negative control. At least three replications were included for each assay. The percentage of calcein released from the vesicles (I%) was calculated by the following formula: $I\% = 100 \times (I_t - I_0)/(I_{max} - I_0)$. I_0, I_t, and I_{max} represent the intensity of the fluorescence of the original calcein-encapsulated lipid solution without assayed protein samples, the calcein-encapsulated lipid solution with an assayed protein sample, and the sample with 0.1% Triton X-100 added, respectively.

Author Contributions: E.S. and L.S. conceived and designed the research. H.H., J.Y., Y.Y., L.O. and X.C. conducted the experiments. E.S., H.H. and J.Y. analyzed the data. E.S., X.P., X.G. and L.S. wrote the manuscript. All authors have read and agreed to the published version of the manuscript.

Funding: This research was funded by the National Natural Science Foundation of China (31772539), the Natural Science Foundation of Fujian Province (2021J01097 and 2023J01455), the Science and Technology Innovation Fund of FAFU (KFB23067), and the Scientific Research Activity Funding of National Talents Program (K86ML106A).

Institutional Review Board Statement: Not applicable.

Informed Consent Statement: Not applicable.

Data Availability Statement: The data presented in this study are available on request from the corresponding author Li Sha (shal@fafu.edu.cn).

Acknowledgments: We thank Ping Wang of Cornell University for his suggestions and comments in this work. We thank the Institute of Applied Ecology (IAE), Fujian Agriculture and Forestry University, for providing us with the Bt-susceptible *P. xylostella* strains.

Conflicts of Interest: The authors declare that they have no conflicts of interest.

References

1. Tabashnik, B.E.; Carrière, Y. Surge in insect resistance to transgenic crops and prospects for sustainability. *Nat. Biotechnol.* **2017**, *35*, 926–935. [CrossRef]
2. Bravo, A.; Likitvivatanavong, S.; Gill, S.S.; Soberón, M. *Bacillus thuringiensis*: A story of a successful bioinsecticide. *Insect Biochem. Mol. Biol.* **2011**, *41*, 423–431. [CrossRef]
3. Kurtz, R.W. A review of Vip3A mode of action and effects on Bt Cry protein-resistant colonies of lepidopteran larvae. *Southwest. Entomol.* **2010**, *35*, 391–394. [CrossRef]
4. Soares Figueiredo, C.; Nunes Lemes, A.R.; Sebastião, I.; Desidério, J.A. Synergism of the *Bacillus thuringiensis* Cry1, Cry2, and Vip3 Proteins in Spodoptera frugiperda Control. *Appl. Biochem. Biotechnol.* **2019**, *188*, 798–809. [CrossRef]
5. Yang, F.; González, J.C.S.; Williams, J.; Cook, D.C.; Gilreath, R.T.; Kerns, D.L. Occurrence and ear damage of Helicoverpa zea on transgenic *Bacillus thuringiensis* maize in the field in Texas, US and its susceptibility to Vip3A protein. *Toxins* **2019**, *11*, 102. [CrossRef]
6. Quan, Y. Studies on the Insecticidal Mechanism of *Bacillus thuringiensis* Vip3A and Cry Proteins. 2022. Available online: https://dialnet.unirioja.es/servlet/dctes?codigo=305161 (accessed on 6 January 2024).
7. Quan, Y.; Lázaro-Berenguer, M.; Hernández-Martínez, P.; Ferré, J.; Rudi, K. Critical Domains in the Specific Binding of Radiolabeled Vip3Af Insecticidal Protein to Brush Border Membrane Vesicles from *Spodoptera* spp. and Cultured Insect Cells. *Appl. Environ. Microbiol.* **2021**, *87*, e01787-21. [CrossRef] [PubMed]
8. Chakrabarty, S.; Jin, M.; Wu, C.; Chakraborty, P.; Xiao, Y. *Bacillus thuringiensis* vegetative insecticidal protein family Vip3A and mode of action against pest Lepidoptera. *Pest Manag. Sci.* **2020**, *76*, 1612–1617. [CrossRef]
9. Chakroun, M.; Banyuls, N.; Bel, Y.; Escriche, B.; Ferré, J. Bacterial vegetative insecticidal proteins (Vip) from entomopathogenic bacteria. *Microbiol. Mol. Biol. Rev.* **2016**, *80*, 329–350. [CrossRef] [PubMed]
10. Bel, Y.; Banyuls, N.; Chakroun, M.; Escriche, B.; Ferré, J. Insights into the structure of the Vip3Aa insecticidal protein by protease digestion analysis. *Toxins* **2017**, *9*, 131. [CrossRef] [PubMed]
11. Quan, Y.; Ferré, J. Structural Domains of the *Bacillus thuringiensis* Vip3Af Protein Unraveled by Tryptic Digestion of Alanine Mutants. *Toxins* **2019**, *11*, 368. [CrossRef]
12. Shao, E.; Zhang, A.; Yan, Y.; Wang, Y.; Jia, X.; Sha, L.; Guan, X.; Wang, P.; Huang, Z. Oligomer Formation and Insecticidal Activity of *Bacillus thuringiensis* Vip3Aa Toxin. *Toxins* **2020**, *12*, 274. [CrossRef]

13. Banyuls, N.; Hernández-Martínez, P.; Quan, Y.; Ferré, J. Artefactual band patterns by SDS-PAGE of the Vip3Af protein in the presence of proteases mask the extremely high stability of this protein. *Int. J. Biol. Macromol.* **2018**, *120*, 59–65. [CrossRef]
14. Núñez-Ramírez, R.; Huesa, J.; Bel, Y.; Ferré, J.; Casino, P.; Arias-Palomo, E. Molecular architecture and activation of the insecticidal protein Vip3Aa from *Bacillus thuringiensis*. *Nat. Commun.* **2020**, *11*, 3974. [CrossRef]
15. Byrne, M.J.; Iadanza, M.G.; Perez, M.A.; Maskell, D.P.; George, R.M.; Hesketh, E.L.; Beales, P.A.; Zack, M.D.; Berry, C.; Thompson, R.F. Cryo-EM structures of an insecticidal Bt toxin reveal its mechanism of action on the membrane. *Nat. Commun.* **2021**, *12*, 2791. [CrossRef]
16. Zheng, M.; Evdokimov, A.G.; Moshiri, F.; Lowder, C.; Haas, J. Crystal structure of a Vip3B family insecticidal protein reveals a new fold and a unique tetrameric assembly. *Protein Sci.* **2020**, *29*, 824–829. [CrossRef]
17. Lázaro-Berenguer, M.; Paredes-Martínez, F.; Bel, Y.; Núñez-Ramírez, R.; Arias-Palomo, E.; Casino, P.; Ferré, J. Structural and functional role of Domain I for the insecticidal activity of the Vip3Aa protein from *Bacillus thuringiensis*. *Microb. Biotechnol.* **2022**, *15*, 2607–2618. [CrossRef]
18. Jiang, K.; Zhang, Y.; Chen, Z.; Wu, D.; Cai, J.; Gao, X. Structural and functional insights into the C-terminal fragment of insecticidal Vip3A toxin of *Bacillus thuringiensis*. *Toxins.* **2020**, *12*, 438. [CrossRef] [PubMed]
19. Li, J.; Carroll, J.; Ellar, D.J. Crystal structure of insecticidal δ-endotoxin from *Bacillus thuringiensis* at 2.5 Å resolution. *Nature* **1991**, *353*, 815–821. [CrossRef] [PubMed]
20. Grochulski, P.; Masson, L.; Borisova, S.; Pusztai-Carey, M.; Schwartz, J.L.; Brousseau, R.; Cygler, M. *Bacillus thuringiensis* CrylA (a) Insecticidal Toxin: Crystal Structure and Channel Formation. *J. Mol. Biol.* **1995**, *254*, 447–464. [CrossRef] [PubMed]
21. Pardo-López, L.; Soberón, M.; Bravo, A. *Bacillus thuringiensis* insecticidal three-domain Cry toxins: Mode of action, insect resistance and consequences for crop protection. *FEMS Microbiol. Rev.* **2013**, *37*, 3–22. [CrossRef] [PubMed]
22. Palma, L.; Muñoz, D.; Berry, C.; Murillo, J.; Caballero, P. *Bacillus thuringiensis* toxins: An overview of their biocidal activity. *Toxins* **2014**, *6*, 3296–3325. [CrossRef]
23. Pacheco, S.; Gómez, I.; Peláez-Aguilar, A.E.; Verduzco-Rosas, L.A.; García-Suárez, R.; do Nascimento, N.A.; Rivera-Nájera, L.Y.; Cantón, P.E.; Soberón, M.; Bravo, A. Structural changes upon membrane insertion of the insecticidal pore-forming toxins produced by *Bacillus thuringiensis*. *Front. Insect Sci.* **2023**, *3*, 1188891. [CrossRef]
24. Zack, M.D.; Sopko, M.S.; Frey, M.L.; Wang, X.; Tan, S.Y.; Arruda, J.M.; Letherer, T.T.; Narva, K.E. Functional characterization of Vip3Ab1 and Vip3Bc1: Two novel insecticidal proteins with differential activity against lepidopteran pests. *Sci. Rep.* **2017**, *7*, 11112. [CrossRef]
25. Selvapandiyan, A.; Arora, N.; Rajagopal, R.; Jalali, S.; Venkatesan, T.; Singh, S.; Bhatnagar, R.K. Toxicity analysis of N-and C-terminus-deleted vegetative insecticidal protein from *Bacillus thuringiensis*. *Appl. Environ. Microbiol.* **2001**, *67*, 5855–5858. [CrossRef] [PubMed]
26. Kim, S.; Lee, S.B. Soluble expression of archaeal proteins in Escherichia coli by using fusion-partners. *Protein Expr. Purif.* **2008**, *62*, 116–119. [CrossRef] [PubMed]
27. Pardo-López, L.; Gómez, I.; Muñoz-Garay, C.; Jiménez-Juarez, N.; Soberón, M.; Bravo, A. Structural and functional analysis of the pre-pore and membrane-inserted pore of Cry1Ab toxin. *J. Invertebr. Pathol.* **2006**, *92*, 172–177. [CrossRef] [PubMed]
28. Lebel, G.; Vachon, V.; Préfontaine, G.; Girard, F.; Masson, L.; Juteau, M.; Bah, A.; Larouche, G.; Vincent, C.; Laprade, R. Mutations in domain I interhelical loops affect the rate of pore formation by the *Bacillus thuringiensis* Cry1Aa toxin in insect midgut brush border membrane vesicles. *Appl. Environ. Microbiol.* **2009**, *75*, 3842–3850. [CrossRef] [PubMed]
29. Yang, X.; Wang, Z.; Geng, L.; Chi, B.; Liu, R.; Li, H.; Gao, J.; Zhang, J. Vip3Aa domain IV and V mutants confer higher insecticidal activity against Spodoptera frugiperda and Helicoverpa armigera. *Pest Manag. Sci.* **2022**, *78*, 2324–2331. [CrossRef] [PubMed]
30. Gomez, I.; Arenas, I.; Benitez, I.; Miranda-Rios, J.; Becerril, B.; Grande, R.; Almagro, J.C.; Bravo, A.; Soberon, M. Specific epitopes of domains II and III of *Bacillus thuringiensis* Cry1Ab toxin involved in the sequential interaction with cadherin and aminopeptidase-N receptors in Manduca sexta. *J. Biol. Chem.* **2006**, *281*, 34032–34039. [CrossRef] [PubMed]
31. Martinez-Solis, M.; Pinos, D.; Endo, H.; Portugal, L.; Sato, R.; Ferre, J.; Herrero, S.; Hernández-Martínez, P. Role of *Bacillus thuringiensis* Cry1A toxins domains in the binding to the ABCC2 receptor from Spodoptera exigua. *Insect Biochem. Mol. Biol.* **2018**, *101*, 47–56. [CrossRef]
32. Herrero, S.; Bel, Y.; Hernández-Martínez, P.; Ferré, J. Susceptibility, mechanisms of response and resistance to *Bacillus thuringiensis* toxins in Spodoptera spp. *Curr. Opin. Insect Sci.* **2016**, *15*, 89–96. [CrossRef]
33. Dong, F.; Shi, R.; Zhang, S.; Zhan, T.; Wu, G.; Shen, J.; Liu, Z. Fusing the vegetative insecticidal protein Vip3Aa7 and the N terminus of Cry9Ca improves toxicity against Plutella xylostella larvae. *Appl. Microbiol. Biotechnol.* **2012**, *96*, 921–929. [CrossRef]
34. Gulzar, A.; Wright, D.J. Sub-lethal effects of Vip3A toxin on survival, development and fecundity of Heliothis virescens and Plutella xylostella. *Ecotoxicology* **2015**, *24*, 1815–1822. [CrossRef]
35. Mosbahi, K.; Walker, D.; Lea, E.; Moore, G.R.; James, R.; Kleanthous, C. Destabilization of the colicin E9 Endonuclease domain by interaction with negatively charged phospholipids: Implications for colicin translocation into bacteria. *J. Biol. Chem.* **2004**, *279*, 22145–22151. [CrossRef]
36. Zakharov, S.D.; Cramer, W.A. Colicin crystal structures: Pathways and mechanisms for colicin insertion into membranes. *Biochim. Biophys. Acta-Biomembr.* **2002**, *1565*, 333–346. [CrossRef]

37. Kunthic, T.; Watanabe, H.; Kawano, R.; Tanaka, Y.; Promdonkoy, B.; Yao, M.; Boonserm, P. pH regulates pore formation of a protease activated Vip3Aa from *Bacillus thuringiensis*. *Biochim. Biophys. Acta-Biomembr.* **2017**, *1859*, 2234–2241. [CrossRef] [PubMed]
38. Kato, T.; Higuchi, M.; Endo, R.; Maruyama, T.; Haginoya, K.; Shitomi, Y.; Hayakawa, T.; Mitsui, T.; Sato, R.; Hori, H. *Bacillus thuringiensis* Cry1Ab, but not Cry1Aa or Cry1Ac, disrupts liposomes. *Pestic. Biochem. Physiol.* **2006**, *84*, 1–9. [CrossRef]
39. Fortea, E.; Lemieux, V.; Potvin, L.; Chikwana, V.; Griffin, S.; Hey, T.; McCaskill, D.; Narva, K.; Tan, S.Y.; Xu, X. Cry6Aa1, a *Bacillus thuringiensis* nematocidal and insecticidal toxin, forms pores in planar lipid bilayers at extremely low concentrations and without the need of proteolytic processing. *J. Biol. Chem.* **2017**, *292*, 13122–13132. [CrossRef] [PubMed]
40. Coux, F.; Vachon, V.; Rang, C.; Moozar, K.; Masson, L.; Royer, M.; Bes, M.; Rivest, S.; Brousseau, R.; Schwartz, J.-L. Role of interdomain salt bridges in the pore-forming ability of the *Bacillus thuringiensis* toxins Cry1Aa and Cry1Ac. *J. Biol. Chem.* **2001**, *276*, 35546–35551. [CrossRef] [PubMed]
41. Sellami, S.; Jemli, S.; Abdelmalek, N.; Cherif, M.; Abdelkefi-Mesrati, L.; Tounsi, S.; Jamoussi, K. A novel Vip3Aa16-Cry1Ac chimera toxin: Enhancement of toxicity against Ephestia kuehniella, structural study and molecular docking. *Int. J. Biol. Macromol.* **2018**, *117*, 752–761. [CrossRef] [PubMed]
42. Song, F.; Chen, C.; Wu, S.; Shao, E.; Li, M.; Guan, X.; Huang, Z. Transcriptional profiling analysis of Spodoptera litura larvae challenged with Vip3Aa toxin and possible involvement of trypsin in the toxin activation. *Sci. Rep.* **2016**, *6*, 23861. [CrossRef]
43. Pogulis, R.J.; Vallejo, A.N.; Pease, L.R. In vitro recombination and mutagenesis by overlap extension PCR. *Vitr. Mutagen. Protoc.* **1996**, *57*, 167–176.
44. Adang, M.J.; Staver, M.J.; Rocheleau, T.A.; Leighton, J.; Barker, R.F.; Thompson, D.V. Characterized full-length and truncated plasmid clones of the crystal protein of *Bacillus thuringiensis* subsp. kurstaki HD-73 and their toxicity to Manduca sexta. *Gene* **1985**, *36*, 289–300. [CrossRef] [PubMed]
45. Chen, W.; Amir, M.B.; Liao, Y.; Yu, H.; He, W.; Lu, Z. New insights into the Plutella xylostella detoxifying enzymes: Sequence evolution, structural similarity, functional diversity, and application prospects of glucosinolate sulfatases. *J. Agric. Food Chem.* **2023**, *71*, 10952–10969. [CrossRef] [PubMed]
46. Kain, W.C.; Zhao, J.-Z.; Janmaat, A.F.; Myers, J.; Shelton, A.M.; Wang, P. Inheritance of resistance to *Bacillus thuringiensis* Cry1Ac toxin in a greenhouse-derived strain of cabbage looper (Lepidoptera: Noctuidae). *J. Econ. Entomol.* **2004**, *97*, 2073–2078. [CrossRef] [PubMed]
47. Wang, P.; Zhao, J.-Z.; Rodrigo-Simón, A.; Kain, W.; Janmaat, A.F.; Shelton, A.M.; Ferré, J.; Myers, J. Mechanism of resistance to *Bacillus thuringiensis* toxin Cry1Ac in a greenhouse population of the cabbage looper, Trichoplusia ni. *Appl. Environ. Microbiol.* **2007**, *73*, 1199–1207. [CrossRef]
48. Pan, Z.-Z.; Xu, L.; Liu, B.; Chen, Q.-X.; Zhu, Y.-J. Key residues of *Bacillus thuringiensis* Cry2Ab for oligomerization and pore-formation activity. *AMB Express* **2021**, *11*, 112. [CrossRef]

toxins

MDPI

Article

Domain III β4–β5 Loop and β14–β15 Loop of *Bacillus thuringiensis* Vip3Aa Are Involved in Receptor Binding and Toxicity

Xiaoyue Hou [1,2,3,4,5], Mengjiao Li [4], Chengjuan Mao [4], Lei Jiang [4], Wen Zhang [4], Mengying Li [4], Xiaomeng Geng [4], Xin Li [4], Shu Liu [1,2,3,4], Guang Yang [1,2,3,4], Jing Zhou [6], Yaowei Fang [1,2,3,4,*] and Jun Cai [5,*]

1 Jiangsu Key Laboratory of Marine Bioresources and Environment, Jiangsu Ocean University, Lianyungang 222005, China; 2020000057@jou.edu.cn (X.H.); 2007000028@jou.edu.cn (S.L.); 2020000050@jou.edu.cn (G.Y.)

2 Co–Innovation Center of Jiangsu Marine Bio–Industry Technology, Jiangsu Ocean University, Lianyungang 222005, China

3 Jiangsu Marine Resources Development Research Institute, Jiangsu Ocean University, Lianyungang 222005, China

4 College of Marine Food and Bioengineering, Jiangsu Ocean University, Lianyungang 222005, China; xydxhl@163.com (M.L.); mcj15055202109@163.com (C.M.); jl17892521856@163.com (L.J.); zw19975066385@163.com (W.Z.); lmy13775446376@163.com (M.L.); gxm18913965276@163.com (X.G.); 2021122077@jou.edu.cn (X.L.)

5 College of Life Sciences, Nankai University, Tianjin 300071, China

6 Lianyungang City Quality Technology Comprehensive Inspection and Quality Inspection Center, Lianyungang 222346, China; zj13179573851@163.com

* Correspondence: 2007000027@jou.edu.cn (Y.F.); caijun@nankai.edu.cn (J.C.)

Abstract: Vip3Aa, secreted by *Bacillus thuringiensis*, is effective at controlling major agricultural pests such as *Spodoptera frugiperda*. However, to control Vip3Aa resistance evolved in the field by different lepidoptera species, an in–depth study of sequence—structure—activity relationships is necessary to design new Vip3Aa variants. In this study, the four specific loops (β4–β5 loop, β9–β10 loop, β12–β13 loop, and β14–β15 loop) in domain III were selected and four loop mutants were constructed by replacing all residues in each specific loop with alanine. We obtained soluble proteins for three of the loop mutants, excluding the β9–β10 loop. These loop mutants have been characterized by toxicity bioassays against *S. frugiperda*, proteolytic processing, and receptor binding. These results indicate that the β4–β5 loop and β14–β15 loop are involved in receptor binding and Vip3Aa toxicity. Based on this, we constructed numerous mutants and obtained three single mutants (Vip3Aa–S366T, Vip3Aa–S366L, and Vip3Aa–R501A) that exhibited significantly increased toxicity of 2.61–fold, 3.39–fold, and 2.51–fold, respectively. Compared to Vip3Aa, the receptor affinity of Vip3Aa–S366T and Vip3Aa–S366L was significantly enhanced. Furthermore, we also analyzed and aligned the three–dimensional structures of the mutants and Vip3Aa. In summary, these results indicate that the loops in domain III have the potential to be targeted to enhance the insecticidal toxicity of the Vip3Aa protein.

Keywords: microbial pesticides; Vip3Aa; domain III; binding affinity; *Spodoptera frugiperda*

Key Contribution: Vip3Aa domain III β4–β5 loop and β14–β15 loop are involved in receptor binding. Vip3Aa domain III β9–β10 loop is essential for the proper protein folding of Vip3Aa. Vip3Aa–S366T, Vip3Aa–S366L, and Vip3Aa–R501A exhibited significantly increased toxicity of 2.61–fold, 3.39–fold, and 2.51–fold, respectively.

Citation: Hou, X.; Li, M.; Mao, C.; Jiang, L.; Zhang, W.; Li, M.; Geng, X.; Li, X.; Liu, S.; Yang, G.; et al. Domain III β4–β5 Loop and β14–β15 Loop of *Bacillus thuringiensis* Vip3Aa Are Involved in Receptor Binding and Toxicity. *Toxins* **2024**, *16*, 23. https://doi.org/10.3390/toxins16010023

Received: 17 November 2023
Revised: 23 December 2023
Accepted: 30 December 2023
Published: 1 January 2024

1. Introduction

Spodoptera frugiperda (Lepidoptera, Noctuidae) is native to the tropics and subtropics of the Americas and has become a major global invasive pest in the past decade [1]. Planting crops that express insecticidal proteins is a green pest control strategy, as these crops exhibit specific toxicity to different pests [2]. Currently, widely used insecticidal proteins are generally derived from *Bacillus thuringiensis* and are categorized into insecticidal crystal proteins (ICPs) and vegetative insecticidal proteins (Vips). Given that some lepidopteran pests, such as *S. frugiperda*, have developed resistance to certain ICPs, it has become urgent to study Vips that do not share any amino acid sequence homology with ICPs [3].

Based on the protein structure, Vips are divided into three categories: Vip3, Vpa, and Vpb [4]. Currently, the focus of research and application is mainly on Vip3 proteins, which can not only specifically control lepidopteran pests such as *S. frugiperda*, *Agrotis ipsilon*, and *Spodoptera exigua* but are also different from ICPs in terms of their receptor binding sites [5]. The amino acid similarity among Vip3Aa protein members is up to 95% and the average molecular weight is approximately 88 kDa. Vip3A can be hydrolyzed by trypsin or insect midgut juice into two fragments, one of 22 kDa (corresponding to the first 198 amino acids) and the other of 66 kDa (corresponding to the amino acids after position 198). The 66 kDa band was initially believed to be a fragment of toxic activation. With the in–depth study of the mechanism of action of Vip3Aa protein, researchers discovered that after being cleaved by trypsin, the 22 kDa fragment and the 66 kDa fragment still remain bound together, forming the activated toxin [6,7].

Trypsin treatment of alanine mutants revealed that the Vip3Af protein consists of five domains, of which domains I to III are required for protein tetramerization [8]. Domain I consists of highly curved α–helices and may be responsible for the ability of the Vip3A protein to insert into the cell membrane; domain I also plays a key role in maintaining the stability of Vip3Aa in the presence of midgut proteases [9,10]. In addition, domains II and III of the Vip3Aa protein are the core regions responsible for its binding with the brush border membrane vesicles (BBMVs) of *S. frugiperda*, with domain II potentially playing a role in stabilizing Vip3 oligomers [7,10]. Domain III was first shown to play a major role in the binding between Vip3Aa and Sf9 cells and recent studies have shown that Vip3Aa truncated variants lacking domain II and domain III or containing only one of them, fail to bind to target cells, suggesting that domain II and domain III together are the receptor–binding domains of the Vip3Aa protein [10,11]. The Vip3 proteins contain carbohydrate–binding modules in domains IV and V [7]. These modules potentially interact with the peritrophic membrane to facilitate the passage of the Vip3 protein through it, enabling contact with the insect midgut epithelium [10].

After determining the 3D structure of the Vip3 protein, there has been a gradual increase in the number of studies investigating its structure—-effect relationship. Trypsin cleaves the Vip3Aa protein from its protoxin form to an activated toxin, which is a prerequisite for the insecticidal activity of the Vip3Aa protein. Helix $\alpha1$ plays an important role in restricting the conformation of domain I in the Vip3Aa protoxin and ensuring the insecticidal activity of the Vip3Aa activated toxin [12]. There are multiple protease cleavage sites in the loop region between domain I and domain II and it is advantageous to increase the insecticidal activity of Vip3Aa by adding more cleavage sites [13]. In the 3D protein structure, loops connecting secondary structures play crucial roles in the function of the protein. Domain III contains three antiparallel β–sheets and is similar to domain II of Cry3A [7,14]. Domain II of the Cry proteins has receptor recognition and receptor binding functions; in which the loops play an important role [15,16]. Therefore, loops in domain III, especially those exposed on the protein structure surface, may be involved in the binding process between the Vip3Aa protein and its receptor. However, studies on the loops in domain III of the Vip3Aa protein are not yet available, which limits studies related to Vip3Aa protein receptor recognition and receptor binding.

In the present study, we focused on four specific loops ($\beta4$–$\beta5$ loop, $\beta9$–$\beta10$ loop, $\beta12$–$\beta13$ loop, and $\beta14$–$\beta15$ loop) in domain III of Vip3Aa and replaced all residues in each of the loops with alanine, respectively, to generate loop alanine mutants. Compared to the wild–type protein Vip3Aa, these mutants exhibited decreased insecticidal activity against *S. frugiperda*, possibly due to reduced stability in the midgut juice or impaired binding to receptors on midgut epithelial cells. In addition, we constructed single mutants and multiple mutants in the $\beta4$–$\beta5$ loop and $\beta14$–$\beta15$ loop of domain III and found that the mutants with improved insecticidal activity exhibited stronger binding ability to BBMVs of *S. frugiperda*. Finally, we analyzed the potential roles of the $\beta4$–$\beta5$ loop and $\beta14$–$\beta15$ loop in the insecticidal activity of Vip3Aa using the predicted 3D protein structure. In summary, these results demonstrate that the loops in domain III are involved in receptor binding and toxicity.

2. Results

2.1. Substituting the Sequence of the Loop in Domain III Reduces the Insecticidal Activity of Vip3Aa

Vip3Aa contains five domains and domain III, composed of three antiparallel β–sheets, plays an important role in the interactions between Vip3Aa and the receptors in *S. frugiperda* BBMVs [8,10]. Four special loops ($\beta4$–$\beta5$ loop, $\beta9$–$\beta10$ loop, $\beta12$–$\beta13$ loop, and $\beta14$–$\beta15$ loop) in domain III were selected and mutated in this study to further elucidate the insecticidal mechanism of Vip3Aa (Figure 1A). The $\beta4$–$\beta5$ loop ($D^{365}SI^{367}$), $\beta9$–$\beta10$ loop ($K^{429}KMKTL^{434}$), and $\beta14$–$\beta15$ loop ($E^{498}NSR^{501}$) are from different β–sheets in domain III and are protruding loops. The $\beta12$–$\beta13$ loop ($S^{468}ANDDG^{473}$) is the least conserved loop in Vip3 protein domain III (Figure S1). Moreover, these four loops are exposed to the surface of the Vip3Aa protein (Figure S2). Four loop mutants (Vip3Aa–loop4–5A, Vip3Aa–loop9–10A, Vip3Aa–loop12–13A, and Vip3Aa–loop14–15A) were constructed by replacing all residues in each specific loop with alanine.

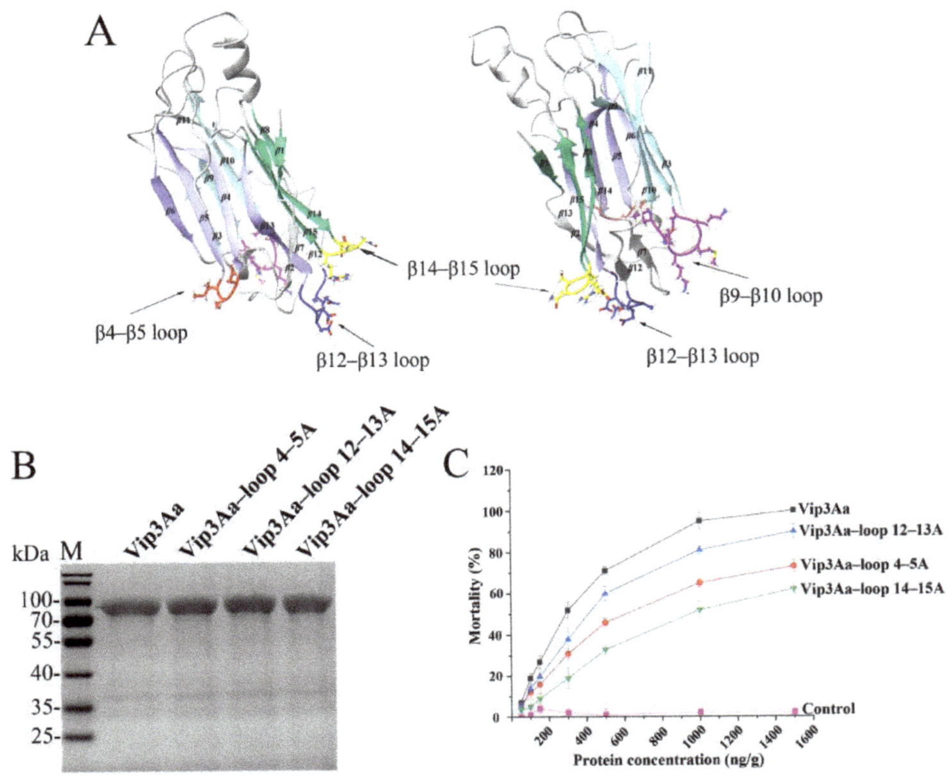

Figure 1. The Vip3Aa loop mutants showed a decrease in toxicity against *S. frugiperda* neonate larvae. (**A**) The spatial location of the four special loops in domain III. (**B**) SDS–PAGE results of Vip3Aa and its mutants. (**C**) The insecticidal activity of Vip3Aa and its mutants.

Unfortunately, we did not obtain the soluble protein for mutant Vip3Aa–loop9–10A (mutation of residues 428–KKMKTL–435 to 428–AAAAAA–435). As shown in Figure 1B, we obtained mutant proteins with a molecular mass of approximately 88 kDa. These mutant proteins exhibited expression levels similar to those of Vip3Aa. To explore the effect of the mutations in domain III loops, the three–loop residue substitution mutants were bioassayed against neonate larvae of the insect species *S. frugiperda*. The mortality rate of *S. frugiperda* reached 95% when treated with 1000 ng/g Vip3Aa but the lethality rates for Vip3Aa–loop4–5A, Vip3Aa–loop12–13A, and Vip3Aa–loop14–15A at the same concentration were 65%, 81%, and 52%, respectively (Figure 1C). The wild–type protein Vip3Aa showed an LC_{50} value of 251 (226–307) ng/g; however, the mutant proteins Vip3Aa–loop4–5A, Vip3Aa–loop12–13A, and Vip3Aa–loop14–15A showed decreased toxicity compared to Vip3Aa (Table 1).

Table 1. Toxicity of the Vip3Aa and loop mutant proteins against *S. frugiperda* neonate larvae.

Protein	Mutation Description	LC_{50} (ng/g) (95% Fiducial Limits)	Slope ± SE	χ^2	DF
Vip3Aa	Wild type, no mutation	251 (226–307)	2.42 ± 0.15	8.97	5
Vip3Aa–loop 4–5A	Mutation of residues 364–DSI–368 to 364–AAA–368	594 (506–713)	1.56 ± 0.12	0.38	5
Vip3Aa–loop12–13A	Mutation of residues 467–SANDDG–474 to 467–AAAAAA–474	375 (330–427)	2.00 ± 0.13	1.55	5
Vip3Aa–loop14–15A	Mutation of residues 497–ENSR–502 to 497–AAAA–502	970 (805–1219)	1.59 ± 0.13	1.58	5

2.2. Substituting the Sequence of the Loop in Domain III Affects the Stability of Vip3Aa

Domain III is essential to the tetramerization of Vip3 [7,9]. To test the effect of the domain III loop mutation on the proteolytic pattern of Vip3Aa, trypsin and *S. frugiperda* midgut juice were used to treat the mutant proteins. As shown in Figure 2A–D, the mutant proteins and Vip3Aa showed similar proteolytic patterns, with the major fragment being approximately 66 kDa. With the prolongation of trypsin hydrolysis time or the increase in midgut juice concentration, the proteolytic patterns did not change. However, compared to Vip3Aa, the activated band ratio of mutant proteins decreased to varying degrees, especially Vip3Aa–loop12–13A (Figure 2E,F). When the mutant proteins were treated with trypsin for 60 min, there was a significant decrease in the percentage of 66 kDa fragments in all three mutant proteins. Combined with the results of the mutant proteins treated with *S. frugiperda* midgut juice, we conclude that the β12–β13 loop may have a greater effect on the stability of the Vip3Aa protein than the β4–β5 loop and β14–β15 loop.

2.3. Substituting the Sequence of the Loop in Domain III Affects the Binding of Vip3Aa to *S. frugiperda* BBMVs

The interaction between Vip3Aa and the target pest midgut BBMVs is widely perceived as the vital step for insecticidal activity [17,18]. To analyze the effect of domain III loop residue substitution with alanine on binding between Vip3Aa and its receptors, we first performed ELISA binding saturation assays of the toxins with biotin tag to *S. frugiperda* BBMVs. As shown in Figure 3A–D, unlike Vip3Aa–loop 12–13A, the equilibrium dissociation constants (K_d) of Vip3Aa–loop 4–5A (128.73 ± 11.75 nM) or Vip3Aa–loop 14–15A (144.25 ± 18.18 nM) binding to BBMVs were significantly larger than that of Vip3Aa (94.04 ± 8.93 nM). Furthermore, when the residues in both β4–β5 loop and β14–β15 loop were replaced by alanine, the equilibrium dissociation constant of Vip3Aa–loop4–5A&14–15A binding to BBMVs was elevated even more (Figure 3E). In addition, it was shown that Vip3Aa enters Sf9 cells via receptor–mediated endocytosis and that the amount of Vip3Aa entering the cells correlates with cytotoxicity [19]. To investigate whether the β4–β5 loop and β14–β15 loop in domain III are involved in the binding of Vip3Aa to its internalization-related receptors, we constructed the fusion protein composed of Vip3Aa (or Vip3Aa–loop 4–5A or Vip3Aa–loop 14–15A) and red fluorescence protein (RFP). A greater number of red fluorescent dots could be observed in Vip3Aa–RFP–treated cells than in Vip3Aa–loop 4–5A–RFP or Vip3Aa–loop 14–15A–RPF–treated cells; Vip3Aa–loop 4–5A and Vip3Aa–loop 14–15A were also less cytotoxic than Vip3Aa (Figure 3F and Figure S3). These results suggest that the β4–β5 loop or β14–β15 loop in domain III should be involved in the interaction between Vip3Aa and its receptors.

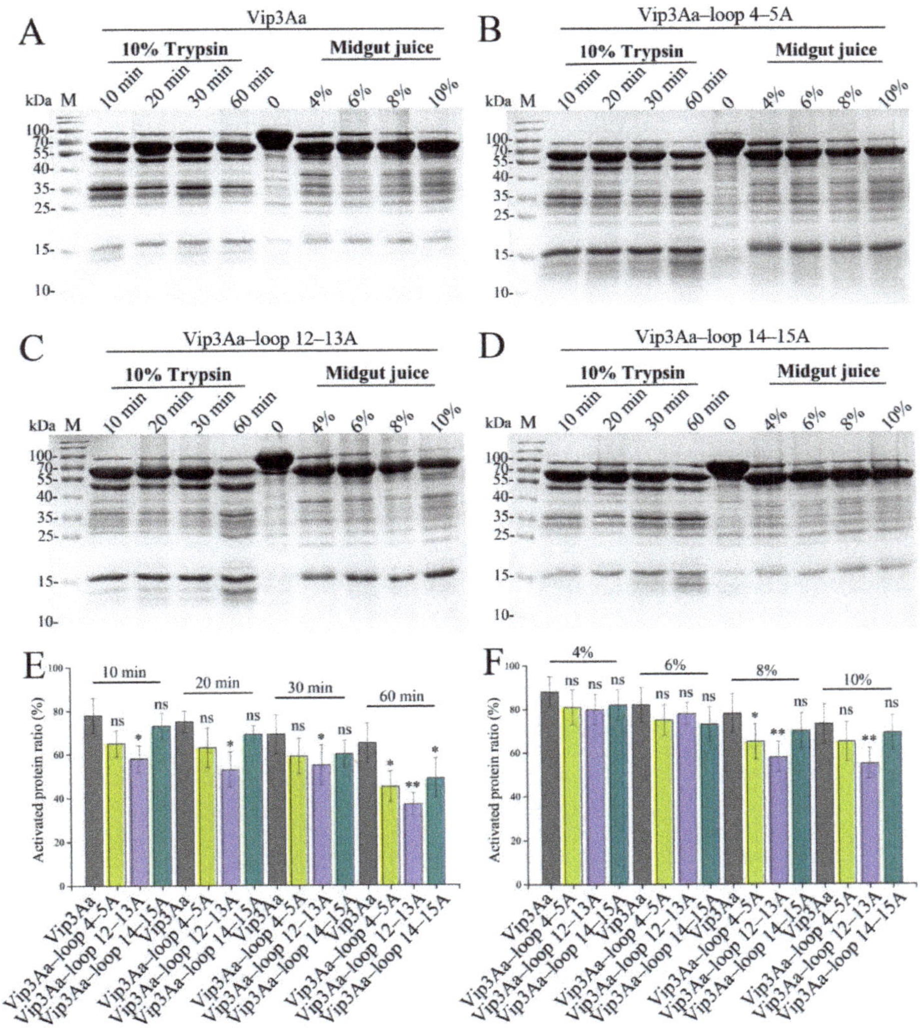

Figure 2. Proteolytic processing of Vip3Aa (**A**), Vip3Aa–loop 4–5A (**B**), Vip3Aa–loop 12–13A (**C**), and Vip3Aa–loop 14–15A (**D**). The cleaved fragments were separated via SDS–PAGE. (**E**) Percentage of 66 kDa fragment after trypsin treatment in Vip3Aa and its mutants. (**F**) Percentage of 66 kDa fragment after *S. frugiperda* midgut juice treatment in Vip3Aa and its mutants. Significant differences from the controls are shown as * $p < 0.05$ and ** $p < 0.01$. ns: no significant.

2.4. Selective Modification of the Loops in Domain III Contributes to the Insecticidal Activity of Vip3Aa

To better understand the role of residues in the β4–β5 loop and β14–β15 loop in the insecticidal activity of Vip3Aa, these residues were modified by site–directed mutagenesis. As indicated in Figure 4A, the insecticidal activity against *S. frugiperda* was significantly impaired when the three residues (D365, S366, and I367) in the β4–β5 loop were replaced by alanine, especially S366 and I367. Unlike the mutants of the β4–β5 loop, the mutants of the β14–β15 loop showed widely varying toxicity. The toxicity of mutant Vip3Aa–E498A was significantly impaired, while the mutants Vip3Aa–S500A (LC50: 185 (162–217) ng/g) and Vip3Aa–R501A (LC50: 100 (82–117) ng/g) showed enhanced toxicity compared to the Vip3Aa protein (LC50: 251 (226–307) ng/g).

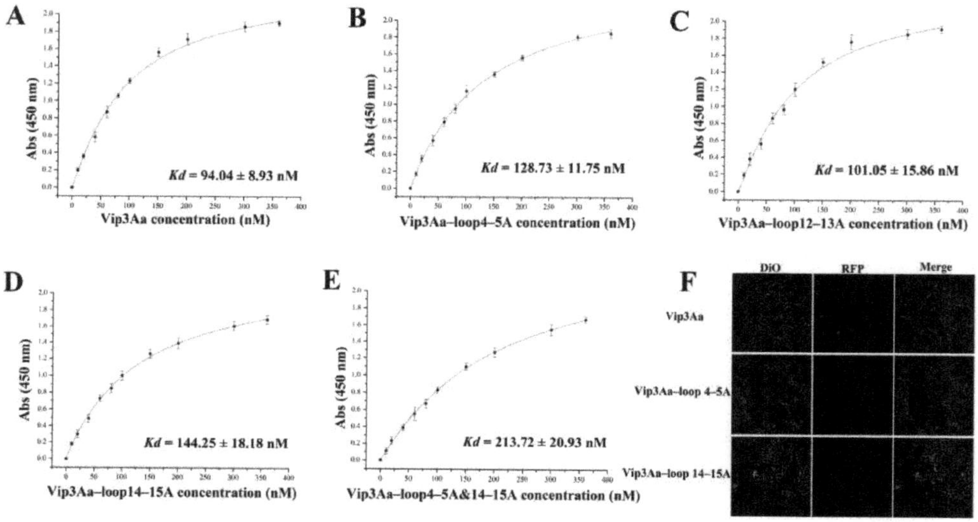

Figure 3. Binding analysis of Vip3Aa and its mutants to *S. frugiperda* BBMVs. Saturation binding of biotinylated Vip3Aa (**A**), Vip3Aa–loop 4–5A (**B**), Vip3Aa–loop 12–13A (**C**), or Vip3Aa–loop 14–15A (**D**) to *S. frugiperda* BBMVs. (**E**) Saturation binding of biotinylated Vip3Aa–loop4–5A&14–15A. (**F**) Confocal microscopy analysis of the entry of Vip3Aa or its mutants into Sf9 cells.

In addition, the residues in the β4–β5 loop and β14–β15 loop were also substituted with amino acids that differed in the multiple sequence alignments of the Vip3 proteins (Figure S1). However, only five (Vip3Aa–S366R, Vip3Aa–E498K, Vip3Aa–N499R, Vip3Aa–R501K, and Vip3Aa–R501Q) of the 14 mutants displayed toxicity similar to that of the Vip3Aa protein. The remaining 10 mutants either had severely impaired insecticidal toxicity or were unable to be obtained as soluble proteins.

Studies have shown that the mutants (S9N, S193T, and S194L) displayed an approximately two–fold insecticidal activity against *Helicoverpa armigera* larvae compared with Vip3Aa11 [20]. To obtain mutant proteins with enhanced insecticidal activity against *S. frugiperda*, more mutants were generated by replacing serine at position 366 with asparagine, leucine, threonine, and charged polar amino acids (aspartic acid, glutamic acid, and lysine), respectively (Figure 4). The LC_{50} values of Vip3Aa–S366T and Vip3Aa–S366L were 96 (75–122) and 74 (56–111) ng/g, respectively, indicating that these two mutants had approximately 2.6– and 3.4–fold higher activity against *S. frugiperda* than Vip3Aa. However, the insecticidal activity of other mutants was lower than that of Vip3Aa, especially Vip3Aa–S500N and Vip3Aa–S500T (Figure 4A).

The replacement of glutamic acid (E, negative) with lysine (K, positive) at position 498 had little influence on the insecticidal activity of Vip3Aa against *S. frugiperda* (Figure 4B). In a subunit of the Vip3Aa protein tetramer, residue N499 does not have any contact with other residues (Table S2). Based on the difference in hydrophilicity of the amino acids, asparagine at position 499 was substituted with aspartic acid and threonine to obtain mutants Vip3Aa–N499D and Vip3Aa–N499T, respectively. We did not obtain soluble protein for Vip3Aa–N499T. However, Vip3Aa–N499D (LC_{50}: 184 (139–219) ng/g) showed slightly increased insecticidal activity over Vip3Aa (LC_{50}: 251 (226–307) ng/g).

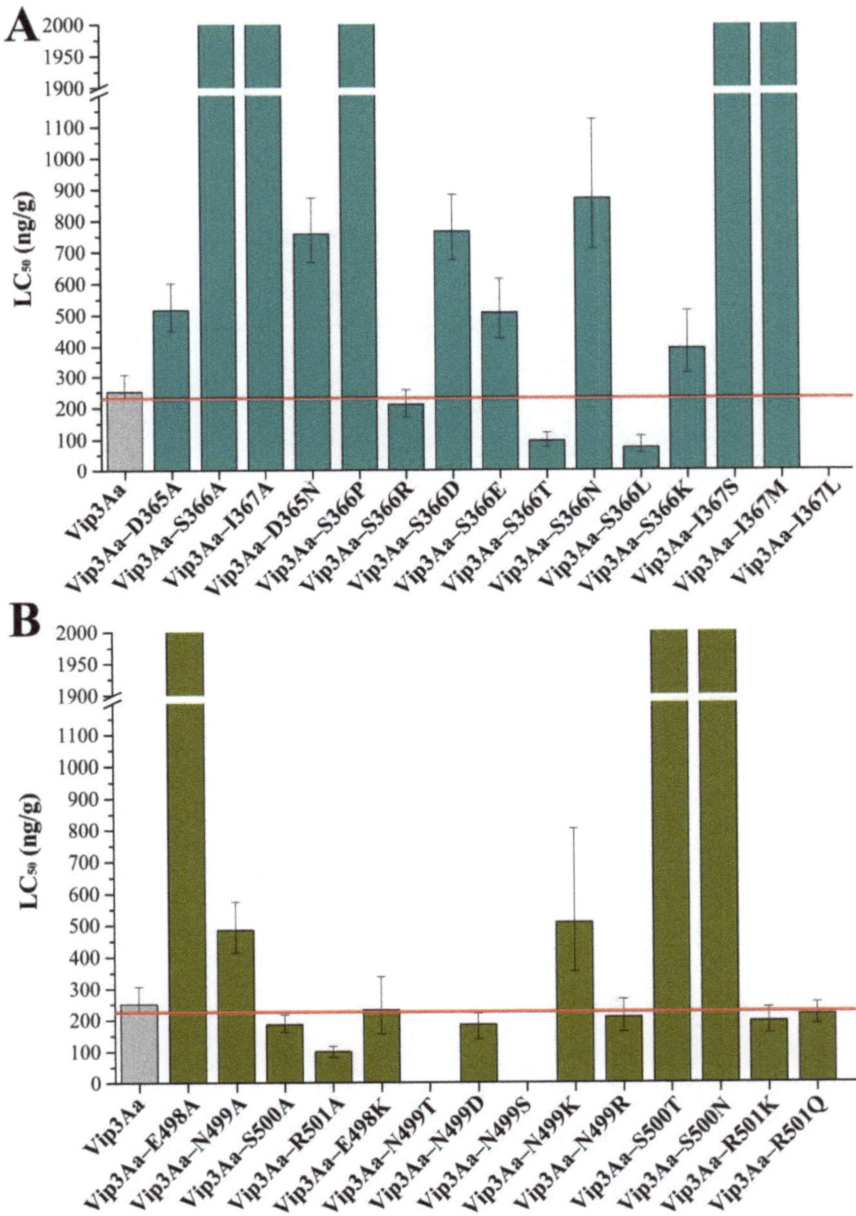

Figure 4. Toxicity of Vip3Aa single mutants against *S. frugiperda* neonate larvae. (**A**) Mutants in the β4–β5 loop ($D^{365}SI^{367}$); (**B**) Mutants in the β14–β15 loop ($E^{498}NSR^{501}$). Vip3Aa–I367L, Vip3Aa–N499T, and Vip3Aa–N499S did not obtain soluble protein.

Thus far, we have obtained five single mutants (Vip3Aa–S366L, Vip3Aa–S366T, Vip3Aa–R501A, Vip3Aa–S500A, and Vip3Aa–N499D) with enhanced insecticidal activity compared to Vip3Aa but Vip3Aa–S500A and Vip3Aa–N499D had less toxicity than the other three single mutants. In addition, compared to Vip3Aa–loop4–5A and Vip3Aa–loop14–15A, the mutant Vip3Aa–loop12–13A showed less impaired insecticidal activity, so we constructed only one mutant (Vip3Aa–N470K) in the β12–β13 loop, which replaced the asparagine at position 470 with lysine. Mutant Vip3Aa–N470K (LC_{50}: 123 (107–141) ng/g) showed enhanced toxicity compared to Vip3Aa (Table 2). Subsequently, multiple mutants were constructed, including Vip3Aa–S366T/R501A, Vip3Aa–S366L/R501A, Vip3Aa–S366T/N470K, Vip3Aa–S366L/N470K, Vip3Aa–N470K/R501A, Vip3Aa–S366T/N470K/R501A, and Vip3Aa–S366L/N470K/R501A. Among these multiple mutants, only Vip3Aa–S366L/R501A did not show enhanced toxicity compared to Vip3Aa. However, only the multiple mutants Vip3Aa–S366T/N470K and Vip3Aa–S366L/N470K showed higher toxicity than Vip3Aa–S366L (Table 2 and Figure 4A). These results suggest that selective modification of loops in domain III can increase the toxicity of Vip3Aa.

Table 2. Toxicity of modified Vip3Aa proteins against *S. frugiperda* neonate larvae.

Protein	Position	LC_{50} (ng/g) (95% Fiducial Limits)
Vip3Aa	-	251 (226–307)
Vip3Aa–N470K	β12–β13 loop	123 (107–141)
Vip3Aa–S366T/R501A	β4–β5 loop and β14–β15 loop	145 (122–168)
Vip3Aa–S366L/R501A		128 (104–172)
Vip3Aa–S366T/N470K	β4–β5 loop and β12–β13 loop	68 (51–86)
Vip3Aa–S366L/N470K		56 (39–77)
Vip3Aa–N470K/R501A	β12–β13 loop and β14–β15 loop	113 (106–131)
Vip3Aa–S366T/N470K/R501A	β4–β5 loop, β12–β13 loop and β14–β15 loop	108 (92–122)
Vip3Aa–S366L/N470K/R501A		106 (86–125)

2.5. *Toxicity–Enhanced Vip3Aa Mutants Bind More Strongly to S. frugiperda BBMVs*

Our results suggested that the β4–β5 loop or β14–β15 loop in domain III is involved in the interaction between Vip3Aa and its receptors. Therefore, the binding of toxicity–enhanced Vip3Aa mutants to *S. frugiperda* BBMVs was measured by ELISA binding saturation assays. As shown in Figure 5, the equilibrium dissociation constants (K_d) of Vip3Aa binding to BBMVs, Vip3Aa–S366L, Vip3Aa–S366T, and Vip3Aa–R501A were 91.89 ± 10.02 nM, 59.34 ± 3.85 nM, 64.83 ± 5.12 nM, and 76.93 ± 7.11 nM, respectively. These results indicate that the toxicity–enhanced Vip3Aa mutants (except Vip3Aa–R501A) have greater binding with *S. frugiperda* BBMVs than Vip3Aa.

2.6. *Three–Dimensional Structural Analysis of Vip3Aa Mutants*

Based on the results of the insecticidal activity of the mutants against *S. frugiperda* neonate larvae, we categorized the single mutants into four groups: (i) the insecticidal activity was significantly improved; (ii) the insecticidal activity was severely impaired; (iii) the soluble protein was not available; and (iv) the insecticidal activity was affected. We used the Phyre2 server to predict the 3D structure of the mutants except for mutant (iv).

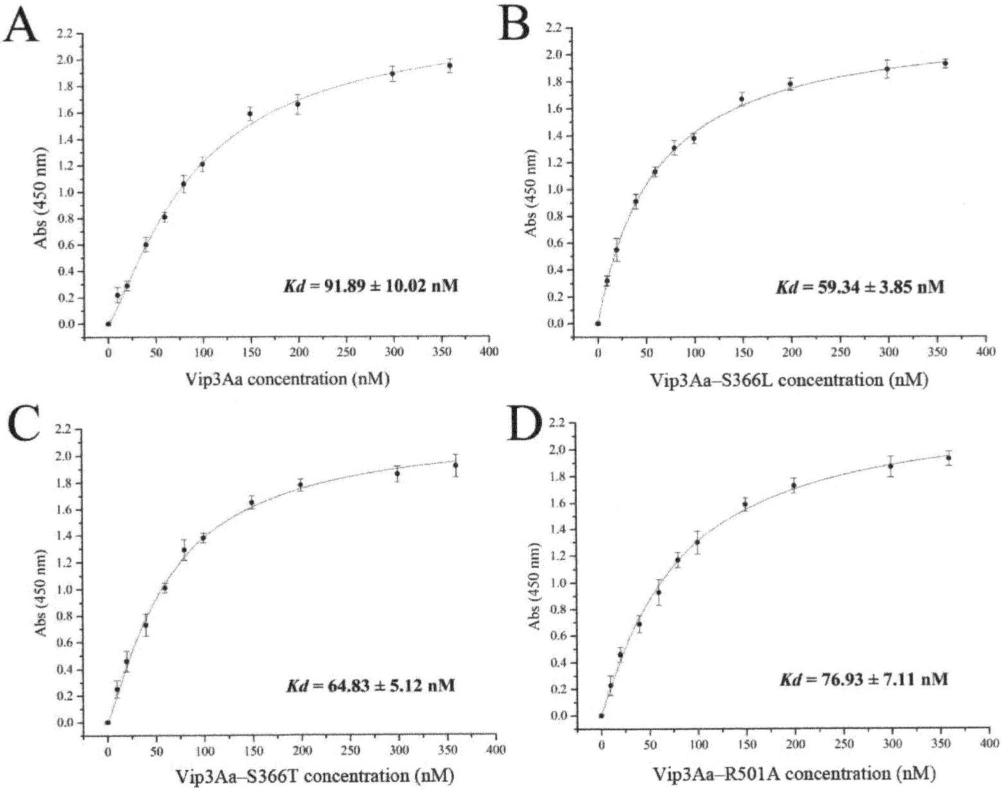

Figure 5. Saturation binding of biotinylated Vip3Aa (**A**), Vip3Aa–S366L (**B**), Vip3Aa–S366T (**C**), or Vip3Aa–R501A (**D**) to *S. frugiperda* BBMVs.

The contacts program in the software Chimera X v.1.8 was used to de–predict the contacts of the mutated residues with the surrounding residues in the 3D structures of the mutants. As shown in Figure 6A and Table S2, when serine at position 366 was replaced with alanine, proline, or threonine, 3D structural analysis of the mutants showed that the mutation not only reduced the contacts of the mutated residue with other residues but also significantly reduced the contacts of residues in the $\beta4$–$\beta5$ loop with other residues. Moreover, mutants at position 367 exhibited the same phenomenon (Figure 6B and Table S2). However, the 3D structure of Vip3Aa–S366L shows increased contacts between L366 and other residues (Figure 6A and Table S2). Additionally, we discovered that residue N633 in the 3D structure of the Vip3Aa protein forms nine contacts (four contacts with S366 and five contacts with I367) with residues in the $\beta4$–$\beta5$ loop ($D^{365}SI^{367}$) (Figure 6). Except for Vip3Aa–S366T and Vip3Aa–S366L, the contacts between residue N633 and residue at position 367 in the 3D structures of other mutants were no less than the contacts with residue at position 366.

Unlike the mutant Vip3Aa–R501A, it was not difficult to observe that the mutants with severely impaired toxicity (or insoluble mutant proteins) all contacted with N470, potentially restricting the flexibility of both the $\beta14$–$\beta15$ loop and $\beta12$–$\beta13$ loop (Figure 7 and Table S2). In addition, 3D structural analysis of the toxicity–enhanced mutant Vip3Aa–N470K suggests some contacts between K470 and R501 (Table S2). However, lysine and arginine are both positively charged amino acids and repel each other. This analysis shows that the $\beta14$–$\beta15$ loop and $\beta12$–$\beta13$ loop staying away from each other may assist in the toxicity of Vip3Aa.

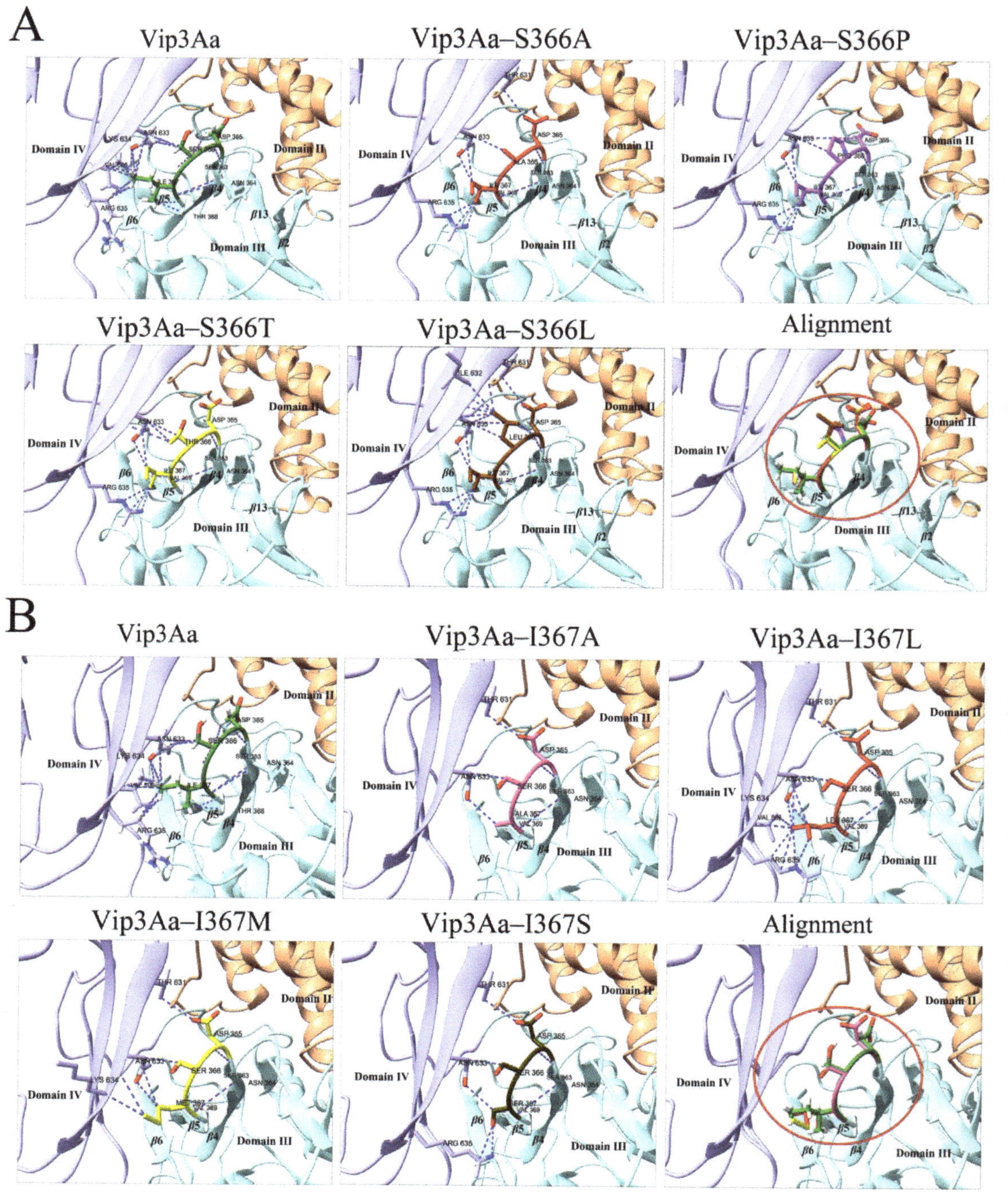

Figure 6. Contact analysis between residues by chimera X. (**A**) Vip3Aa and mutants of S366. (**B**) Vip3Aa and mutants of S367. Alignment indicates the result of comparison of β4–β5 loop in the structure of the Vip3Aa protein with that of the mutant proteins. The β4–β5 loop in Vip3Aa and its mutants were labeled with different colors. Blue: domain III; Purple: domain IV; Red circle: β4–β5 loop.

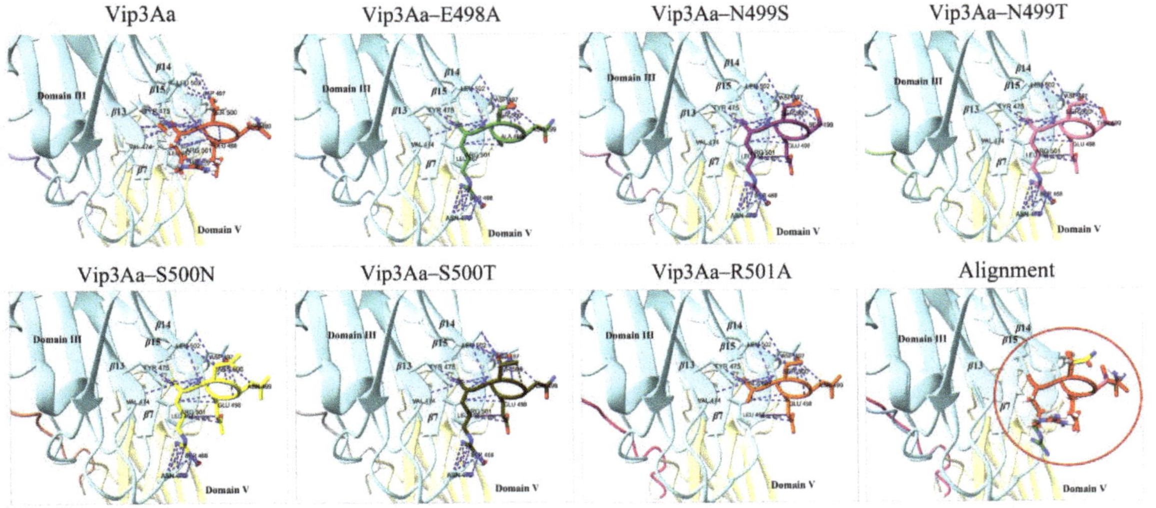

Figure 7. Contact analysis between residues in Vip3Aa (or mutants in the β14–β15 loop) by Chimera X. Alignment indicates the result of the comparison of the β14–β15 loop in the structure of the Vip3Aa protein with that of the mutant proteins. The β14–β15 loop in Vip3Aa and its mutants were labeled with different colors. Blue: domain III; Yellow: domain V; Red circle: β14–β15 loop.

3. Discussion

Vip3Aa, a soluble protein secreted by *B. thuringiensis*, can effectively control lepidopteran pests such as *S. frugiperda*. However, the lack of relevant research on the structure-function relationship of the Vip3Aa protein greatly hinders its application in pest control. The receptor binding function of domain III has been demonstrated but the specific regions involved in binding are still unclear. Therefore, we selected four specific loops (β4–β5 loop, β9–β10 loop, β12–β13 loop, and β14–β15 loop) in domain III as our research targets and constructed various mutants to investigate their insecticidal activity, proteolytic activity, and receptor binding ability. Additionally, we analyzed the impact of these loop mutations on the conformation of the Vip3Aa protein in an attempt to correlate structural features with insecticidal activity.

The correct folding is crucial for the Vip3Aa protein to exhibit insecticidal activity [8,21]. When the residues in the β9–β10 loop were all replaced with alanine, the mutant failed to form soluble protein, suggesting that the β9–β10 loop may be involved in the proper folding of Vip3Aa. To further analyze the effect of alanine substitution on Vip3Aa–loop9–10A, we predicted the 3D structure of Vip3Aa–loop9–10A using the Phyre2 server. The residues ($K^{429}KMKTL^{434}$) in the β9–β10 loop have 156 contacts with other residues but after replacing the residues in the β9–β10 loop with alanine, only 47 contacts are made (Figure S4). Additionally, there is a change in the size of the loop region before and after the mutation. In the structure of the Vip3Aa protein, most of the residues interacting with residues in the β9–β10 loop are located in domain III and some are located in domain V (Figure S4A).

The results of treating the mutants with trypsin or *S. frugiperda* midgut juice indicated that the stability of the other three mutants (Vip3Aa–loop4–5A, Vip3Aa–loop12–13A, and Vip3Aa–loop14–15A) was also affected but the impact on Vip3Aa–loop4–5A and Vip3Aa–loop14–15A was relatively minor compared to that on Vip3Aa–loop12–13A (Figure 2). This may be related to the larger regional span of the β12–β13 loop (Figure 1A). Even so, the percentage of 66 kDa fragments in Vip3Aa–loop12–13A was still higher than that of 75% of the activated fragments in Vip3Aa (Figure 2E,F). The affinity of Vip3Aa–loop4–5A, Vip3Aa–loop14–15A, or Vip3Aa–loop4–5A&14–15A for *S. frugiperda* BBMVs was significantly lower than that of Vip3Aa (Figure 3). Therefore, we speculate that the reason for the significant decrease in insecticidal activity of Vip3Aa–loop4–5A or Vip3Aa–loop14–15A is mainly related to the weakened affinity of the mutants for *S. frugiperda* BBMVs.

Currently, there is a lack of studies on the function of loops in the Vip3Aa protein. However, there are more reports on the roles of loops in Cry proteins. Loop 1 in domain II of the Cry2A toxin is involved in receptor recognition [22]. The Cry1Aa protein domain II loops contain binding sites for two functional receptors in the midgut of *Bombyx mori* and these binding sites overlap with each other [23]. Domain II loop 3 of Cry1Ab toxin is involved in a "ping pong" binding mechanism with *Manduca sexta* aminopeptidase–N and cadherin receptors [16]. Loops α–8 and 2 in domain II of Cry1Ab toxin interact with the Bt–R1 receptor in the midgut of *Manduca sexta* [24]. In this study, we found that Vip3Aa toxicity may be enhanced when N633 interacts more strongly with the residue at position 366 than with the residue at position 367 (β4–β5 loop: 364–DSI–368). However, N633 is located in domain IV, so the effect of these speculative residue interactions on Vip3Aa protein activity needs to be further investigated. In addition, ensuring the respective flexibility of both the β12–β13 loop and β14–β15 loop is indispensable in the insecticidal activity of Vip3Aa. However, we did not explore the binding of these loops to the identified Vip3Aa–interacting receptors (Sf–SR–C and Sf–FGFR) in Sf9 cells. Sf–SR–C and Sf–FGFR were identified as receptors for the action of Vip3Aa in Sf9 cells and were associated with cytotoxicity [19]. However, further investigation is needed to determine whether they are the receptors for Vip3Aa in the midgut of *S. frugiperda*. Our data (Figure 5) demonstrate that the affinity between the enhanced toxic mutants and *S. frugiperda* BBMVs is increased, indicating that the β4–β5 loop and β14–β15 loop in domain III are involved in the binding of Vip3Aa to its receptors. In addition, the affinity of Vip3Aa–loop 4–5A and Vip3Aa–loop 14–15A for *S. frugiperda* BBMVs, although decreased, was not lost and thus it is possible that other regions in domain III may be involved in the binding of Vip3Aa protein to the receptors. Mutants Vip3Aa–S366T and Vip3Aa–S366L showed significantly increased insecticidal toxicity compared to Vip3Aa. In the 3D structure of Vip3Aa–S366T or Vip3Aa–S366L, it was observed that N633 (located in domain IV) had more interactions with residue 366 rather than residue 367 in comparison to other β4–β5 mutants and Vip3Aa. Mutant Vip3Aa–N470K also shows enhanced insecticidal toxicity and the 3D structure of this mutant revealed a repulsion between residue K470 and residue R501. Furthermore, the β4–β5 loop, β12–β13 loop, and β14–β15 loop are all situated on one right side of domain III. Therefore, we propose that the loops on the right side of domain III play a role in the binding process of Vip3Aa to its receptors and that an appropriate increase in the distance between these prominent loops in domain III could facilitate the binding of Vip3Aa to its receptors.

After being ingested by insects, the Vip3Aa protein is activated by the intestinal juice and then binds to specific receptors, thereby exerting its insecticidal activity [7,10,25]. Therefore, increasing the activation efficiency of the Vip3Aa protein or enhancing its affinity for the receptors in the midgut may contribute to the insecticidal activity. Elucidation of the conformational relationship of the Vip3Aa protein is necessary to efficiently obtain Vip3Aa protein mutants with enhanced insecticidal activity. The increase in proteolytic cleavage sites between domain I and domain II resulted in the generation of a mutant ($Vip3Aa^{SS193RA/197RA}$) with enhanced insecticidal activity against *S. frugiperda* [13]. Replacement of M34 in domain I with leucine increased the insecticidal activity against *S. exigua* and *S. littoralis* [12,26]. Mutants in domains IV and V showed improved structural stability and affinity for BBMVs and the insecticidal activity of these mutants against *S. frugiperda* and *Helicoverpa armigera* was also significantly improved [27]. In this study, we obtained single mutants that exhibited higher affinity for *S. frugiperda* BBMVs and showed increased insecticidal activity against *S. frugiperda* (Figure 2). However, except for Vip3Aa–S366T/N470K and Vip3Aa–S366L/N470K, the multiple mutants did not show improvement in insecticidal activity compared to the single mutants and even tended to decrease (Table 2 and Figure 4). Further investigation is needed to determine the specific reasons for this phenomenon. In addition, we also observed a curious phenomenon in our results, in which the replacement of individual residues with alanine in β4–β5 loop or β14–β15 loop largely attenuated the insecticidal toxicity of the Vip3Aa protein (S366A, I367A, and E498A), whereas the replacement of all the residues in β4–β5 loop or β14–β15 loop with alanine only reduced the toxicity to a certain extent (about 2–4 –fold). To explore the possible reasons for this phenomenon, we analyzed the contacts of residues in Vip3Aa and Vip3Aa mutants (Vip3Aa–loop 4–5A, Vip3Aa–S366A, Vip3Aa–I367A, Vip3Aa–loop 14–15A, and Vip3Aa–E498A) using Chimera X. As shown in Figure S5A, mutants Vip3Aa–S366A and Vip3Aa–I367A differ from Vip3Aa as well as Vip3Aa–loop 4–5A by increasing the contact of residue Asp365 with residue Thr631 (Figure S5B). However, as to whether these contacts are detrimental to the toxicity of Vip3Aa needs to be further investigated, since we can also observe the contacts of residue Asp365 with residue Thr631 in the mutant Vip3Aa–S366L.

4. Conclusions

In summary, we investigated the potential roles of four specific loops in domain III in the insecticidal activity of the Vip3Aa protein. The β9–β10 loop may be involved in the proper folding of Vip3Aa, while the other three loops (β4–β5 loop, β12–β13 loop, and β14–β15 loop) may be involved in the stability and insecticidal activity of Vip3Aa. Additionally, the β4–β5 loop and β14–β15 loop are involved in the receptor binding and insecticidal activity of Vip3Aa. Furthermore, maintaining the flexibility of the β14–β15 loop or appropriately increasing the spatial distance between the β14–β15 loop and β12–β13 loop also facilitates the enhancement of the Vip3Aa toxicity. Our findings are important for optimizing the insecticidal activity of Vip3Aa protein and shed light on the role of domain III in its functionality.

5. Materials and Methods

5.1. Plasmid Construction

The plasmid pET28a–Vip3Aa, which carries Vip3Aa11 (NCBI accession No. AAR36859), was used as the template to generate the expression plasmids for Vip3Aa mutants. The Vip3Aa protein structure (PDB: 6TFJ) reported by Núñez–Ramírez et al. was used as a reference for selecting amino acid mutation sites [7]. Two kinds of mutant vectors (loop replacement mutations and point mutations) were constructed in this study. Take the β4–β5 loop of Domain III in Vip3Aa as an example to expound the plasmid construction of loop replacement mutations. In brief, the loop sequence substitution (mutation of residues 364–DSI–368 to 364–AAA–368) was obtained by PCR amplification using the primers loop 4–5A–F/loop 4–5A–R. T4 ligase (TransGen Biotech, Beijing, China) was then used to ligate the phosphorylated PCR products to obtain the β4–β5 loop replacement mutant plasmid pET28a–Vip3Aa–loop 4–5A. All point mutation vectors were obtained using the Fast MultiSite Mutagenesis System (TransGen Biotech, Beijing, China). All primers used for plasmid construction in this study are listed in Table S1.

5.2. Protein Expression and Purification

Protein expression and purification were performed using the previously described method [28]. Briefly, *Escherichia coli* BL21 (DE3) cells harboring the mutant vectors were grown in LB broth medium supplemented with 50 µg kanamycin/mL at 37 °C with shaking until the OD600 reached 0.8–1.0. Then, the cell cultures were treated with 0.5 mM isopropyl–β–D–1–thiogalactopyranoside (IPTG) at 16 °C for 16–20 h. The target proteins were released from the cells by sonication and further purified with a Ni Sepharose™ affinity column. The proteins were dialyzed in a buffer containing 25 mM Tris–HCl (pH 7.4) and 150 mM NaCl at 4 °C. The protein concentration was measured by the BCA Protein Quantitation Kit (Solarbio Science and Technology, Beijing, China).

5.3. Bioassay

S. frugiperda eggs and feed were purchased from Keyun Biological Pesticide (Zhengzhou, China). The toxicity of Vip3Aa mutant proteins to *S. frugiperda* neonate was tested by feeding with various concentrations (50, 100, 150, 300, 500, 1000, and 1500 ng/g) in a rearing chamber under a 16:8 h dark/light photocycle at 28 °C, with 50% relative humidity. The following are details about the major steps of the bioassay. First, the diet containing the toxin was divided into 24–well plates (~1 g per well) and the plates were placed at room temperature for approximately 2–3 h to dry the diet completely. Then, one *S. frugiperda* neonate larva was placed per well and the plates with larvae were cultured in the rearing chamber for seven days. Finally, the results of at least three independent trials were used to evaluate the mortality rate at each toxin dose. GraphPad Prism v.8.0 (GraphPad, San Diego, USA) was used to calculate the lethal concentration (LC_{50}) values [27]. The toxicity of the proteins was quantified by probit analysis using SPSS Statistics (IBM, Chicago, IL, USA).

5.4. Proteolysis Assay

The preparation of midgut juice from *S. frugiperda* was carried out according to a protocol described elsewhere [29]. The protein concentration in midgut juice was determined by the BCA method. The trypsin used in this study was purchased from Solarbio Life Sciences (Beijing, China).

The purified protein (30 μg) and 10% trypsin were incubated at 37 °C for different durations (10 min, 20 min, 30 min, and 60 min). In addition, the purified proteins were treated with varying proportions (4%, 6%, 8%, and 10%) of the *S. frugiperda* midgut juice at 37 °C for 1 h. The protease inhibitor PMSF (1 mM) was used to terminate the hydrolysis reaction. Protein hydrolysis fragments were analyzed using SDS-PAGE.

5.5. ELISA Binding Assays

A saturation binding assay of protein to *S. frugiperda* BBMVs was performed as described by Yang et al. [27]. *S. frugiperda* BBMVs were prepared using the differential magnesium precipitation method [30]. The obtained BBMVs were rapidly frozen in liquid nitrogen and stored at −80 °C.

Briefly, a fixed amount of *S. frugiperda* BBMVs (1 μg) was added to each well of an ELISA plate (a 96–well plate) and incubated overnight at 4 °C. PBS buffer was used to remove the BBMVs that were not immobilized on the plate. After blocking with 2% bovine serum albumin (BSA), proteins labeled with biotin were added to the wells coated with BBMVs and incubated at 37 °C for 2 h. PBS buffer was used to remove the excess biotinylated protein. Streptavidin–horseradish peroxidase (1:20,000) was added to the wells and the plates were incubated at 37 °C for 1 h. Then, 3,3′,5,5′–tetramethylbenzidine substrate solution was added to each well and the plate was incubated for 10 min at 37 °C in the dark. Finally, the reaction was terminated by the addition of HCl (2 M) to each well. The results were measured at 450 nm using a microplate reader. The relative binding affinities were analyzed via Scatchard analysis with SigmaPlot v.14.0.

PBS buffer, BSA, streptavidin–horseradish peroxidase, and 3,3′,5,5′–tetramethylbenzidine were purchased from Solarbio Life Sciences (Beijing, China).

5.6. Vip3Aa or Mutant Subcellular Localization in Sf9 Cells

The subcellular localization of Vip3Aa or mutants in Sf9 cells was described in our previous study [28]. Briefly, Vip3Aa (or mutant proteins) with red fluorescent protein (RFP) was used to treat Sf9 cells for 6 h. After washing with PBS buffer three times, the Sf9 cells were stained with Dio dye (Beyotime, Shanghai, China) at 28 °C in the dark for 45 min. Then, the cells were imaged with a Zeiss LSM710 fluorescence microscope.

5.7. Protein Structure Modeling and Analysis

The Phyre2 server was used to predict the three–dimensional (3D) structure of the mutants [31]. Chimera X was used to analyze the contacts of residues [32].

5.8. Statistical Analysis

The significance was tested using one–way analysis of variance using Student's t–test. If the p value was ≤ 0.05, the results were considered significant. All experiments were performed with at least three biological replicates and technical replicates.

Supplementary Materials: The following supporting information can be downloaded at https://www.mdpi.com/article/10.3390/toxins16010023/s1. Table S1: Primers used in this study; Table S2: Analysis of residue interactions; Table S3. Contact residues analysis with residues in β9–β10 loop; Figure S1: Alignment of the amino acid sequences in domain III; Figure S2: The four special loops exposed on the surface of the Vip3Aa protein domain III; Figure S3: Western blot analysis of the binding between Vip3Aa or its mutants and *S. frugiperda* BBMVs; Figure S4: Interaction analysis between residues by chimera X. Figure S5: Contacts analysis between residues by chimera X. (A) Vip3Aa and β4–β5 loop mutants. (B) Vip3Aa and β14–β15 loop mutants. Green: domain III, pink: domain IV.

Author Contributions: Data curation, L.J., X.G. and X.L.; Formal analysis, S.L., G.Y. and J.Z.; Investigation, W.Z. and M.L. (Mengying Li); Methodology, M.L. (Mengjiao Li) and C.M.; Writing—original draft, X.H.; Writing—review and editing, Y.F. and J.C. All authors have read and agreed to the published version of the manuscript.

Funding: This work was supported by the National Natural Science Foundation of China (32102270 and 32272610), the Natural Science Foundation of Jiangsu Province (BK20210923), the China Postdoctoral Science Foundation (2023M731398), the National Natural Science Foundation of Jiangsu Ocean University (KQ20052), and the Postgraduate Research and Practice Innovation Program of Jiangsu Province (SJCX22_1660).

Institutional Review Board Statement: Not applicable.

Informed Consent Statement: Not applicable.

Data Availability Statement: All original research data supporting reported results can be made available upon request.

Acknowledgments: We thank the Jiangsu Key Laboratory of Marine Bioresources and Environment for the facility of the equipment used in this study.

Conflicts of Interest: The authors declare no conflict of interest.

References

1. Tay, W.T.; Meagher, R.L., Jr.; Czepak, C.; Groot, A.T. *Spodoptera frugiperda*: Ecology, Evolution, and Management Options of an Invasive Species. *Annu. Rev. Entomol.* **2023**, *68*, 299–317. [CrossRef] [PubMed]
2. Li, G.; Feng, H.; Ji, T.; Huang, J.; Tian, C. What type of Bt corn is suitable for a region with diverse lepidopteran pests: A laboratory evaluation. *GM Crops Food* **2021**, *12*, 115–124. [CrossRef] [PubMed]
3. Badran, A.H.; Guzov, V.M.; Huai, Q.; Kemp, M.M.; Vishwanath, P.; Kain, W.; Nance, A.M.; Evdokimov, A.; Moshiri, F.; Turner, K.H.; et al. Continuous evolution of *Bacillus thuringiensis* toxins overcomes insect resistance. *Nature* **2016**, *533*, 58–63. [CrossRef] [PubMed]
4. Crickmore, N.; Berry, C.; Panneerselvam, S.; Mishra, R.; Connor, T.R.; Bonning, B.C. A structure–based nomenclature for *Bacillus thuringiensis* and other bacteria–derived pesticidal proteins. *J. Invertebr. Pathol.* **2021**, *186*, 107438. [CrossRef] [PubMed]
5. Chakroun, M.; Banyuls, N.; Bel, Y.; Escriche, B.; Ferré, J. Bacterial Vegetative Insecticidal Proteins (Vip) from Entomopathogenic Bacteria. *Microbiol. Mol. Biol. Rev.* **2016**, *80*, 329–350. [CrossRef] [PubMed]
6. Bel, Y.; Banyuls, N.; Chakroun, M.; Escriche, B.; Ferré, J. Insights into the Structure of the Vip3Aa Insecticidal Protein by Protease Digestion Analysis. *Toxins* **2017**, *9*, 131. [CrossRef] [PubMed]
7. Núñez–Ramírez, R.; Huesa, J.; Bel, Y.; Ferré, J.; Casino, P.; Arias–Palomo, E. Molecular architecture and activation of the insecticidal protein Vip3Aa from *Bacillus thuringiensis*. *Nat. Commun.* **2020**, *11*, 3974. [CrossRef]
8. Quan, Y.; Ferré, J. Structural Domains of the *Bacillus thuringiensis* Vip3Af Protein Unraveled by Tryptic Digestion of Alanine Mutants. *Toxins* **2019**, *11*, 368. [CrossRef]
9. Byrne, M.J.; Iadanza, M.G.; Perez, M.A.; Maskell, D.P.; George, R.M.; Hesketh, E.L.; Beales, P.A.; Zack, M.D.; Berry, C.; Thompson, R.F. Cryo–EM structures of an insecticidal Bt toxin reveal its mechanism of action on the membrane. *Nat. Commun.* **2021**, *12*, 2791. [CrossRef]
10. Jiang, K.; Chen, Z.; Zang, Y.; Shi, Y.; Shang, C.; Jiao, X.; Cai, J.; Gao, X. Functional characterization of Vip3Aa from *Bacillus thuringiensis* reveals the contributions of specific domains to its insecticidal activity. *J. Biol. Chem.* **2023**, *299*, 103000. [CrossRef]

11. Quan, Y.; Lázaro–Berenguer, M.; Hernández–Martínez, P.; Ferré, J. Critical Domains in the Specific Binding of Radiolabeled Vip3Af Insecticidal Protein to Brush Border Membrane Vesicles from *Spodoptera* spp. and Cultured Insect Cells. *Appl. Env. Microbiol.* **2021**, *87*, e0178721. [CrossRef] [PubMed]
12. Lázaro–Berenguer, M.; Paredes–Martínez, F.; Bel, Y.; Núñez–Ramírez, R.; Arias–Palomo, E.; Casino, P.; Ferré, J. Structural and functional role of Domain I for the insecticidal activity of the Vip3Aa protein from *Bacillus thuringiensis*. *Microb. Biotechnol.* **2022**, *15*, 2607–2618. [CrossRef] [PubMed]
13. Jiang, K.; Chen, Z.; Shi, Y.; Zang, Y.; Shang, C.; Huang, X.; Zang, J.; Bai, Z.; Jiao, X.; Cai, J.; et al. A strategy to enhance the insecticidal potency of Vip3Aa by introducing additional cleavage sites to increase its proteolytic activation efficiency. *Eng. Microbiol.* **2023**, *3*, 100083. [CrossRef]
14. Li, J.D.; Carroll, J.; Ellar, D.J. Crystal structure of insecticidal delta–endotoxin from *Bacillus thuringiensis* at 2.5 A resolution. *Nature* **1991**, *353*, 815–821. [CrossRef] [PubMed]
15. Shao, E.; Lin, L.; Chen, C.; Chen, H.; Zhuang, H.; Wu, S.; Sha, L.; Guan, X.; Huang, Z. Loop replacements with gut–binding peptides in Cry1Ab domain II enhanced toxicity against the brown planthopper, *Nilaparvata lugens* (Stål). *Sci. Rep.* **2016**, *6*, 20106. [CrossRef] [PubMed]
16. Pacheco, S.; Gómez, I.; Arenas, I.; Saab–Rincon, G.; Rodríguez–Almazán, C.; Gill, S.S.; Bravo, A.; Soberón, M. Domain II loop 3 of *Bacillus thuringiensis* Cry1Ab toxin is involved in a "ping pong" binding mechanism with *Manduca sexta* aminopeptidase–N and cadherin receptors. *J. Biol. Chem.* **2009**, *284*, 32750–32757. [CrossRef]
17. Liu, J.G.; Yang, A.Z.; Shen, X.H.; Hua, B.G.; Shi, G.L. Specific binding of activated Vip3Aa10 to *Helicoverpa armigera* brush border membrane vesicles results in pore formation. *J. Invertebr. Pathol.* **2011**, *108*, 92–97. [CrossRef]
18. Chakroun, M.; Ferré, J. In vivo and in vitro binding of Vip3Aa to *Spodoptera frugiperda* midgut and characterization of binding sites by (125)I radiolabeling. *Appl. Env. Microbiol.* **2014**, *80*, 6258–6265. [CrossRef]
19. Jiang, K.; Hou, X.Y.; Tan, T.T.; Cao, Z.L.; Mei, S.Q.; Yan, B.; Chang, J.; Han, L.; Zhao, D.; Cai, J. Scavenger receptor–C acts as a receptor for *Bacillus thuringiensis* vegetative insecticidal protein Vip3Aa and mediates the internalization of Vip3Aa via endocytosis. *PLoS Pathog.* **2018**, *14*, e1007347. [CrossRef]
20. Liu, M.; Liu, R.; Luo, G.; Li, H.; Gao, J. Effects of Site–Mutations Within the 22 kDa No–Core Fragment of the Vip3Aa11 Insecticidal Toxin of *Bacillus thuringiensis*. *Curr. Microbiol.* **2017**, *74*, 655–659. [CrossRef]
21. Palma, L.; Scott, D.J.; Harris, G.; Din, S.U.; Williams, T.L.; Roberts, O.J.; Young, M.T.; Caballero, P.; Berry, C. The Vip3Ag4 Insecticidal Protoxin from *Bacillus thuringiensis* Adopts A Tetrameric Configuration That Is Maintained on Proteolysis. *Toxins* **2017**, *9*, 165. [CrossRef] [PubMed]
22. Gomez, I.; Miranda–Rios, J.; Rudiño–Piñera, E.; Oltean, D.I.; Gill, S.S.; Bravo, A.; Soberón, M. Hydropathic complementarity determines interaction of epitope (869)HITDTNNK(876) in *Manduca sexta* Bt–R(1) receptor with loop 2 of domain II of *Bacillus thuringiensis* Cry1A toxins. *J. Biol. Chem.* **2002**, *277*, 30137–30143. [CrossRef] [PubMed]
23. Adegawa, S.; Nakama, Y.; Endo, H.; Shinkawa, N.; Kikuta, S.; Sato, R. The domain II loops of *Bacillus thuringiensis* Cry1Aa form an overlapping interaction site for two *Bombyx mori* larvae functional receptors, ABC transporter C2 and cadherin–like receptor. *Biochim. Biophys. Acta Proteins Proteom.* **2017**, *1865*, 220–231. [CrossRef] [PubMed]
24. Gómez, I.; Dean, D.H.; Bravo, A.; Soberón, M. Molecular basis for *Bacillus thuringiensis* Cry1Ab toxin specificity: Two structural determinants in the *Manduca sexta* Bt–R1 receptor interact with loops alpha–8 and 2 in domain II of Cy1Ab toxin. *Biochemistry* **2003**, *42*, 10482–10489. [CrossRef] [PubMed]
25. Lee, M.K.; Walters, F.S.; Hart, H.; Palekar, N.; Chen, J.S. The mode of action of the *Bacillus thuringiensis* vegetative insecticidal protein Vip3A differs from that of Cry1Ab delta–endotoxin. *Appl. Env. Microbiol.* **2003**, *69*, 4648–4657. [CrossRef] [PubMed]
26. Banyuls, N.; Quan, Y.; González–Martínez, R.M.; Hernández–Martínez, P.; Ferré, J. Effect of substitutions of key residues on the stability and the insecticidal activity of Vip3Af from *Bacillus thuringiensis*. *J. Invertebr. Pathol.* **2021**, *186*, 107439. [CrossRef] [PubMed]
27. Yang, X.; Wang, Z.; Geng, L.; Chi, B.; Liu, R.; Li, H.; Gao, J.; Zhang, J. Vip3Aa domain IV and V mutants confer higher insecticidal activity against *Spodoptera frugiperda* and *Helicoverpa armigera*. *Pest. Manag. Sci.* **2022**, *78*, 2324–2331. [CrossRef]
28. Hou, X.; Han, L.; An, B.; Zhang, Y.; Cao, Z.; Zhan, Y.; Cai, X.; Yan, B.; Cai, J. Mitochondria and Lysosomes Participate in Vip3Aa–Induced *Spodoptera frugiperda* Sf9 Cell Apoptosis. *Toxins* **2020**, *12*, 116. [CrossRef]
29. Chakroun, M.; Bel, Y.; Caccia, S.; Abdelkefi–Mesrati, L.; Escriche, B.; Ferré, J. Susceptibility of *Spodoptera frugiperda* and *S. exigua* to *Bacillus thuringiensis* Vip3Aa insecticidal protein. *J. Invertebr. Pathol.* **2012**, *110*, 334–339. [CrossRef]
30. Wolfersberger, M.G. Preparation and partial characterization of amino acid transporting brush border membrane vesicles from the larval midgut of the gypsy moth (*Lymantria dispar*). *Arch. Insect Biochem. Physiol.* **1993**, *24*, 139–147. [CrossRef]
31. Kelley, L.A.; Sternberg, M.J. Protein structure prediction on the Web: A case study using the Phyre server. *Nat. Protoc.* **2009**, *4*, 363–371. [CrossRef]
32. Goddard, T.D.; Huang, C.C.; Meng, E.C.; Pettersen, E.F.; Couch, G.S.; Morris, J.H.; Ferrin, T.E. UCSF ChimeraX: Meeting modern challenges in visualization and analysis. *Protein Sci.* **2018**, *27*, 14–25. [CrossRef]

toxins

MDPI

Communication

Bacillus thuringiensis Bt_UNVM-84, a Novel Strain Showing Insecticidal Activity against *Anthonomus grandis* Boheman (Coleoptera: Curculionidae)

Diego Herman Sauka [1,2,†], Cecilia Peralta [1,3,4], Melisa Paula Pérez [2], Antonella Molla [3], Tadeo Fernandez-Göbel [5], Federico Ocampo [5] and Leopoldo Palma [1,3,4,*,†]

1 Consejo Nacional de Investigaciones Científicas y Técnicas (CONICET), Ciudad Autónoma de Buenos Aires C1425FQB, Argentina; sauka.diego@inta.gob.ar (D.H.S.); cecilia.peralta@uv.es (C.P.)

2 Instituto Nacional de Tecnología Agropecuaria (INTA), Instituto de Microbiología y Zoología Agrícola (IMYZA), Hurlingham, Ciudad Autónoma de Buenos Aires 1686, Argentina; perez.melisa@inta.gob.ar

3 Instituto Multidisciplinario de Investigación y Transferencia Agroalimentaria y Biotecnológica (IMITAB), Consejo Nacional de Investigaciones Científicas y Técnicas (CONICET), Universidad Nacional de Villa María (UNVM), Villa María 1555, Argentina; antonellabmolla@hotmail.com

4 Laboratorio de Control Biotecnológico de Plagas, Instituto BIOTECMED, Departamento de Genética, Universitat de València, 46100 València, Spain

5 Elytron Biotech S.A., 275 Ing. Enrique Butty Street, Ciudad Autónoma de Buenos Aires C1001, Argentina; tfernandezgobel@elytronbiotech.com (T.F.-G.); focampo@elytronbiotech.com (F.O.)

* Correspondence: leopoldo.palma@uv.es; Tel.: +34-963544297

† These authors contributed equally to this work.

Abstract: *Bacillus thuringiensis* is a Gram-positive bacterium known for its insecticidal proteins effective against various insect pests. However, limited strains and proteins target coleopteran pests like *Anthonomous grandis* Boheman, causing substantial economic losses in the cotton industry. This study focuses on characterizing a *Bacillus* sp. strain, isolated from Oncativo (Argentina), which exhibits ovoid to amorphous parasporal crystals and was designated Bt_UNVM-84. Its genome encodes genes for the production of two pairs of binary Vpb1/Vpa2 proteins and three Cry-like proteins showing similarity with different Cry8 proteins. Interestingly, this gene content was found to be conserved in a previously characterized Argentine isolate of *B. thuringiensis* designated INTA Fr7-4. SDS-PAGE analysis revealed a major band of 130 kDa that is proteolytically processed to an approximately 66-kDa protein fragment by trypsin. Bioassays performed with spore-crystal mixtures demonstrated an interesting insecticidal activity against the cotton boll weevil *A. grandis* neonate larvae, resulting in 91% mortality. Strain Bt_UNVM-84 is, therefore, an interesting candidate for the efficient biological control of this species, causing significant economic losses in the cotton industry in the Americas.

Keywords: *Bacillus thuringiensis*; insect pests; insecticidal proteins; Vpb1/Vpa2 proteins; Cry8 proteins; biological control; bioinsecticides

Key Contribution: We describe a novel *Bacillus thuringiensis* strain showing a potential set of coleoptericidal-encoded proteins and insecticidal activity against *Anthonomus grandis* Boheman (Coleoptera: Curculionidae), a harmful cotton pest causing significant losses to the cotton industry in the Americas.

Citation: Sauka, D.H.; Peralta, C.; Pérez, M.P.; Molla, A.; Fernandez-Göbel, T.; Ocampo, F.; Palma, L. *Bacillus thuringiensis* Bt_UNVM-84, a Novel Strain Showing Insecticidal Activity against *Anthonomus grandis* Boheman (Coleoptera: Curculionidae). *Toxins* **2024**, *16*, 4. https://doi.org/10.3390/toxins16010004

Received: 14 November 2023
Revised: 16 December 2023
Accepted: 18 December 2023
Published: 20 December 2023

1. Introduction

Bacillus thuringiensis is a ubiquitous Gram-positive, sporulated bacterium well-known for its ability to produce proteins with toxic activity against insect pests, human-disease vectors (mosquitoes), and nematodes [1]. Different strains of this bacterium have been successfully used for decades in sprayable products for the biological control of insect pests (e.g., *B. thuringiensis* svar. *kurstaki*), ranging from small domestic vegetable gardens to crop fields. Moreover, the genes encoding such insecticidal proteins have been introduced into crops (Bt crops), conferring specific resistance to insect pests and promoting the reduction in the use of synthetic insecticides in integrated pest management (IPM) programs, toward sustainable agricultural practices [2]. Synthetic insecticides not only pollute the environment but also pose harm to humans, animals, and other non-target insects such as natural enemies of pests and pollinators [3].

B. thuringiensis produces its insecticidal proteins during both the vegetative growth (vegetative insecticidal proteins) and the stationary growth phases (delta-endotoxins) [4]. The vegetative insecticidal proteins include Vpb1/Vpa2 (formerly Vip1/Vip2) with activity against coleopterans, Vip (formerly Vip3) with activity against lepidopterans [5], and Vpb4 (formerly Vip4) with activity against *Diabrotica virgifera virgifera* Chevrolat (Coleoptera: Chrysomelidae) [6]. The delta-endotoxins include Cry (crystal) proteins with activity against several orders of insects and some nematodes, plus Cyt (cytolytic) proteins with activity against dipterans (mosquitoes and black flies) [7]. In addition, *B. thuringiensis* can produce Thuringiensin, also known as β-exotoxin, a non-proteinic, thermostable, and secretable secondary metabolite showing toxicity against a wide range of insects and some nematodes. Nowadays, Thuringiensin is considered an adenine nucleoside oligosaccharide analog that may interfere with RNA synthesis [8], and, as such, it has been banned from public use due to its potential toxicity against mammals and its high persistence in the environment [9].

However, only a few *B. thuringiensis* proteins have been described for their activity against coleopteran pests, providing a limited number of options for the biological control of these insects [10]. Therefore, finding novel strains showing coleoptericidal activity is paramount to achieving efficient crop protection against coleopteran herbivorous insects. A notable example is the cotton boll weevil, *Anthonomus grandis* Boheman (Coleoptera: Curculionidae), which is considered the most damaging pest, causing substantial economic losses to the cotton industry in the Americas [11]. Its behavior, protecting larvae from insecticides, is significant, since fertilized females lay eggs (around 200 per female) on cotton flower buds, where the main damage occurs during egg laying and larval feeding [12]. In addition, some populations in Brazil have shown resistance to certain insecticides, such as beta-cyfluthrin [13].

In this work, we report the molecular and insecticidal characterization of a novel *Bacillus* sp. strain isolated from Oncativo (Argentina), showing ovoid to amorphous parasporal crystals, designated as Bt_UNVM-84 strain. Its genome sequence harbors genes for the production of proteins with potential insecticidal activity against Coleoptera such as binary Vpb1/Vpa2 homolog proteins plus proteins with potential dual activity against Coleoptera and Lepidoptera, including three Cry8-like homolog proteins. Bioassays performed with spore-crystal mixtures demonstrated insecticidal activity against *A. grandis*, resulting in 91% mortality. Interestingly, a previously characterized *B. thuringiensis* strain designated INTA Fr7-4 exhibited a conserved insecticidal-gene content with a lower insecticidal activity against *A. grandis* [14]. The results provided by this study demonstrate that strain Bt_UNVM-84 emerges as a compelling candidate for the biological control of this insect species, which is causing significant economic losses in the cotton industry in the Americas.

2. Results

2.1. Strain Isolation and Identification

The new *Bacillus* sp. isolate exhibited typical *B. thuringiensis* morphology in the bacterial colony (flat, dry, matt-white color with uneven borders). Under the light microscope, it also showed Coomassie-blue-stained parasporal crystals with an ovoid to amorphous shape (Figure 1A), later confirmed by SEM examination (Figure 1B).

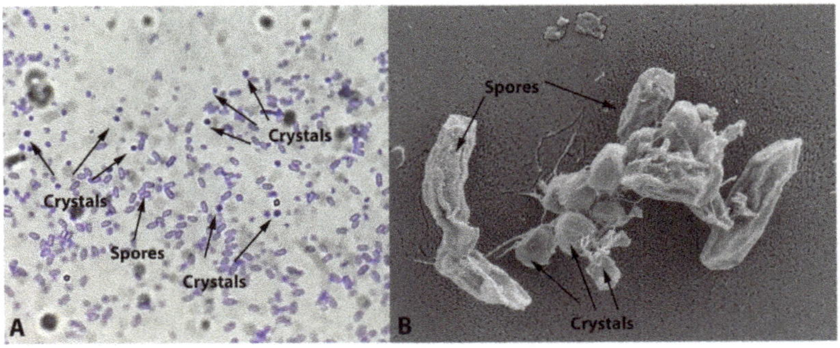

Figure 1. Microscopic analysis of the isolate. (**A**) Ovoid to amorphous parasporal crystals stained with Coomassie blue stain [15] and (**B**) parasporal crystals' examination using SEM microscope.

The screening of *cry* genes by PCR produced an amplicon of approximately 1.5 kb, slightly larger than the amplicon produced by *B. thuringiensis* svar. *kurstaki* HD-1 strain used as positive control (Figure 2A). SDS-PAGE analysis showed a main band of approximately 130 kDa, comparable to the size of the band from the control HD-1 strain. In addition, the band was digested by trypsin, exhibiting proteolytically digested products and a smaller band that may correspond to an activated Cry protein of approximately 66 kDa (Figure 2B). Considering these findings, the isolate was preliminary designated as *B. thuringiensis* strain Bt_UNVM-84.

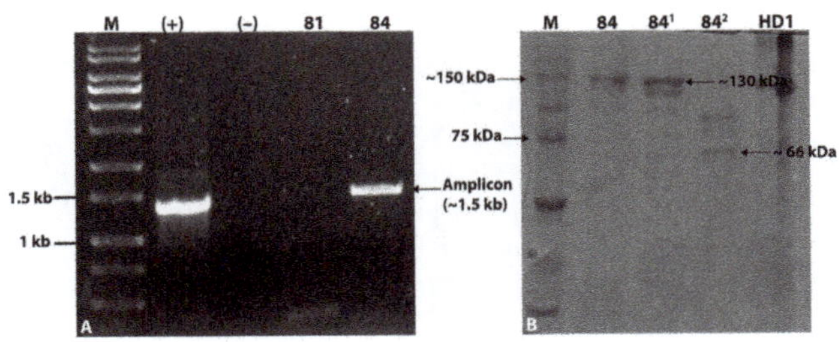

Figure 2. Detection of *cry* genes and proteins from the parasporal crystals. (**A**) Detection of *cry* genes by PCR with degenerate primers [16]. Amplicons were electrophoresed in 1% agarose gel: M molecular weight marker 1 kb, (+) *B. thuringiensis* strain HD1 control, (−) negative control with water; 81 is a negative control strain whereas 84 is the new isolate and (**B**) SDS-PAGE analysis: M molecular weight marker (Precision Plus Proteins Dual Color), 84 dried Bt_UNVM-84 biomass, 84^1 solubilized Bt_UNVM-84 biomass, and 84^2 solubilized Bt_UNVM-84 biomass digested (potentially activated) with the enzyme trypsin.

2.2. Genome Sequencing and Annotation

Genome sequencing produced 12,299,332 million Illumina pair-end (raw) reads, which were trimmed and assembled into 81 contig sequences, resulting in a genome size of 6,081,079 bp, with a G+C content of 34.7% with 6249 predicted protein-coding genes (CDS) and 103 RNAs. These values were consistent in both size and G + C% with other sequenced *B. thuringiensis* genomes [17,18].

Phylogenetic analysis using the Type Strain Genome Server (TYGS) showed that the new isolate branched into a unique cluster along with type strains *Bacillus cereus* ATCC 14579 and *B. thuringiensis* ATCC 10792 (Figure 3).

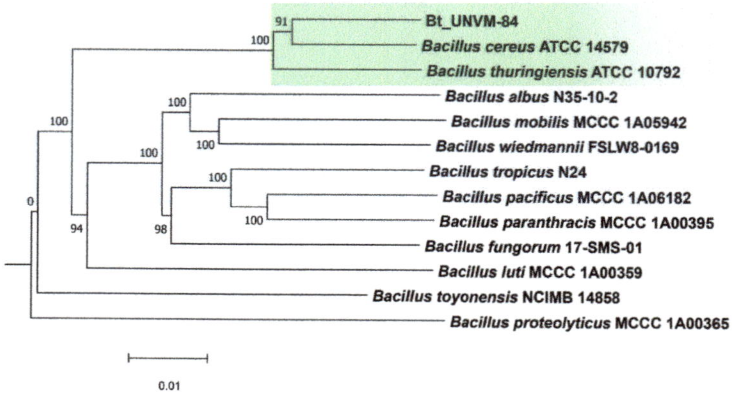

Figure 3. GBDP tree (whole-genome sequence-based) using TYGS server (average branch support 98.5%) [19]. Green color was used to highlight Bt_UNVM-84 strain clustered along with type strains.

In addition, mapped reads over the draft genome sequence of the related *B. thuringiensis* strain INTA Fr7-4 (Acc. no. MSFC00000000) showed 98.5% pairwise nucleotide identity covering 92.2% of the reference genomic sequence.

The draft genome sequence from the Bt-UNVM_84 was also searched for putative insecticidal proteins and other virulence factors that may have a role in insect pathogenesis. The genome harbors seven CDSs showing significant BlastX [20] similarity with Vpa1/Vpa2 proteins and Cry8 crystal proteins.

The repertoire of insecticidal CDSs was found to be highly conserved between the INTA Fr7-4 and Bt_UNVM-84 strains. The seven homologous genes were located in the pFR260 (Acc. no. KX258624) megaplasmid harbored by strain INTA Fr7-4 strain [21]. This megaplasmid, 260,595 bp in size, encodes Cry8Qa2, Cry8Kb3, and Cry8Pa3 proteins, along with two pairs of binary proteins, namely, Vpb1Ea1/Vpa2Ah1 and Vpb1Ea2/VpaAh2. Mapping analysis using Bt_UNVM-84 Illumina reads on the pFR260 plasmid sequence showed 96.1% pairwise identity, covering 88.1% of the plasmid (used as the reference sequence). The encoded CDSs in the plasmid pFR260 showed more than 97% pairwise nucleotide identity with mapped reads from the Bt_UNVM-84 strain (Table 1).

Table 1. Insecticidal CDSs comparison by mapping Bt_UNVM-84 reads over pFR260 megaplasmid.

CDSs [a]	% Pairwise Nucleotide Identity	Contig	% Ref-Seq Coverage	Gene Length (bp)
vpa2Ah1	98.8	NCA	93.9	1338
vpb1Ea1	99.0	52	99.7	2625
vpa2Ah2	98.9	61	100	1338
vpb1Ea2	98.7	61	100	2619
cry8Kb3	96.3	61	42.5	3510
cry8Pa3	97.7	39	52.8	3531
cry8Qa2	97.7	75	98.7	3555

[a] Closest homolog. NCA = no contig assigned, detected by read-mapping over plasmid pFR206 [21].

The Bt_UNVM-84 genome also harbors three CDSs showing 36%, 34%, and 37% pairwise amino similarity to the proteins Mpp4Aa1, Mpp46Ab1, and Xpp22Ba1, respectively. Additionally, strain INTA Fr7-4 also harbors CDS coding for Mpp4Aa1 and Xpp22Ba1, with pairwise amino similarities of 38% and 36%, respectively, but lacks the CDS encoding the Mpp46Ab1 homolog. Strain Bt_UNVM-84 also exhibited other CDSs encoding two putative chitinases and three chitin-binding proteins, whereas strain INTA Fr7-4 harbors two putative chitinases along with five chitin-binding proteins. Furthermore, the gene *thuE*, involved in the thuringiensin synthesis pathway [8,9], was not detected either by RAST server, custom BlastX analyses, or PCR amplification, following the methodology described by Sauka et al. [9].

2.3. Insect Bioassays

Mixed spore-crystal suspensions of the strain exhibited insecticidal activity against *A. grandis* and no toxicity against *Alphitobius diaperinus* and *Cydia pomonella* (Table 2).

Table 2. Bioassays with insects were conducted with the whole strain (spore-crystal mixtures).

Species	% Average Mortality ± SD	% Corrected Mortality ± SD
C. pomonella	25.0 ± 5.9 [a]	5.3 ± 7.5
A. diaperinus	21.7 ± 12.3 [b]	7.2 ± 14.6
A. grandis	91.7 ± 5.9 [b]	91.1 ± 6.3

[a] 5 μg/mL diet. [b] 1 mg/mL diet.

In consistency with both the lack of the PCR amplification of the *thuE* gene and the absence of its coding sequence in the Bt_UNVM-84 genome, no teratological effects were detected during the pupal emergence in adults of *Musca domestica* Linnaeus (Diptera: Muscidae).

3. Discussion

B. thuringiensis is the most used bacterium for controlling invertebrate pests, either through the use of spray formulations (e.g., Dipel, Xentary, etc.) or by expressing its insecticidal proteins in transgenic crops [22]. However, its effectiveness has been compromised, as some insect populations have evolved to become resistant through different mechanisms [23] to both spray formulations [24] and the most commonly used *B. thuringiensis* proteins (e.g., Cry1Ac) [25]. In addition to these problems, only a few types of insecticidal proteins from *B. thuringiensis* have been found to be effective against coleopteran pests [10,26], including those capable of controlling the cotton boll weevil *A. grandis*. This species is one of the most important pests, causing significant economic losses in the cotton industry in the Americas [26]. While chemical insecticides are efficient in controlling this pest in cotton, they are harmful to non-target organism, polluting the environment and increasing farmers' expenses during the growing seasons [26]. In addition, the life cycle and behavior of this insect may limit its contact with synthetic insecticides, thereby enhancing survival and causing damage to cotton plants in cultivated areas. For these reasons, finding novel genes with insecticidal activity against coleopteran pests is crucial for improving biological control strategies in integrated pest management programs through the construction of genetically modified cotton with insect resistance against *A. grandis*.

Here, we report the molecular and insecticidal characterization of a novel *Bacillus* sp. strain isolated from a soil sample obtained at Oncativo, Córdoba (Argentina). This strain exhibited ovoid to amorphous parasporal crystals, briefly described in a previous work [27]. These crystal shapes bear resemblance to those produced by strain INTA Fr7-4 and other strains showing insecticidal activity against *A. grandis* [14]. Moreover, similar parasporal-crystal shapes were identified through the cloning and expression of the Cry8Qa2 gene from strain INTA Fr7-4 into an acrystalliferous *B. thuringiensis* strain [28]. However, spore-crystal mixtures from this recombinant strain were not tested against *A. grandis*, and the mortality rate against *Anticarsia gemmatalis* Hübner (Lepidoptera: Noctuidae) was only 13.8% [28].

The genome sequence from strain Bt-UNVM-84 harbors seven insecticidal-like CDSs, showing similarity to well-known insecticidal proteins, including two pairs of binary Vpb1/Vpa2 proteins plus three Cry proteins showing similarity to Cry8 proteins (Table 1). Analysis by reads mapping has shown that the insecticidal CDSs are highly conserved with those already described in plasmid pFR206 [21]. Strain Bt_UNVM-84 also encodes CDSs showing similarity to the Mpp4Aa1, Mpp46Ab1, and Xpp22Ba1 proteins, with the last one absent in strain INTA Fr7-4, which has been described to exhibit insecticidal activity against *A. grandis* and *D. virgifera virgifera* [29]. Furthermore, Bt_UNVM-84 lacks the genes for toxins Cry1Ba and the binary Mpp23Aa/Xpp37Aa, which are highly active against *A.*

grandis [26]. Although the novel strain was more closely related to *B. cereus* ATCC 14579 strain, *B. thuringiensis* is only differentiated from *B. cereus* by the production of typical parasporal crystal proteins [30] and was, therefore, designated here as *Bacillus thuringiensis* strain Bt_UNVM-84 instead of *B. cereus* sensu stricto biovar Thuringiensis by following the taxonomic nomenclature proposal for the *Bacillus cereus* group [31].

Insect bioassays using spore-crystal mixtures showed an interesting insecticidal activity, resulting in a 91% mean mortality for *A. grandis*. This suggests that the parasporal crystals may contain an active protein or proteins to control this pest. However, no toxicity has been observed against *C. pomonella* and *A. diaperinus*. The closely related INTA Fr7-4 strain has also been described to exhibit insecticidal activity against *A. grandis*, showing a lower percentage of mean mortality (32.5%). We hypothesize that this difference could be due to either the minor differences found in the coding nucleotide sequences (Table 1), which may be producing more active proteins, or the more efficient expression of some of them in the Bt_UNVM-84 strain. The results of this work suggest, at least at this stage, an exclusive coleoptericidal activity of strain Bt_UNVM-84. The identification of proteins in the crystals using liquid chromatography coupled with mass spectrometry should shed light on crystal composition and their role in the demonstrated activity against *A. grandis*. In addition, the absence of *thuE* coding sequences, confirmed by reads mapping, PCR amplification, and bioassays with *M. domestica*, suggests that this strain should be tested for the development of a sprayable formulation against other coleopteran pests. However, more experiments are necessary to elucidate which encoded gene or genes are responsible for the toxic activity against *A. grandis* and to unravel the full insecticidal potential of this novel and interesting *B. thuringiensis* strain.

4. Conclusions

A novel *Bacillus* sp. strain showing insecticidal activity against *A. grandis* was isolated from Oncativo, Argentina, designated as *B. thuringiensis* strain Bt_UNVM-84 and characterized at the molecular and insecticidal levels. The genomic sequence exhibited a set of genes showing significant similarity with several insecticidal proteins active against coleopteran and lepidopteran pests. Strain INTA Fr7-4 was closely related to strains showing the same insecticidal-gene configuration but lower insecticidal activity against *A. grandis*. Although more studies are necessary to describe the full insecticidal potential of the strain and its encoded proteins, the results obtained in this work indicate that strain Bt_UNVM-84 is an interesting candidate to provide novel tools for the biological control of, at least, the cotton boll weevil *A. grandis*.

5. Materials and Methods

5.1. Strain Isolation and Characterization

Soil samples were collected using a soil sampling tube from Establecimiento Norma Lucía farm in Oncativo, Córdoba province (Argentina), where *Medicago sativa* L. was planted. The final sample was a composite of five random sub-samples, totaling approximately 200 g of soil. This sample was stored at 4 °C in zip-lock bags until processed for bacteria isolation. Bacterial colonies, originating from sporulated species, were obtained as previously described by Iriarte et al. (1998) [32]. Colonies with a flat, dry, matt-white color and uneven borders (*B. thuringiensis*-like phenotype) were analyzed through PCR using degenerate primers (forward 5′-TATGCWCAAGCWGCCAATYTWCATYT-3′ and reverse 5′-GGRATAAATTCAATTYKRTCWA-3′) for the detection of three-domain *cry* genes [16]. In order to detect type I β-exotoxin production from strain Bt_UNVM-84, a qualitative PCR-based method for the detection of the *thuE* gene was performed as described previously by Sauka et al. [9] with the following forward 5′-GCGGCAGCCGTTTATTCAAA-3′ and reverse 5′-CCCCTTCCCATGGAGAAACA-3′ BEF primers, which produce 406-bp amplicons. The presence of parasporal crystals was examined using a light microscope, following the methodology described by Ammons et al. (2002) [15]. The production of parasporal crystals and their morphology were later confirmed by Scanning Electron Microscopy

(SEM) at the Comprehensive Center for Electron Microscopy (CIME-CONICET, Argentina). The purified (axenic) sporulated colony was then stored at our bacterial collection in 15% glycerol at −80 °C. The composition of parasporal crystals and trypsin activation was determined by SDS-PAGE, following the procedure described by Pérez et al. [4], with *B. thuringiensis* svar. *kurstaki* HD-1strain used as reference.

5.2. Genome Sequencing and Annotation

Total DNA containing chromosome and plasmids was obtained using the Wizard genomic DNA purification kit (Promega, Madison, WI, USA), following the manufacturer's instructions for the purification of DNA from Gram-positive bacteria. DNA was electrophoresed in 1% agarose gels stained with SYBR Safe (ThermoFisher Scientific, Waltham, MA, USA) and quantified using a PICODROP PICO 100 µL spectrophotometer. The purified DNA was then utilized to construct a pooled Illumina library, which was sequenced at the Genomics Unit from the National Institute of Agricultural Technology (INTA, Argentina) by using high-throughput Illumina sequencing technology.

The obtained Illumina (raw) reads were analyzed, and trimmed regions were deleted before being assembled into contigs using Velvet plug in Geneious version R11 software suite (www.geneious.com, accessed on 10 November 2023), with the de novo assembly tool and default parameters. The resulting contigs were then analyzed with BLAST [20] using a customized personal non-redundant insecticidal protein database (2023 update). Genome annotation was initially performed with the NCBI Prokaryotic Genome Annotation Pipeline (2023 release) and the RAST server [33]. Species delimitation was performed with the Type (Strain) Genome Server [19]. The analysis of % pairwise identity by reads mapping over pFR260 plasmid and the genome of strain INTA Fr7-4 was performed using the *map to reference* tool embodied in Geneious R11 software.

5.3. Insect Bioassays

The insecticidal activity of the strain was qualitatively evaluated against three insect species, including second instars of *Alphitobius diaperinus* Panzer (Coleoptera: Tenebrionidae), neonates of *Anthonomus grandis* Boheman (Coleoptera: Curculionidae), and neonates of *Cydia pomonella* Linnaeus (Lepidoptera: Tortricidae). Coleopteran and lepidopteran larvae were obtained from colonies reared at the Institute of Microbiology and Agricultural Zoology (IMYZA-INTA, Argentina).

For insect bioassays, strain Bt_UNVM-84 was grown in 100 mL of BM medium (2.5 g NaCl, 1 g KH_2PO_4, 2.5 g K_2HPO_4, 0.25 g $MgSO_4{\cdot}7H_2O$, 0.1 g $MnSO_4{\cdot}H_2O$, 5 g glucose, 2.5 g starch and 4 g yeast extract for 1 litter and with the pH adjusted to 7.2) at 340 rpm and 30 °C during approximately 72 h until complete autolysis was observed under the microscope. Spore-crystal mixtures were obtained by centrifugation at 12,000× *g* and 4 °C for 15 min; then, the pellets were freeze-dried, and the resultant powder composed of spore and crystals was kept at −20 °C until use. Each spore crystal mixture (final concentration of 5–1000 µg/mL) was incorporated into polypropylene conical tubes containing the corresponding artificial diet for each species (maintained at 60 °C) and poured into each well of a 24-well plate (Nunc 143982) [14]. Twenty-four coleopteran and lepidopteran larvae were used, and the bioassays were repeated twice. Mortality was recorded after 15 days at 29 °C for *A. diaperinus* and *A. grandis*, whereas 5 days were spent for *C. pomonella* at 29 °C. Larvae were accounted as dead if they did not respond to gentle probing. Distilled (sterile) water was used instead of spore crystal mixtures as mortality controls. Schneider-Orelli's formula was used to calculate corrected mortality compared to the untreated control. The InfoStat software (Universidad Nacional de Córdoba, version 2014 was used for the statistical analysis, and the statistical significance was set at $p < 0.05$.

Strain Bt_UNVM-84 was also analyzed for its capability to synthesize type I β-exotoxin by counting the number of emerged *M. domestica* adults, following the methodology described by Sauka et al. [9] and using the β-exotoxin producer strain HD-2 as positive control.

Author Contributions: Conceptualization, L.P. and D.H.S.; methodology, L.P., C.P., M.P.P. and D.H.S.; validation, L.P. and D.H.S.; formal analysis, L.P. and D.H.S.; investigation, L.P., C.P., M.P.P., A.M. and D.H.S.; resources, L.P. and D.H.S.; data curation, L.P. and D.H.S.; writing—original draft preparation, L.P.; writing—review and editing, L.P., D.H.S., T.F.-G. and F.O.; visualization, L.P.; supervision, L.P. and D.H.S.; project administration, L.P.; funding acquisition, L.P. All authors have read and agreed to the published version of the manuscript.

Funding: This research was funded by FONCYT, Plan Argentina Innovadora 2020, grant number PICT 2017-0087, by the Consejo Nacional de Investigaciones Científicas y Técnicas (CONICET), grant number PIP 2017 GRUPO INVES, and by Universidad Nacional de Villa María-Instituto de Investigación, grant number PIC 2017-2019, as well as, by a Research I+D Agreement EX-2021-117239788-APN-GVT#CONICET between CONICET and Elytron Biotech S.A. (www.elytronbiotech.com, accessed on 14 November 2023).

Institutional Review Board Statement: Not applicable.

Informed Consent Statement: Not applicable.

Data Availability Statement: This whole-genome shotgun project has been deposited in DDBJ/ENA/GenBank under the accession number JAWUAH000000000. The version described in this paper is the first version, JAWUAH010000000. Illumina paired-end (raw) reads were deposited at Sequence Read Archive (SRA) database under accession PRJNA1035773.

Acknowledgments: We thank Hernán Esquivel and Luciano Martínez from CIME-CONICET (Argentina) and Andrea Puebla from Unidad de Genómica at INTA Castelar (Argentina) for their technical assistance in this work.

Conflicts of Interest: Authors Tadeo Fernandez-Göbel and Federico Ocampo were employed by the company Elytron Biotech S.A. The remaining authors declare that the research was conducted in the absence of any commercial or financial relationships that could be construed as a potential conflict of interest.

References

1. Melo, A.L.d.A.; Soccol, V.T.; Soccol, C.R. *Bacillus thuringiensis*: Mechanism of action, resistance, and new applications: A review. *Crit. Rev. Biotechnol.* **2016**, *36*, 317–326. [CrossRef] [PubMed]
2. Kumar, S.; Chandra, A.; Pandey, K.C. *Bacillus thuringiensis* (Bt) transgenic crop: An environment friendly insect-pest management strategy. *J. Environ. Biol.* **2008**, *29*, 641–653. [PubMed]
3. Pathak, V.M.; Verma, V.K.; Rawat, B.S.; Kaur, B.; Babu, N.; Sharma, A.; Dewali, S.; Yadav, M.; Kumari, R.; Singh, S.; et al. Current status of pesticide effects on environment, human health and it's eco-friendly management as bioremediation: A comprehensive review. *Front. Microbiol.* **2022**, *13*, 962619. [CrossRef] [PubMed]
4. Crickmore, N.; Berry, C.; Panneerselvam, S.; Mishra, R.; Connor, T.R.; Bonning, B.C. A structure-based nomenclature for *Bacillus thuringiensis* and other bacteria-derived pesticidal proteins. *J. Invertebr. Pathol.* **2021**, *186*, 107438. [CrossRef]
5. Chakroun, M.; Banyuls, N.; Bel, Y.; Escriche, B.; Ferré, J. Bacterial Vegetative Insecticidal Proteins (Vip) from Entomopathogenic Bacteria. *Microbiol. Mol. Biol. Rev.* **2016**, *80*, 329–350. [CrossRef] [PubMed]
6. Yin, Y.; Flasinski, S.; Moar, W.; Bowen, D.; Chay, C.; Milligan, J.; Kouadio, J.L.; Pan, A.; Werner, B.; Buckman, K.; et al. A new *Bacillus thuringiensis* protein for Western corn rootworm control. *PLoS ONE* **2020**, *15*, e0242791. [CrossRef] [PubMed]
7. Schnepf, E.; Crickmore, N.; Van Rie, J.; Lereclus, D.; Baum, J.; Feitelson, J.; Zeigler, D.R.; Dean, D.H. *Bacillus thuringiensis* and Its Pesticidal Crystal Proteins. *Microbiol. Mol. Biol. Rev.* **1998**, *62*, 775–806. [CrossRef]
8. Liu, X.; Ruan, L.; Peng, D.; Li, L.; Sun, M.; Yu, Z. Thuringiensin: A thermostable secondary metabolite from *Bacillus thuringiensis* with insecticidal activity against a wide range of insects. *Toxins* **2014**, *6*, 2229–2238. [CrossRef]
9. Sauka, D.H.; Pérez, M.P.; López, N.N.; Onco, M.I.; Berretta, M.F.; Benintende, G.B. PCR-based prediction of type I β-exotoxin production in *Bacillus thuringiensis* strains. *J. Invertebr. Pathol.* **2014**, *122*, 28–31. [CrossRef]
10. Dominguez-Arrizabalaga, M.; Villanueva, M.; Fernandez, A.B.; Caballero, P. A Strain of *Bacillus thuringiensis* Containing a Novel *cry7Aa2* Gene that Is Toxic to *Leptinotarsa decemlineata* (Say) (Coleoptera: Chrysomelidae). *Insects* **2019**, *10*, 259. [CrossRef]
11. Sánchez-Reyes, U.J.; Jones, R.W.; Raszick, T.J.; Ruiz-Arce, R.; Sword, G.A. Potential Distribution of Wild Host Plants of the Boll Weevil (*Anthonomus grandis*) in the United States and Mexico. *Insects* **2022**, *13*, 337. [CrossRef] [PubMed]
12. Health, E.P.o.P.; Jeger, M.; Bragard, C.; Caffier, D.; Candresse, T.; Chatzivassiliou, E.; Dehnen-Schmutz, K.; Gilioli, G.; Gregoire, J.C.; Jaques Miret, J.A. Pest categorisation of *Anthonomus grandis*. *EFSA J.* **2017**, *15*, e05074.
13. Rolim, G.G.; Coelho, R.R.; Antonino, J.D.; Arruda, L.S.; Rodrigues, A.S.; Barros, E.M.; Torres, J.B. Field-evolved resistance to beta-cyfluthrin in the boll weevil: Detection and characterization. *Pest. Manag. Sci.* **2021**, *77*, 4400–4410. [CrossRef] [PubMed]

14. Pérez, M.P.; Sauka, D.H.; Onco, M.I.; Berretta, M.F.; Benintende, G.B. Selection of *Bacillus thuringiensis* strains toxic to cotton boll weevil (*Anthonomus grandis*, Coleoptera: Curculionidae) larvae. *Rev. Argent. Microbiol.* **2017**, *49*, 264–272. [CrossRef] [PubMed]
15. Ammons, D.; Rampersad, J.; Khan, A. Usefulness of staining parasporal bodies when screening for *Bacillus thuringiensis. J. Invertebr. Pathol.* **2002**, *79*, 203–204. [CrossRef] [PubMed]
16. Noguera, P.A.; Ibarra, J.E. Detection of new *cry* genes of *Bacillus thuringiensis* by use of a novel PCR primer system. *Appl. Environ. Microbiol.* **2010**, *76*, 6150–6155. [CrossRef] [PubMed]
17. Fang, Y.; Li, Z.; Liu, J.; Shu, C.; Wang, X.; Zhang, X.; Yu, X.; Zhao, D.; Liu, G.; Hu, S.; et al. A pangenomic study of *Bacillus thuringiensis. J. Genet. Genom.* **2011**, *38*, 567–576. [CrossRef]
18. Ibrahim, M.A.; Griko, N.; Junker, M.; Bulla, L.A. Bacillus thuringiensis. *Bioeng. Bugs.* **2010**, *1*, 31–50. [CrossRef]
19. Meier-Kolthoff, J.P.; Göker, M. TYGS is an automated high-throughput platform for state-of-the-art genome-based taxonomy. *Nat. Commun.* **2019**, *10*, 2182. [CrossRef]
20. Altschul, S.F.; Gish, W.; Miller, W.; Myers, E.W.; Lipman, D.J. Basic local alignment search tool. *J. Mol. Biol.* **1990**, *215*, 403–410. [CrossRef]
21. Navas, L.E.; Amadio, A.F.; Ortiz, E.M.; Sauka, D.H.; Benintende, G.B.; Berretta, M.F.; Zandomeni, R.O. Complete Sequence and Organization of pFR260, the *Bacillus thuringiensis* INTA Fr7-4 Plasmid Harboring Insecticidal Genes. *J. Mol. Microbiol. Biotechnol.* **2017**, *27*, 43–54. [CrossRef] [PubMed]
22. Romeis, J.; Naranjo, S.E.; Meissle, M.; Shelton, A.M. Genetically engineered crops help support conservation biological control. *Biol. Control* **2019**, *130*, 136–154. [CrossRef]
23. Jurat-Fuentes, J.L.; Heckel, D.G.; Ferré, J. Mechanisms of Resistance to Insecticidal Proteins from *Bacillus thuringiensis. Annu. Rev. Entomol.* **2021**, *66*, 121–140. [CrossRef] [PubMed]
24. Zago, H.B.; Siqueira, H.; Pereira, E.J.; Picanço, M.C.; Barros, R. Resistance and behavioural response of *Plutella xylostella* (Lepidoptera: Plutellidae) populations to *Bacillus thuringiensis* formulations. *Pest. Manag. Sci.* **2014**, *70*, 488–495. [CrossRef] [PubMed]
25. Liu, Z.; Fu, S.; Ma, X.; Baxter, S.W.; Vasseur, L.; Xiong, L.; Huang, Y.; Yang, G.; You, S.; You, M. Resistance to *Bacillus thuringiensis* Cry1Ac toxin requires mutations in two *Plutella xylostella* ATP-binding cassette transporter paralogs. *PLoS Pathog.* **2020**, *16*, e1008697. [CrossRef] [PubMed]
26. De Oliveira, J.A.; Negri, B.F.; Hernández-Martínez, P.; Basso, M.F.; Escriche, B. Mpp23Aa/Xpp37Aa Insecticidal Proteins from *Bacillus thuringiensis* (Bacillales: Bacillaceae) Are Highly Toxic to *Anthonomus grandis* (Coleoptera: Curculionidae) Larvae. *Toxins* **2023**, *15*, 55. [CrossRef] [PubMed]
27. Peralta, C.; Sauka, D.H.; Marozzi, A.; Del Valle, E.E.; Palma, L. Argentinean *Bacillus thuringiensis* strains exhibiting distinct morphology of their parasporal crystals. *Rev. Argent. Microbiol.* **2021**, *53*, 378–379. [CrossRef]
28. Amadio, A.F.; Navas, L.E.; Sauka, D.H.; Berretta, M.F.; Benintende, G.B.; Zandomeni, R.O. Identification, cloning and expression of an insecticide *cry8* gene from *Bacillus thuringiensis* INTA Fr7-4. *J. Mol. Microbiol. Biotechnol.* **2013**, *23*, 401–409. [CrossRef]
29. Isaac, B.; Krieger, E.K.; Light, A.-M.; Farhad, M.; Sivasupramanian, S. Polypeptide Compositions Toxic to Anthonomus Insects, and Methods of Use. U.S. Patent 6,541,448 B2, 22 November 2001.
30. Helgason, E.; Okstad, O.A.; Caugant, D.A.; Johansen, H.A.; Fouet, A.; Mock, M.; Hegna, I.; Kolstø, A.B. *Bacillus anthracis, Bacillus cereus,* and *Bacillus thuringiensis*—One species on the basis of genetic evidence. *Appl. Environ. Microbiol.* **2000**, *66*, 2627–2630. [CrossRef]
31. Carroll, L.M.; Wiedmann, M.; Kovac, J. Proposal of a Taxonomic Nomenclature for the *Bacillus cereus* Group Which Reconciles Genomic Definitions of Bacterial Species with Clinical and Industrial Phenotypes. *mBio* **2020**, *11*, e00034-20. [CrossRef]
32. Iriarte, J.; Bel, Y.; Ferrandis, M.D.; Andrew, R.; Murillo, J.; Ferré, J.; Caballero, P. Environmental distribution and diversity of *Bacillus thuringiensis* in Spain. *Syst. Appl. Microbiol.* **1998**, *21*, 97–106. [CrossRef] [PubMed]
33. Aziz, R.K.; Bartels, D.; Best, A.A.; DeJongh, M.; Disz, T.; Edwards, R.A.; Formsma, K.; Gerdes, S.; Glass, E.M.; Kubal, M.; et al. The RAST Server: Rapid annotations using subsystems technology. *BMC Genom.* **2008**, *9*, 75. [CrossRef] [PubMed]

toxins

MDPI

Article

Field Evaluation of Cotton Expressing Mpp51Aa2 as a Management Tool for Cotton Fleahoppers, *Pseudatomoscelis seriatus* (Reuter)

Brady P. Arthur [1], Charles P. Suh [2], Benjamin M. McKnight [3], Megha N. Parajulee [4], Fei Yang [5] and David L. Kerns [1,*]

1 Department of Entomology, Texas A&M University, College Station, TX 77843, USA; bparthur@tamu.edu
2 USDA-ARS Southern Plains Agricultural Research Center, College Station, TX 77845, USA; charles.suh@usda.gov
3 Department of Soil and Crop Sciences, Texas A&M University, College Station, TX 77843, USA; benjamin.mcknight@ag.tamu.edu
4 AgriLife Research and Extension Center, Texas A&M University, Lubbock, TX 79403, USA; megha.parajulee@ag.tamu.edu
5 Department of Entomology, University of Minnesota, St. Paul, MN 55108, USA; yang8905@umn.edu
* Correspondence: david.kerns@ag.tamu.edu

Abstract: The cotton fleahopper (*Pseudatomoscelis seriatus* Reuter) is considered a highly economically damaging pest of cotton (*Gossypium hirsutum* L.) in Texas and Oklahoma. Current control methods rely heavily on the use of foliar-applied chemical insecticides, but considering the cost of insecticides and the critical timeliness of applications, chemical control methods are often not optimized to reduce potential yield losses from this pest. The *Bacillus thuringiensis* (*Bt*) Mpp51Aa2 (formerly Cry51Aa2.834_16) protein has proven effective against thrips and plant bugs with piercing and sucking feeding behaviors, but the impact of this toxin on cotton fleahoppers has not been investigated. To evaluate the Mpp51Aa2 trait effectiveness towards the cotton fleahopper, field trials were conducted in 2019, 2020, and 2021, comparing a cotton cultivar containing the Mpp51Aa2 trait to a non-traited isoline cultivar under insecticide-treated and untreated conditions. Populations of cotton fleahopper nymphs and adults were estimated weekly by visually inspecting cotton terminals. Square retention was also assessed during the first week of bloom to provide some insight on how the *Bt* trait may influence yield. While cotton fleahopper population differences between the traited and non-traited plants were not consistently noted during the pre-bloom squaring period, there was a consistent increase in square retention in cotton expressing Mpp51Aa2 relative to non-traited cotton. Additionally, cotton expressing Mpp51Aa2 offered similar square protection relative to non-traited cotton treated with insecticides for the cotton fleahopper. These findings indicate that the Mpp51Aa2 protein should provide benefits of delayed nymphal growth, population suppression, and increased square retention.

Keywords: Mpp51Aa2; ThryvOn; cotton fleahopper; *Psuedatomoscelis seriatus*; *Gossypium hirsutum*

Key Contribution: Cotton expressing Mpp51Aa2 protein has proven efficacious against the cotton fleahopper. Benefits of the transgenic trait include delayed cotton fleahopper infestations and increased fruit retention.

Citation: Arthur, B.P.; Suh, C.P.; McKnight, B.M.; Parajulee, M.N.; Yang, F.; Kerns, D.L. Field Evaluation of Cotton Expressing Mpp51Aa2 as a Management Tool for Cotton Fleahoppers, *Pseudatomoscelis seriatus* (Reuter). *Toxins* **2023**, *15*, 644. https://doi.org/10.3390/toxins15110644

Received: 20 October 2023
Revised: 2 November 2023
Accepted: 3 November 2023
Published: 5 November 2023

1. Introduction

For decades, the cotton fleahopper (*Pseudatomoscelis seriatus* Reuter) has been considered one of the top five most damaging insect pests of cotton in Texas, resulting in an average reduction of 100,577 bales annually from 2010 to 2021 [1–5]. In 2021 alone, this pest reportedly infested 98% of cotton acres across the state, resulting in a loss of 494,000 bales of cotton. Estimated crop losses from this reduction in bale production was equivalent to

USD 266.5 million, 61% of the total lost cotton revenue in Texas [4]. While some efforts have shown promise in selecting for natural resistance mechanisms, the primary strategy for managing this persistent pest has been through insecticidal control [6–8]. However, the average cost of foliar treatments for this pest has risen from USD 4.72 per acre in 2010 to USD 15.19 per acre in 2021 [4,5]. Consequently, there has been a growing interest in developing a transgenic trait to control the cotton fleahopper [9].

The cotton fleahopper is a generalist feeder with a host range of over 160 species of plants. The adults move from one species of host to another, based on the time of year [10]. Despite the wide host range, cotton fleahoppers have preferred host species, including woolly croton (*Croton capitatus* Michx), horsemint (*Monarda punctata* L.), parthenium ragweed (*Parthenium hysterophorus* L.), silverleaf nightshade (*Solanum elaeagnifolium* Cav.) and others. The cotton fleahopper is initially observed and monitored in cotton during the early portion of the cotton growing season, when their preferred wild host species are less abundant [11–13]. Cotton fleahoppers generally feed on the newly formed pre-floral structures known as squares, as well as the apical meristem. Feeding on the squares can cause abscission of squares and ultimately the loss of the fruit, while feeding on the plant terminal can shorten internodes, cause abnormal node formation, and potentially delay maturity [11]. Like most Hemipteran: Miridae pests, the cotton fleahopper feeds using a piercing and sucking stylet which allows penetration of the plant tissues and ingestion of the sap [10]. After stylet insertion, the cotton fleahopper injects saliva into the plant tissues. This saliva contains enzymes that digest the plant cell walls and aid in initiating the flow of plant fluids [14,15].

Since 1996, the use of plant-incorporated insecticidal proteins derived from soil bacteria *Bacillus thuringiensis* (Berliner) (*Bt*) has provided excellent control for insect pest species in cotton and corn, reducing the necessity for foliar applications of insecticides [16]. However, these *Bt* traits have exclusively been introduced to target coleopteran and lepidopteran species with no activity on hemipterans [17]. With Bayer CropScience's (St. Louis, MO, USA) discovery of the *Bt* protein Mpp51Aa2 (formerly referred to as Cry51Aa2) [18] and its variants, there is potential to utilize a plant-incorporated protein to control hemipteran and thysanopteran pests in cotton [9,19]. Similar to other Cry proteins, Mpp51Aa2 protein is a β pore-forming toxin acting on the epithelium of the insect midgut [17]. Utilizing an artificial diet containing the Mpp51Aa2.834_16 variant of the protein, Gowda et al. [17] determined that the lethal concentrations of the toxin towards the western tarnished plant bug, *Lygus hesperus* (Knight) and the tarnished plant bug, *Lygus lineolaris* (Palisor de Beauvois), were $LC_{50} = 0.30$ μg mg^{-1} and $LC_{50} = 0.853$ μg mg^{-1}, respectively. In another study using an artificial diet, Graham et al. [20] showed that the presence of Mpp51Aa2 protein did not affect the feeding of first- or third-instar tarnished plant bug nymphs. However, the adults preferred a diet that did not contain the Mpp51Aa2 protein. This preference for non-*Bt* was further confirmed by Graham et al. [20], in observations that tarnished plant bug adults favored non-*Bt* excised squares and laid significantly more eggs in non-*Bt* diet packs. In field trials, the Mpp51Aa2 cotton variant, termed as ThryvOn (Bayer CropScience, St. Louis, MO, USA), has been shown to reduce the population density of tarnished plant bug adults by 23% relative to non-traited cotton [21]. Similarly, a 19-fold reduction in tarnished plant bug large nymphs (fourth and fifth instar) was observed in Mpp51Aa2 cotton relative to a non-traited cultivar [22]. Furthermore, cotton that contained the Mpp51Aa2 trait had higher levels of square retention relative to a non-traited isoline when exposed to tarnished plant bugs. When integrated into a pest management system using foliar insecticides, treatments containing the Mpp51Aa2 trait required 50% fewer insecticide applications relative to their non-traited counterparts [23]. Because the cotton fleahopper shares similarities in biology and feeding strategies with the tarnished plant bug and the western tarnished plant bug [24], it is plausible that cotton expressing the Mpp51Aa2 protein may exhibit similar activity against the cotton fleahopper. In a recent study by Bachman et al. [22], they did not observe any differences in the number of large cotton fleahopper nymphs between Mpp51Aa2 and non-traited cotton, but noted that Mpp51Aa2 had a 1.7-fold decrease in adult cotton fleahoppers in the subsequent generation. The objective of this study was to determine the efficacy of Mpp51Aa2-traited cotton on cotton fleahopper populations and explore the potential integration of this new *Bt* technology into a comprehensive pest management system.

2. Results

2.1. Cotton Fleahopper Populations

In 2019, the percentage of infested terminals was not significant for any life stage of cotton fleahopper during the first week of squaring (Table 1). The insecticide application following the first week of squaring significantly reduced populations of both small nymphs ($p = 0.042$) and large nymphs ($p = 0.026$), leading to differences in total cotton fleahoppers ($p = 0.008$) during the second week of squaring. During the third week of squaring, a significant interaction between the spray and trait effects was noted for small nymphs ($p = 0.024$) and large nymphs ($p = 0.011$), while only a spray effect was evident for adults ($p = 0.001$) and the total cotton fleahopper population ($p < 0.001$). When examining populations in the fourth week of squaring, a significant trait effect was observed for the number of large nymphs ($p = 0.013$). Additionally, a significant interaction between the spray treatment and trait was noted for adults ($p < 0.001$) and total cotton fleahoppers ($p < 0.001$), with the non-traited untreated treatment combination revealing significantly higher cotton fleahopper densities compared to all other treatment combinations.

In 2020, there was a significant trait effect on the number of adults ($p = 0.004$) and total cotton fleahoppers ($p = 0.009$) during the first week of squaring (Table 2). Similarly, during the second week of squaring, significant trait effects were observed for the number of adults ($p = 0.031$) and total cotton fleahoppers ($p = 0.025$). Following the first insecticide application during the second week of squaring, the insecticide treatment significantly reduced the numbers of large nymphs ($p = 0.016$), adults ($p = 0.001$), and total cotton fleahoppers ($p = 0.003$) during the third week of squaring. Cotton fleahopper counts for the fourth week of squaring showed that there was a significant spray effect on populations of small nymphs ($p < 0.001$), large nymphs ($p < 0.001$), and the total cotton fleahoppers ($p < 0.001$). There was also a significant interaction of the spray and trait effects on adults ($p = 0.040$). However, there were no significant treatment effects on the number of adults during the fifth week of squaring, although a significant spray treatment effect was observed for small nymphs ($p = 0.001$), large nymphs ($p < 0.001$), and total cotton fleahoppers ($p < 0.001$) during that week.

In 2021, significant trait effects were initially observed during the first week of squaring for the number of small nymphs ($p = 0.024$), adults ($p = 0.005$), and total cotton fleahoppers ($p = 0.004$) (Table 3). The first insecticide application made during the first week of squaring significantly reduced populations of all cotton fleahopper life stages during the second week of squaring; small nymphs ($p = 0.001$), large nymphs ($p = 0.005$), adults ($p = 0.002$), and total cotton fleahoppers ($p = 0.001$). During the third week of squaring, significant insecticide effects were observed on the abundance of small nymphs ($p < 0.0001$), large nymphs ($p = 0.012$), and total cotton fleahoppers ($p = 0.001$), but no significant effect was detected for adults ($p = 0.349$). In the fourth week of squaring, a significant interaction between trait and insecticide treatments was evident for all life stages; small nymphs ($p = 0.034$), large nymphs ($p = 0.005$), adults ($p = 0.044$), and total cotton fleahoppers ($p = 0.004$). Total cotton fleahoppers in the non-traited untreated treatment significantly exceeded that in all other treatment combinations.

Table 1. Percent (mean ± SEM) cotton fleahopper infested terminals by insect stage and week of squaring during 2019.

Treatment	First Week of Squaring			
	Small nymphs	Large nymphs	Adults	Total
Non-traited Treated	52.00 ± 13.56	24.00 ± 7.48	24.00 ± 11.66	100.00 ± 20.98
Non-traited Untreated	40.00 ± 6.32	12.00 ± 4.90	40.00 ± 8.94	92.00 ± 17.44
ThryvOn Treated	44.00 ± 7.48	12.00 ± 8.00	24.00 ± 14.70	80.00 ± 14.14
ThryvOn Untreated	36.00 ± 11.66	8.00 ± 4.90	16.00 ± 7.48	60.00 ± 17.89
Trait *p value*	0.565	0.235	0.293	0.163
Insecticide *p value*	0.341	0.235	0.722	0.443
Trait * Insecticide *p value*	0.857	0.546	0.293	0.740
	Second week of squaring			
	Small nymphs **	Large nymphs **	Adults	Total **
Non-traited Treated	4.00 ± 4.00	0.00 ± 0.00	36.00 ± 11.66	40.00 ± 0.00
Non-traited Untreated	16.00 ± 7.48	32.00 ± 13.56	52.00 ± 18.55	100.00 ± 13.56
ThryvOn Treated	12.00 ± 4.90	4.00 ± 4.00	16.00 ± 7.48	32.00 ± 4.00
ThryvOn Untreated	28.00 ± 8.00	12.00 ± 8.00	48.00 ± 14.97	88.00 ± 8.00
Trait *p value*	0.133	0.339	0.397	0.611
Insecticide *p value*	0.042	0.026	0.101	0.008
Trait * Insecticide *p value*	0.756	0.159	0.570	0.919
	Third week of squaring			
	Small nymphs ***	Large nymphs ***	Adults **	Total **
Non-traited Treated	8.00 ± 4.90 bc	0.00 ± 0.00 c	8.00 ± 8.00	16.00 ± 7.48
Non-traited Untreated	32.00 ± 4.90 ab	64.00 ± 9.80 a	44.00 ± 7.48	140.00 ± 17.89
ThryvOn Treated	0.00 ± 0.00 c	0.00 ± 0.00 c	4.00 ± 4.00	4.00 ± 4.00
ThryvOn Untreated	60.00 ± 12.65 a	28.00 ± 8.00 b	36.00 ± 11.66	124.00 ± 18.33
Trait *p value*	0.185	0.011	0.477	0.315
Insecticide *p value*	0.0001	0.0001	0.001	0.0001
Trait * Insecticide *p value*	0.024	0.011	0.811	0.884
	Fourth week of squaring			
	Small nymphs ***	Large nymphs *	Adults ***	Total ***
Non-traited Treated	12.00 ± 4.90 a	16.00 ± 4.00	44.00 ± 14.7 b	72.00 ± 18.55 b
Non-traited Untreated	40.00 ± 10.95 a	56.00 ± 17.20	140.00 ± 14.14 a	236.00 ± 32.50 a
ThryvOn Treated	28.00 ± 8.00 a	8.00 ± 8.00	48.00 ± 4.90 b	84.00 ± 9.80 b
ThryvOn Untreated	24.00 ± 4.00 a	8.00 ± 4.90	80.00 ± 14.14 b	112.00 ± 20.59 b
Trait *p value*	1.00	0.013	0.042	0.021
Insecticide *p value*	0.128	0.063	0.0001	0.001
Trait * Insecticide *p value*	0.048	0.063	0.022	0.001

Means in a column followed by a common letter are not significantly different (Tukey's HSD, $p < 0.05$). * Denotes that there is only a significant trait effect. ** Denotes that there is only a significant spray effect. *** Denotes a significant interaction of insecticide treatments and trait effects.

Across all three years and sample dates for the untreated treatments, the ratio of small nymphs (≤third instar) to large nymphs (fourth and fifth instar) was 2.6:1 (small nymphs:large nymphs) in the ThryvOn treatment and 1.1:1 in the non-traited treatment (Figure 1). A significant chi-square analysis ($\chi^2 = 15.45$, df = 1, $p < 0.0001$) revealed the ThryvOn cultivar had a significantly lower proportion of large nymphs relative to small nymphs compared to the non-traited cultivar.

Table 2. Percent (mean ± SEM) cotton fleahopper infested terminals by insect stage and week of squaring during 2020.

Treatment	First Week of Squaring			
	Small nymphs	Large nymphs	Adults *	Total *
Non-traited Treated	1.00 ± 0.58	0.00 ± 0.00	0.33 ± 0.33	1.33 ± 0.67
Non-traited Untreated	0.00 ± 0.00	0.00 ± 0.00	0.33 ± 0.33	0.33 ± 0.33
ThryvOn Treated	1.67 ± 0.33	0.00 ± 0.00	0.33 ± 1.20	5.00 ± 1.53
ThryvOn Untreated	2.00 ± 1.15	0.00 ± 0.00	3.00 ± 0.58	5.00 ± 1.73
Trait *p value*	0.081	1.000	0.004	0.009
Insecticide *p value*	0.631	1.000	0.820	0.691
Trait * Insecticide *p value*	0.347	1.000	0.820	0.691
	Second week of squaring			
	Small nymphs	Large nymphs	Adults *	Total *
Non-traited Treated	3.33 ± 1.2	0.00 ± 0.00	6.67 ± 2.03	10.00 ± 3.20
Non-traited Untreated	2.67 ± 0.67	0.67 ± 0.67	5.33 ± 1.2	8.67 ± 2.33
ThryvOn Treated	5.33 ± 1.67	1.33 ± 0.88	10.67 ± 1.33	17.33 ± 1.76
ThryvOn Untreated	6.33 ± 1.76	1.00 ± 0.58	12 ± 3.06	19.33 ± 4.91
Trait *p value*	0.0766	0.218	0.031	0.025
Insecticide *p value*	0.908	0.796	1.000	0.922
Trait * Insecticide *p value*	0.567	0.446	0.532	0.625
	Third week of squaring			
	Small nymphs	Large nymphs **	Adults **	Total **
Non-traited Treated	0.00 ± 0.00	0 ± 0	0.33 ± 0.33	0.33 ± 0.33
Non-traited Untreated	1.67 ± 0.33	0.67 ± 0.67	16 ± 4.04	18.33 ± 4.26
ThryvOn Treated	0.33 ± 0.33	0.00 ± 0.00	0.00 ± 0.00	0.33 ± 0.33
ThryvOn Untreated	6.00 ± 5.51	3.00 ± 1.00	14.33 ± 4.67	23.33 ± 8.82
Trait *p value*	0.423	0.088	0.755	0.624
Insecticide *p value*	0.221	0.016	0.001	0.003
Trait * Insecticide *p value*	0.490	0.088	0.834	0.624
	Fourth week of squaring			
	Small nymphs **	Large nymphs **	Adults ***	Total **
Non-traited Treated	0.33 ± 0.33	0.00 ± 0.00	7.00 ± 1.50 b	7.33 ± 1.45
Non-traited Untreated	15.67 ± 2.33	5.67 ± 1.45	19.33 ± 2.60 a	40.67 ± 4.81
ThryvOn Treated	0.33 ± 0.33	0.00 ± 0.00	6.33 ± 2.33 b	6.67 ± 2.40
ThryvOn Untreated	12.33 ± 2.4	7.00 ± 0.00	8.67 ± 1.45 b	28.00 ± 3.79
Trait *p value*	0.353	0.386	0.024	0.083
Insecticide *p value*	<0.0001	0.0001	0.007	<0.0001
Trait * Insecticide *p value*	0.353	0.386	0.040	0.113
	Fifth week of squaring			
	Small nymphs **	Large nymphs **	Adults	Total **
Non-traited Treated	1.33 ± 0.88	0.67 ± 0.33	4.33 ± 1.2	6.33 ± 1.45
Non-traited Untreated	17.00 ± 3.79	11.67 ± 1.76	5.00 ± 1.53	33.67 ± 4.84
ThryvOn Treated	2.00 ± 1.15	0.33 ± 0.33	2.67 ± 1.45	5.00 ± 2.64
ThryvOn Untreated	18.67 ± 4.09	8.67 ± 2.85	5.33 ± 0.88	32.67 ± 4.81
Trait *p value*	0.696	0.353	0.620	0.763
Insecticide *p value*	0.001	0.0004	0.233	<0.0001
Trait * Insecticide *p value*	0.866	0.453	0.491	0.966

Means in a column followed by a common letter are not significantly different (Tukey's HSD, $p < 0.05$). * Denotes that there is only a significant trait effect. ** Denotes that there is only a significant spray effect. *** Denotes a significant interaction of insecticide treatments and trait effects.

Table 3. Percent (mean ± SEM) cotton fleahopper infested terminals by insect stage and week of squaring during 2021.

Treatment	First Week of Squaring			
	Small nymphs *	Large nymphs	Adults *	Total *
Non-traited Treated	1.13 ± 0.61	0.00 ± 0.00	1.63 ± 0.38	2.75 ± 0.88
Non-traited Untreated	0.25 ± 0.16	0.00 ± 0.00	0.75 ± 0.31	1.00 ± 0.38
ThryvOn Treated	1.75 ± 0.31	0.00 ± 0.00	2.38 ± 0.50	4.12 ± 0.74
ThryvOn Untreated	2.25 ± 0.84	0.00 ± 0.00	2.75 ± 0.59	5.00 ± 1.22
Trait *p value*	0.024	1.000	0.005	0.004
Insecticide *p value*	0.575	1.000	0.589	0.616
Trait * Insecticide *p value*	0.317	1.000	0.182	0.139
	Second week of squaring			
	Small nymphs **	Large nymphs **	Adults **	Total **
Non-traited Treated	0.00 ± 0.00	0.00 ± 0.00	0.13 ± 0.38	0.13 ± 0.13
Non-traited Untreated	0.75 ± 0.37	0.75 ± 0.37	1.75 ± 0.31	3.25 ± 0.86
ThryvOn Treated	0.13 ± 0.13	0.13 ± 0.13	0.50 ± 0.50	0.75 ± 0.37
ThryvOn Untreated	1.75 ± 0.49	0.88 ± 0.3	1.38 ± 0.59	4.00 ± 0.57
Trait *p value*	0.087	0.611	1.000	0.222
Insecticide *p value*	0.001	0.005	0.002	0.001
Trait * Insecticide *p value*	0.418	1.000	0.300	0.911
	Third week of squaring			
	Small nymphs **	Large nymphs **	Adults	Total **
Non-traited Treated	0.13 ± 0.13	0.88 ± 0.23	1.00 ± 0.33	2.00 ± 0.5
Non-traited Untreated	2.00 ± 0.57	2.13 ± 0.69	2.00 ± 0.33	6.13 ± 0.79
ThryvOn Treated	0.50 ± 0.27	0.50 ± 0.13	1.38 ± 0.46	2.00 ± 0.57
ThryvOn Untreated	3.50 ± 0.98	0.88 ± 0.49	1.13 ± 0.44	5.88 ± 0.83
Trait *p value*	0.121	0.078	0.532	0.881
Insecticide *p value*	0.0003	0.012	0.349	0.001
Trait * Insecticide *p value*	0.345	0.889	0.124	0.183
	Fourth week of squaring			
	Small nymphs ***	Large nymphs ***	Adults ***	Total ***
Non-traited Treated	0.88 ± 0.35 b	0.38 ± 0.18 b	0.88 ± 0.35 b	2.13 ± 0.61 bc
Non-traited Untreated	4.75 ± 1.05 a	4.63 ± 0.92 a	3.13 ± 0.64 a	12.50 ± 2.02 a
ThryvOn Treated	0.25 ± 0.16 b	0.00 ± 0.00 b	0.13 ± 0.13 b	0.38 ± 0.26 c
ThryvOn Untreated	2.50 ± 0.63 ab	1.50 ± 0.38 b	1.13 ± 0.3 b	5.13 ± 1.04 b
Trait *p value*	0.049	0.003	0.003	0.002
Insecticide *p value*	0.003	0.0001	0.003	0.0001
Trait * Insecticide *p value*	0.034	0.005	0.044	0.004

Means in a column followed by a common letter are not significantly different (Tukey's HSD, $p < 0.05$). * Denotes that there is only a significant trait effect. ** Denotes that there is only a significant spray effect. *** Denotes a significant interaction of insecticide treatments and trait effects.

As there were significant insecticide effects on cotton fleahopper populations, our analysis of cumulative insect days (CIDs) considered only the untreated treatments. A one-way ANOVA, followed by post hoc Student's *t*-tests, was used to determine differences in CIDs between ThryvOn and the non-traited for each of the life stages (small nymphs, large nymphs, and adults). There were no significant differences in CIDs between the non-traited and ThryvOn for small nymphs ($p = 0.109$) or across all cotton fleahopper life stages ($p = 0.081$) (Figure 2A,D). However, CIDs were significantly reduced for large nymphs ($p = 0.026$) and adults ($p = 0.017$) in ThryvOn compared to the non-traited cultivar (Figure 2B,C).

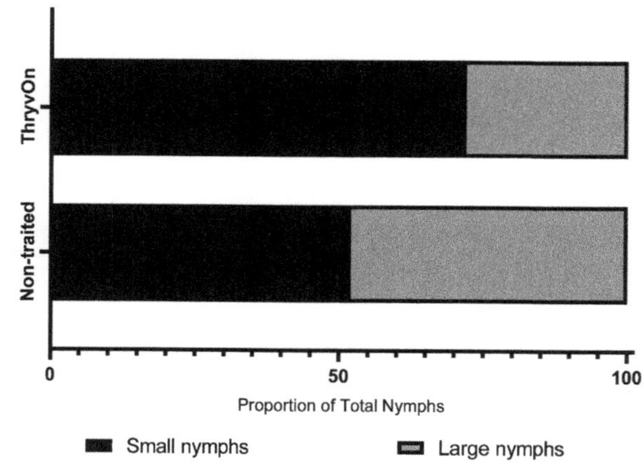

Figure 1. Chi-square analysis of the proportion of small nymphs (≤3rd instar) and large nymphs (4th and 5th instars) between non-traited untreated and ThryvOn untreated treatments across all years and sample dates ($n = 376$, $\chi^2 = 15.45$, $p < 0.0001$).

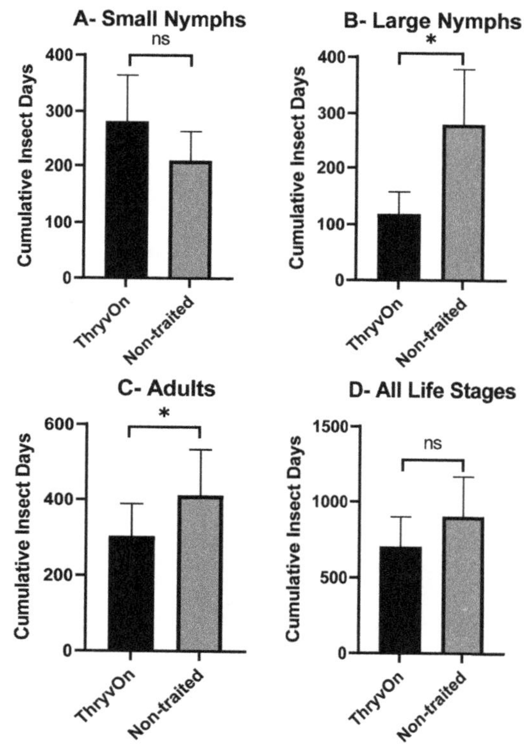

Figure 2. Cumulative insect days (mean ± SEM) of non-insecticide treated plots across all three years by cotton fleahopper life stage. (**A**) Small nymphs ≤ 3rd instar. (**B**) Large nymphs 4th and 5th instar. (**C**) Adult cotton fleahoppers. (**D**) All life stages combined. Significant differences between traits are denoted by "*" (ANOVA, Student's *t*-test, $p < 0.05$).

2.2. Square Retention

Square retention values estimated at first bloom consistently showed the non-traited, untreated treatment had the lowest square retention across all three years (Table 4). In 2019, there were significant trait ($p < 0.0001$) and insecticide treatment ($p < 0.0001$) effects, but no significant interaction ($p = 0.898$). Square retention for the ThryvOn cultivar was significantly higher at 66% compared to 46% for the non-traited ($p < 0.0001$). In 2020, there was no significant trait effect ($p = 0.266$), but significant insecticide ($p = 0.020$) and interaction ($p = 0.047$) effects were detected. Closer examination of the significant interaction revealed that square retention of the non-traited, untreated treatment (69%) was significantly reduced in comparison to the non-traited, treated treatment (86%). Similarly, in 2021, there was no significant trait effect ($p = 0.215$), but both insecticide treatment ($p = 0.002$) and interaction ($p = 0.013$) effects were significant. Square retention in the non-traited, untreated treatment (87.63%) was significantly lower than those observed for the non-traited treated, ThryvOn treated and ThryvOn untreated treatments. The ThryvOn untreated treatment was statistically similar to the non-traited treated and ThryvOn treated treatments.

Table 4. Percent fruit retention (mean ± SEM) during the 4th week of squaring by year.

	2019 * **	2020 ***	2021 ***
Non-traited Treated	61.26 ± 2.98	86.39 ± 2.59 a	91.81 ± 0.52 a
Non-traited Untreated	30.53 ± 3.71	69.08 ± 3.81 b	87.63 ± 0.64 b
ThryvOn Treated	81.5 ± 3.62	82.61 ± 0.14 ab	90.77 ± 0.88 a
ThryvOn Untreated	51.70 ± 3.70	80.75 ± 4.72 ab	90.05 ± 0.40 a
Trait effect *p value*	<0.001	0.266	0.215
Insecticide effect *p value*	<0.001	0.020	0.002
Trait * Insecticide effect *p value*	0.898	0.047	0.013

Means in a column followed by a common letter are not significantly different (Tukey's HSD, $p < 0.05$). * Denotes that there is only a significant trait effect. ** Denotes that there is only a significant insecticide spray effect. *** Denotes a significant interaction of spray and trait effects.

Regression analyses of the cumulative insect days (Figure 2D) against square retention in the untreated treatments produced significant relationships (Figure 3). The regression model for ThryvOn ($R^2 = 0.8732$) was ($SR = 66.53 \times e^{-0.00051*CID} + 24.52$), where SR represents square retention and CID indicates cumulative insect days. The model for the non-traited ($R^2 = 0.9300$) was ($SR = 70.01 \times e^{-0.001*CID} + 25.57$). The two models were significantly different, based on the analysis of the extra-sum-of-squares F-test ($F = 3.769$, df = 3, 26, $p = 0.022$). The difference in the span (difference between the Y intercept and the plateau) of the two models suggests the ThryvOn treatment would shed fewer squares compared to non-traited treatments as the number of cumulative insect days increases. The half-life of the ThryvOn and non-traited models was 1366 and 669.7 CIDs, respectively, suggesting that ThryvOn requires a higher number of cumulative insect days to inflict a similar magnitude of square abscission.

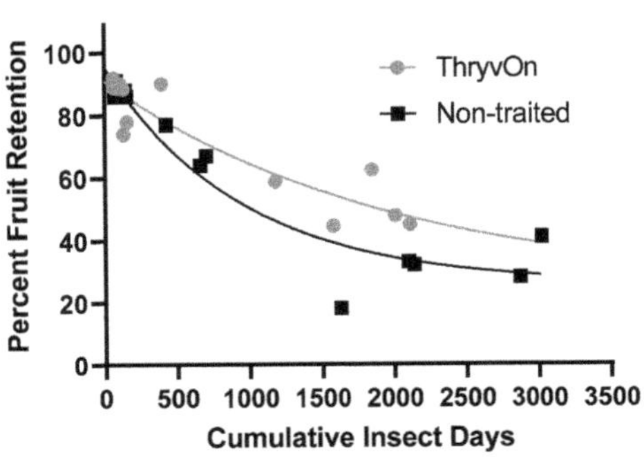

Figure 3. Relationship of cumulative insect days of cotton fleahoppers, across life stages and years, and fruit retention on untreated ThryvOn and non-traited cotton cultivars. An extra-sum-of-squares F-test was used to determine that the single-phase decay model shown that best fits each of the cultivars were significantly different ($p = 0.02$). ThryvOn ($y = 66.53 \times e^{-0.00051*x} + 24.52$) ($R^2 = 0.8732$), non-traited ($y = 70.01 \times e^{-0.001*x} + 25.57$) ($R^2 = 0.9300$).

3. Discussion

The population counts for 2019, 2020, and 2021 showed varying levels of infestation, characterized as relatively high in 2019, moderate in 2020, and relatively low in 2021. As the season progressed in each year of the study, differences in nymphal and adult population levels in the non-sprayed plots became more evident between the ThryvOn and non-traited (Tables 1–3). The high nymphal and adult populations observed during the squaring period in 2019 indicated that ThryvOn alone does not prevent cotton fleahopper colonization. Nevertheless, there is a notable benefit even in high populations when ThryvOn is coupled with insecticides, as cotton fleahopper populations increased at a significantly slower rate compared with those in the non-traited, treated cultivar, thus limiting the number of large nymphs and adults at the end of the squaring period.

Compared to 2019, infestations in 2020 were moderate, but still reached threshold levels on a consistent basis. Early in 2020, before cotton fleahopper populations reached spray treatment thresholds, there was an apparent trait effect on the number of adults. Interestingly, adult populations in ThryvOn treatments were significantly higher than those observed in the non-traited treatments. Possible explanations for this unexpected observation may be related to initial dispersal behaviors of cotton fleahoppers into the cotton field from the weeds along the field margins. Parajulee et al. [25] indicated that adult populations of cotton fleahoppers can exhibit a clumped distribution during the initial stages of infestation but that adults tend to disperse throughout the field and exhibit a more uniform distribution as the season progresses. Nevertheless, following the initial application of insecticide and as the squaring season progressed, there was a significant interaction of trait and spray effects on adult populations, where the non-traited, untreated treatment had a 2.2-fold increase in adults compared to the ThryvOn, untreated treatment.

The low levels of infestation throughout the 2021 growing season can be attributed to the diverse non-cotton vegetation in field margins. Ample rainfall during the growing season not only suppressed populations by washing smaller nymphs from the terminal [26], but also potentially increased the diversity and intensity of weedy species adjacent to the cotton field, particularly horsemint and woolly croton, not only serving as a source of cotton fleahoppers but also acting as a trap crop [27]. Although populations of cotton fleahoppers in our test plots were considerably lower in 2021 compared to the previous two years, as the season progressed, dispersal into the field indicated a significant interaction of the trait and spray effects, resulting in higher populations in the non-traited, untreated. In our study, the non-traited, untreated treatment possessed the highest numbers of cotton fleahoppers by the end of the squaring period in each year of the study. This is similar to the results found by Gowda et al. [17], where lower numbers of tarnished plant bugs and western tarnished plant bugs of all life stages were observed in ThryvOn cotton compared to a non-traited control. Likewise, Graham and Stewart [21] and Graham et al. [20] found that ThryvOn had fewer tarnished plant bug adults than the non-traited control, suggesting that the trait had a repellency effect on the pest, with adults preferentially feeding on the non-traited plants. This same result was reported in cotton fleahoppers, where the adult populations in ThryvOn were significantly lower than in the non-traited [28]. In contrast, the adult cotton fleahopper populations at the end of the squaring period in each year of our study suggest the singular *Bt* trait effect did not significantly suppress adult cotton fleahopper populations. Although no differences were observed in adult cotton fleahoppers between the traited and non-traited treatments, a significant interaction occurred between spray treatment and the *Bt* trait. This interaction suggests that there is a benefit to the ThryvOn trait in slowing the development of infestations. This observation aligns with Graham and Stewart [21], who reported a delay in cotton fleahopper reinfestation when ThryvOn cotton was over-sprayed with insecticide, potentially reducing the number of insecticide applications required to manage populations. Similar to the results of Asiimwe et al. [28] populations of nymphs, small or large, did not consistently show significant differences based on trait. However, differences were observed in CIDs for both large nymphs and adults in this study. It has been documented that the presence of *Bt* in the host plant can delay insect development [29], which suggests that the ThryvOn trait was largely responsible for the delayed nymphal development. In ThryvOn, CIDs for large nymphs and adults were reduced by 160 and 111 days, respectively, compared to those estimated for large nymphs and adults in the non-traited. These differences potentially contributed to the increase in square retention in ThryvOn, as large-nymph and adult stages of mirids are known to cause the highest magnitude of damage [22,30]. Given that the ratio of small to large nymphs was 2.6:1 in the ThryvOn treatment but closer to 1:1 in the non-traited cultivar, the rate of nymphal growth appeared to be more rapid and consistent in the non-traited plots. These findings align with those of Whitfield [31], who reported different ratios, small vs. large, of tarnished plant bug nymphs in ThryvOn compared to non-traited. Similarly, Jerga et al. [29] showed that the presence of Mpp51Aa2 protein delayed the development of early-instar tarnished plant bug nymphs.

A potential concern with the results is the presence of other insect pests; however, populations of tarnished plant bugs and western tarnished plant bugs were not observed in the test area for any of the years the field trials were conducted. The presence of late-instar cotton fleahopper nymphs in ThryvOn could potentially point to concerns with the longevity of the technology. As documented for other pest species, low doses of the toxins only eliminate the highly sensitive insects, resulting in potential evolution of resistance [32]. Killing only the homozygous susceptible populations and not the heterozygous resistant will allow for a faster increase in the resistant allele frequency within the pest population [33]. Failure to meet the requirements of a high-dose resistance management strategy is one of the main factors in field-evolved resistance to *Bt* toxins [34,35].

4. Conclusions

In summary, the incorporation of a *Bt* protein into cotton, with activity against sap-feeding insects like the cotton fleahopper, addresses a long-standing need in pest management. The Mpp51Aa2 protein provides benefits for the control of mirid pests of cotton, including delayed field colonization, nymphal development, population reduction, decreased dependance on foliar insecticides, and mitigation of insect-induced fruit abscission. The results of this study confirm that the ThryvOn trait suppresses populations of cotton fleahoppers, which is consistent with the suppression observed in previous work with other piercing–sucking insect pest species. Implementation of this new *Bt* technology, as part of an integrated pest management system for cotton fleahoppers, will potentially reduce the number of insecticide applications, provide producers with more flexibility to make timely applications of insecticides, and limit potential yield losses associated with cotton fleahopper feeding damage.

5. Materials and Methods

5.1. In-Season Data Collection

A three-year (2019–2021) field study was conducted in the Brazos Valley, Texas, to determine if Mpp51Aa2-traited cotton has an impact on cotton fleahopper populations and square retention. The experiment utilized near-isolines of Mpp51Aa2-traited and non-traited cotton with or without insecticidal control of the cotton fleahopper. The Mpp51Aa2-traited cotton cultivars, hereafter referred to as ThryvOn, used in the study were 18R445B3XF, 19R326LB3XF, and DeltaPine 2131B3TXF in 2019, 2020 and 2021, respectively. The non-traited cotton cultivars included an unspecified isoline in 2019, and the near-isoline, DeltaPine 2055B3XF, in 2020 and 2021. All seeds were sourced from Bayer CropScience, had been treated with fungicides and insecticides to prevent early-season pathogens and pre-fruiting pests (Acceleron Standard: Bayer CropScience, St. Louis, MO, USA), and contained the lepidopteran active *Bt* traits Cry1Ac, Cry2Ab2, and Vip3a19. These *Bt* proteins expressed in both cultivars are presumed to have negligible influence on our results, as it is documented that Cry1Ac, Cry2Ab and Vip3Aa are active on lepidopteran pests but not active on mirids such as cotton fleahoppers [19]. The experiment employed a randomized complete block design with all combinations of the two main effects of trait,

ThryvOn and non-traited, and an insecticide regimen, treated and untreated. Treatment combinations included ThryvOn treated, non-traited treated, ThryvOn untreated, and non-traited untreated with five, three, and eight replicates in 2019, 2020, and 2021, respectively. Insecticidal treatments were applied when cotton fleahopper populations of the treated plots reached or exceeded 5 percent infested terminals; however, precipitation and equipment failure prevented timely application in some instances. Treatments designated as 'treated' were sprayed using a ground-driven high-clearance sprayer with commercially available insecticides that had proven effectiveness in cotton fleahopper management, as indicated by prior research conducted at Texas A&M AgriLife Extension [36]. Foliar insecticide treatments were a tank mix of 44.7 g ai per ha of acephate (Orthene 97, AMVAC Chemical Corporation, Newport Beach, CA, USA) and 60.5 g ai per ha of imidacloprid (Admire Pro, Bayer Crop Science, Raleigh, NC, USA). Each experimental plot was approximately 0.2 ha with a row spacing of 1.02 m. The seeding rate was approximately 13 seeds per meter. Applications of fertilizers, herbicides, growth regulators, and insecticides to manage cotton, weeds and other insect pests followed the recommendations of Texas A&M AgriLife Extension [36–38].

Cotton terminal inspections for cotton fleahoppers began at first week of squaring and continued on a weekly basis until first bloom. In 2019, 25 terminals were inspected per treatment plot. In 2020 and 2021, 100 terminals were inspected in each plot to reduce data variability. The total numbers of cotton fleahopper adults, large nymphs (fourth and fifth instar) and small nymphs ($\leq$third instar) in each plot were recorded on each sample date. Cotton fleahopper instars were determined by the size of the head capsule and relative development of wing pads; first, second, and third instars have a narrower head and no presence of wing pads; fourth-instar nymphs have short wing pads that only extend to the base of the second abdominal segment, whereas fifth-instar nymphs have a head equal or greater in width relative to the abdomen, with long wingpads extending to at least the fourth abdominal segment [39,40]. All cotton fleahopper counts for each plot were converted to percent infested terminals by dividing the number of cotton fleahoppers encountered by the number of terminals inspected. Cumulative insect days (CIDs) were calculated using the method described by Ruppel [41], to highlight differences in overall insect pressure across different sample dates.

During the fourth week of squaring, percent fruit retention was estimated by dividing the number of squares present by the total number of fruiting positions on whole plants, with a specific focus on sympodial branches (fruiting branches) [42]. In 2019, five random plants were sampled per plot, whereas twenty random plants per plot were sampled in 2020 and 2021, to reduce variability in the data.

5.2. Statistical Analyses

All statistical analyses were performed utilizing GraphPad Prism 9.3.0 [43]. The percent cotton fleahopper infested terminals and percent fruit retention data were analyzed using a two-way ANOVA with trait, spray regimen, and their interaction as fixed effects. When comparing cumulative insect days by life stage, the main effects of trait, insecticide treatment, and the interaction of trait and insecticide treatment were designated as fixed effects, while year was designated as a random effect. Significant differences between treatment means were separated using Tukey's honest significant difference method ($\alpha = 0.05$). A chi-square test of independence ($\alpha = 0.05$) was performed to examine differences in the small- and large-nymph population dynamics on ThryvOn and non-traited. Regression curves, generated by plotting square retention against cumulative insect days, were compared using an extra sum-of-squares F-test with an $\alpha = 0.05$. The extra-sum-of-squares F-test compared the two regression curves to a simplified regression curve of the entire data set.

Author Contributions: Conceptualization, B.P.A., D.L.K., C.P.S., M.N.P. and B.M.M.; methodology, B.P.A., D.L.K., C.P.S., M.N.P. and B.M.M.; investigation, D.L.K. and B.P.A.; resources, D.L.K.; data curation, D.L.K. and B.P.A.; writing—original draft preparation: B.P.A.; writing—review and editing,

B.P.A., D.L.K., C.P.S., M.N.P., B.M.M. and F.Y.; visualization, B.P.A.; supervision, D.L.K.; funding acquisition, D.L.K. All authors have read and agreed to the published version of the manuscript.

Funding: Funding for this research was provided in-part by Bayer CropScience and Cotton Incorporated.

Institutional Review Board Statement: Not applicable.

Informed Consent Statement: Not applicable.

Data Availability Statement: Not applicable.

Acknowledgments: The authors would like to thank William Coors, Ryan Gilreath and the numerous students who helped with study establishment and data collection.

Conflicts of Interest: The funders had no role in the design of the study; in the collection, analyses, or interpretation of data; in the writing of the manuscript; or in the decision to publish the results.

References

1. Cook, D.R.; Threet, M. Cotton Insect Losses Estimates. In Proceedings of the Beltwide Cotton Conference, Virtual, 5–7 January 2020; pp. 410–465.
2. Cook, D. Cotton insect loss estimates. In Proceedings of the Beltwide Cotton Conference, San Antonio, TX, USA, 3–5 January 2017; pp. 721–780.
3. Williams, M.R. Cotton Insect Losses—2015. In Proceedings of the Beltwide Cotton Conferences, Memphis, TN, USA, 5–7 January 2016; pp. 507–526.
4. Cook, D.R.; Threet, M. Cotton Insect Losses Estimates. In Proceedings of the Beltwide Cotton Conference, San Antonio, TX, USA, 4–6 January 2021.
5. Williams, M.R. Cotton Insect Losses. In Proceedings of the Beltwide Cotton Conference, Atlanta, GA, USA, 4–7 January 2010; pp. 896–940.
6. Knutson, A.E.; Mekala, K.D.; Smith, C.W.; Campos, C. Tolerance to Feeding Damage by Cotton Fleahopper (Hemiptera: Miridae) Among Genotypes Representing Adapted Germplasm Pools of United States Upland Cotton. *J. Econ. Entomol.* **2013**, *106*, 1045–1052. [CrossRef]
7. Parker, R.D.; Knutson, A.; Allen, E.; Biles, S.; Kerns, D.L.; Jungman, M.J. *Managing Cotton Insects in the Southern, Eastern and Blackland Areas of Texas*; Texas AgriLife Extension Service: College Station, TX, USA, 2008.
8. McLoud, L.A.; Hague, S.; Knutson, A.; Wayne Smith, C.; Brewer, M. Cotton Square Morphology Offers New Insights into Host Plant Resistance to Cotton Fleahopper (Hemiptera: Miridae) in Upland Cotton. *J. Econ. Entomol.* **2015**, *109*, 392–398. [CrossRef]
9. Akbar, W.; Gowda, A.; Ahrens, J.E.; Stelzer, J.W.; Brown, R.S.; Bollman, S.L.; Greenplate, J.T.; Gore, J.; Catchot, A.L.; Lorenz, G.; et al. First transgenic trait for control of plant bugs and thrips in cotton. *Pest Manag. Sci.* **2019**, *75*, 867–877. [CrossRef]
10. Esquivel, J.F.; Esquivel, S.V. Identification of Cotton Fleahopper (Hemiptera: Miridae) Host Plants in Central Texas and Compendium of Reported Hosts in the United States. *Environ. Entomol.* **2009**, *38*, 766–780. [CrossRef]
11. Walker, J.K.; Niles, G.A.; Gannaway, J.R.; Robinson, J.V.; Cowan, C.B.; Lukefahr, M.J. Cotton Fleahopper Damage to Cotton Genotypes. *J. Econ. Entomol.* **1974**, *67*, 537–542. [CrossRef]
12. Almand, L.K.; Sterling, W.L.; Green, C.L. *Seasonal Abundance and Dispersal of the Cotton Fleahopper as Related to Host Plant Phenology*; The Texas Agricultural Experiment Station: College Station, TX, USA, 1976.
13. Gaylor, M.J.; Sterling, W.L. Photoperiodic Induction and Seasonal Incidence of Embryonic Diapause in the Cotton Fleahopper, *Pseudatomoscelis seriatus*. *Ann. Entomol. Soc. Am.* **1977**, *70*, 893–897. [CrossRef]
14. Brett, C.H. *The Cotton Flea Hopper in Oklahoma*; Oklahoma Agricultural Experiment Station: Stillwater, OK, USA, 1946.
15. Martin, W.R., Jr.; Morgan, P.W.; Sterling, W.L.; Meola, R.W. Stimulation of Ethylene Production in Cotton by Salivary Enzymes of the Cotton Fleahopper (Heteroptera: Miridae). *Environ. Entomol.* **1988**, *17*, 930–935. [CrossRef]
16. Lu, Y.; Wu, K.; Jiang, Y.; Xia, B.; Li, P.; Feng, H.; Wyckhuys, K.A.; Guo, Y. Mirid bug outbreaks in multiple crops correlated with wide-scale adoption of Bt cotton in China. *Science* **2010**, *328*, 1151–1154. [CrossRef]
17. Gowda, A.; Rydel, T.J.; Wollacott, A.M.; Brown, R.S.; Akbar, W.; Clark, T.L.; Flasinski, S.; Nageotte, J.R.; Read, A.C.; Shi, X.; et al. A transgenic approach for controlling Lygus in cotton. *Nat. Commun.* **2016**, *7*, 12213. [CrossRef]
18. Crickmore, N.; Berry, C.; Panneerselvam, S.; Mishra, R.; Connor, T.R.; Bonning, B.C. A structure-based nomenclature for Bacillus thuringiensis and other bacteria-derived pesticidal proteins. *J. Invertebr. Pathol.* **2021**, *186*, 107438. [CrossRef]
19. Baum, J.A.; Sukuru, U.R.; Penn, S.R.; Meyer, S.E.; Subbarao, S.; Shi, X.; Flasinski, S.; Heck, G.R.; Brown, R.S.; Clark, T.L. Cotton Plants Expressing a Hemipteran-Active Bacillus thuringiensis Crystal Protein Impact the Development and Survival of Lygus hesperus (Hemiptera: Miridae) Nymphs. *J. Econ. Entomol.* **2012**, *105*, 616–624. [CrossRef] [PubMed]
20. Graham, S.H.; Musser, F.M.; Jacobson, A.L.; Chitturi, A.; Catchot, B.; Stewart, S.D. Behavioral Responses of Thrips (Thysanoptera: Thripidae) and Tarnished Plant Bug (Hemiptera: Miridae) to a New Bt Toxin, Cry51Aa2.834_16 in Cotton. *J. Econ. Entomol.* **2019**, *112*, 1695–1704. [CrossRef] [PubMed]
21. Graham, S.H.; Stewart, S.D. Field Study Investigating Cry51Aa2.834_16 in Cotton for Control of Thrips (Thysanoptera: Thripidae) and Tarnished Plant Bugs (Hemiptera: Miridae). *J. Econ. Entomol.* **2018**, *111*, 2717–2726. [CrossRef]

22. Bachman, P.M.; Ahmad, A.; Ahrens, J.E.; Akbar, W.; Baum, J.A.; Brown, S.; Clark, T.L.; Fridley, J.M.; Gowda, A.; Greenplate, J.T.; et al. Characterization of the Activity Spectrum of MON 88702 and the Plant-Incorporated Protectant Cry51Aa2.834_16. *PLoS ONE* **2017**, *12*, e0169409. [CrossRef] [PubMed]
23. Corbin, J.C.; Towles, T.B.; Crow, W.D.; Catchot, A.L.; Cook, D.R.; Dodds, D.M.; Gore, J. Evaluation of Current Tarnished Plant Bug (Hemiptera: Miridae) Thresholds in Transgenic MON 88702 Cotton Expressing the Bt Cry51Aa2.834_16 Trait. *J. Econ. Entomol.* **2020**, *113*, 1816–1822. [CrossRef]
24. Miles, P.W. The Saliva of Hemiptera. In *Advances in Insect Physiology*; Treherne, J.E., Berridge, M.J., Wigglesworth, V.B., Eds.; Academic Press: Cambridge, MA, USA, 1972; Volume 9, pp. 183–255.
25. Parajulee, M.N.; Shrestha, R.B.; Leser, J.F. Sampling Methods, Dispersion Patterns, and Fixed Precision Sequential Sampling Plans for Western Flower Thrips (Thysanoptera: Thripidae) and Cotton Fleahoppers (Hemiptera: Miridae) in Cotton. *J. Econ. Entomol.* **2006**, *99*, 568–577. [CrossRef]
26. Gaylor, M.J.; Sterling, W.L. Simulated rainfall and wind as factors dislodging nymphs of the cotton fleahopper, *Pseudatomoscelis seriatus* (Reuter), from cotton plants. *Texas Agric. Exp. Stn. Prog. Rep.* **1975**, 3356.
27. Hamons, K.; Raszick, T.; Perkin, L.; Sword, G.; Suh, C. Cotton Fleahopper Biology and Ecology Relevant to Development of Insect Resistance Management Strategies. *Southwest. Entomol.* **2021**, *46*, 1–16. [CrossRef]
28. Asiimwe, P.; Brown, C.R.; Ellsworth, P.C.; Reisig, D.D.; Bertho, L.; Jiang, C.; Schapaugh, A.; Head, G.; Burzio, L. Transgenic cotton expressing Mpp51Aa2 does not adversely impact beneficial non-target hemiptera in the field. *Crop Prot.* **2023**, *173*, 106384. [CrossRef]
29. Jerga, A.; Chen, D.; Zhang, C.; Fu, J.; Kouadio, J.-L.K.; Wang, Y.; Duff, S.M.; Howard, J.E.; Rydel, T.J.; Evdokimov, A.G. Mechanistic insights into the first Lygus-active β-pore forming protein. *Arch. Biochem. Biophys.* **2016**, *600*, 1–11. [CrossRef]
30. Cooper, W.R.; Spurgeon, D.W. Feeding injury to cotton caused by *Lygus hesperus* (Hemiptera: Miridae) nymphs and prereproductive adults. *Environ. Entomol.* **2013**, *42*, 967–972. [CrossRef]
31. Whitfield, A. Evaluation of Thresholds, Control, and Behavioral Responses of Tobacco Thrips, *Frankliniella fusca* (Hitch), and Tarnished Plant Bugs, *Lygus lineolaris* (Beauvoris), in ThryvOn Cotton. Master's Thesis, University of Arkansas, Fayetteville, AR, USA, 2023.
32. Gould, F. Sustainability of Transgenic Insecticidal Cultivars: Integrating Pest Genetics and Ecology. *Annu. Rev. Entomol.* **1998**, *43*, 701–726. [CrossRef]
33. Storer, N.P.; Van Duyn, J.W.; Kennedy, G.G. Life History Traits of Helicoverpa zea (Lepidoptera: Noctuidae) on Non-Bt and Bt Transgenic Corn Hybrids in Eastern North Carolina. *J. Econ. Entomol.* **2001**, *94*, 1268–1279. [CrossRef]
34. Tabashnik, B.E.; Brévault, T.; Carrière, Y. Insect resistance to Bt crops: Lessons from the first billion acres. *Nat. Biotechnol.* **2013**, *31*, 510–521. [CrossRef]
35. Bates, S.L.; Zhao, J.-Z.; Roush, R.T.; Shelton, A.M. Insect resistance management in GM crops: Past, present and future. *Nat. Biotechnol.* **2005**, *23*, 57–62. [CrossRef] [PubMed]
36. Vyavhare, S.; Kerns, D.; Allen, C.; Bowling, R.; Brewer, M.; Parajulee, M. *Managing Cotton Insects in Texas*; ENTO-075; Texas A&M Agrilife Extension: College Station, TX, USA, 2018.
37. Lemon, R.; Boman, R.; McFarland, M.; Bean, B.; Provin, T.; Hons, F. *Nitrogen Management in Cotton*; Texas A&M University, Agrilife Extension Service: College Station, TX, USA, 2009.
38. Morgan, G.D.; Baumann, P.A.; Dotray, P.A. *Weed Management*; Texas A&M University, Agrilife Extension Service: College Station, TX, USA, 1999.
39. Reinhard, H.J. *The Cotton Flea Hopper*; Texas Agricultural Experiment Station: College Station, TX, USA, 1926; p. 39.
40. Suh, C.P.-C. Head Capsule Widths of Nymphal Instars of the Cotton Fleahopper. *Southwest. Entomol.* **2007**, *32*, 127–130, 124. [CrossRef]
41. Ruppel, R.F. Cumulative Insect-Days as an Index of Crop Protection1. *J. Econ. Entomol.* **1983**, *76*, 375–377. [CrossRef]
42. Ritchie, G.L.; Bednarz, C.W.; Jost, P.H.; Brown, S.M. *Cotton Growth and Development*; University of Georgia: Athens, GA, USA, 2007.
43. GraphPad Prism. *GraphPad Prism for Windows 64-Bit, 9.3.0*; Dotmatics: Boston, MA, USA, 2021.

toxins

MDPI

Article

Baseline Susceptibility of the Field Populations of *Ostrinia furnacalis* in Indonesia to the Proteins Cry1A.105 and Cry2Ab2 of *Bacillus thuringiensis*

Y. Andi Trisyono [1,*], Valentina E. F. Aryuwandari [1], Teguh Rahayu [1], Samuel Martinelli [2], Graham P. Head [2], Srinivas Parimi [3] and Luis R. Camacho [4]

1 Department of Plant Protection, Faculty of Agriculture, Universitas Gadjah Mada, Yogyakarta 55281, Indonesia
2 Regulatory Science, Bayer Crop Science US, Chesterfield, MO 63017, USA
3 Bayer Crop Science Ltd., Hyderabad 500081, India
4 Bayer (South East Asia) Pte Ltd., 2 Tanjong Katong Road #07-01, Paya Lebar Quarter 3, Singapore 437161, Singapore
* Correspondence: anditrisyono@ugm.ac.id

Abstract: Genetically modified MON 89034 corn (*Zea mays* L.) expressing *Bacillus thuringiensis* (*Bt*) insecticidal proteins, viz. Cry1A.105 and Cry2Ab2, is a biotechnological option being considered for the management of the major corn pest in Indonesia, the Asian corn borer (*Ostrinia furnacalis* (Guenée) (Lepidoptera: Crambidae)). As a part of a proactive resistance-management program for MON 89034 corn in Indonesia, we assessed the baseline susceptibility of field-collected populations of *O. furnacalis* to Cry1A.105 and Cry2Ab2 proteins. Dose–response bioassays using the diet-dipping method indicated that the lethal concentration (LC_{50}) values of Cry1A.105 and Cry2Ab2 in 24 different field populations of *O. furnacalis* ranged from 0.006 to 0.401 μg/mL and from 0.044 to 4.490 μg/mL, respectively, while the LC_{95} values ranged from 0.069 to 15.233 μg/mL for Cry1A.105 and from 3.320 to 277.584 μg/mL for Cry2Ab2. The relative resistance ratios comparing the most tolerant field populations and an unselected laboratory population were 6.0 for Cry1A.105 and 2.0 for Cry2Ab2 based on their LC_{50} values. Some field populations were more susceptible to both proteins than the unselected laboratory population. The LC_{99} and its 95% fiducial limits across the field populations were calculated and proposed as candidate diagnostic concentrations. These data provide a basis for resistance monitoring in *Bt* Corn and further support building resistance-management strategies in Indonesia.

Keywords: Asian corn borer; Cry1A.105; Cry2Ab2; Indonesia; resistance monitoring; susceptibility

Key Contribution: This study provides evidence that the field-collected populations of Asian corn borers in Indonesia are susceptible to Cry1A.105 and Cry2Ab2. This baseline study is an essential element for the resistance management of this insect to the transgenic corn expressing these proteins.

Citation: Trisyono, Y.A.; Aryuwandari, V.E.F.; Rahayu, T.; Martinelli, S.; Head, G.P.; Parimi, S.; Camacho, L.R. Baseline Susceptibility of the Field Populations of *Ostrinia furnacalis* in Indonesia to the Proteins Cry1A.105 and Cry2Ab2 of *Bacillus thuringiensis*. *Toxins* **2023**, *15*, 602. https://doi.org/10.3390/toxins15100602

Received: 18 August 2023
Revised: 6 October 2023
Accepted: 6 October 2023
Published: 7 October 2023

1. Introduction

The Asian corn borer, *Ostrinia furnacalis* (Guenée), is one of the major pests of corn (*Zea mays* Linnaeus) in Southeast Asia, including the Indonesian archipelago, Vietnam, the Philippines, and China [1]. Areekul [2] and Camarao [3] reported two generations of *O. furnacalis* infestation in each growing season in tropical areas with 24–30 total development days for each generation. Damage from *O. furnacalis* occurs not only in the stem of corn plants but also in the whorl, tassel, and ears [4,5]. In vegetative stage plants, the newly hatched larvae feed on young leaf whorls, resulting in holes in the leaves that can widen as the larvae grow and feed more vigorously. In the reproductive stage, the larvae feed on tassel, then bore into the stem, making tunnels, and might continue to feed on the ears [2].

Da-Lopez et al. [6] reported that the presence of *O. furnacalis* egg masses was as high as nine egg masses per plant in fields in Sleman, Yogyakarta. A recent study (2018–2019) in Lampung and Central Java showed that *O. furnacalis* infestations were present in as many as 95% of corn plants with an average of four holes per stalk and gallery lengths of 4–6 cm [7]. This level of damage was above the economic threshold for this insect, estimated as one larva (hole) per stalk [8,9]. One *O. furnacalis* larva boring per plant during V10, R1, or R2 resulted in grain yield losses of 4.94%, 4.56%, or 3.76%, respectively [9].

The economic damage on corn indicates the need for effective and ecologically sound management practices for *O. furnacalis* in Indonesia and other countries in Southeast Asia [10,11], where growers have similar challenges due to this species attacking corn plants [5,12,13]. Chemical control using insecticides is the most commonly practiced control measure in Indonesia for the management of *O. furnacalis* [7]. Genetically modified corn plants expressing *Bacillus thuringiensis* (*Bt*) insecticidal proteins have been commercialized and are widely grown in corn-growing countries, such as the United States, Brazil, Argentina, Canada, South Africa, the Philippines, and Vietnam for the successful management of corn stalk borers, corn ear feeders, and fall armyworm (*Spodoptera frugiperda* (J.E. Smith)) [14–16]. Several *Bt* corn technologies are currently in the process of registration in Indonesia [17], including the MON 89034 corn. MON 89034 is a pyramided transgenic corn event expressing two *Bt* proteins, Cry1A.105 and Cry2Ab2, that are highly effective against key lepidopteran corn pests [18–20]. MON 89034 has been approved for commercialization in the Philippines since 2010 and in Vietnam since 2015 as a viable alternative for the sustainable control of *O. furnacalis* and other corn lepidopteran pests [15,21,22]. In September 2021, MON 89034 was approved as a registered product to control *S. frugiperda* in the Philippines [23].

The development of resistance in populations of the target pests poses the risk to the sustainability of *Bt* crops [24–26]. In Indonesia, the risk of resistance development could be higher due to the year-round cultivation of corn which may result in continuous pressure of selection once MON 89304 is commercialized. The planting of refuges and the adoption of technologies expressing multiple *Bt* proteins with an independent mode of action significantly reduces the risk of resistance development in populations of the target pests [27,28]. Because MON 89034 corn expresses two *Bt* proteins of different mechanisms of action targeting *O. furnacalis*, its inherent resistance risk is likely lower than that for single-gene *Bt* plants [29]. Nevertheless, it is important to monitor the susceptibility of *O. furnacalis* populations to Cry1A.105 and Cry2Ab2 proteins after the introduction of MON 89034 in Indonesia. Establishing the baseline susceptibility of *O. furnacalis* to Cry1A.105 and Cry2Ab2 is important for resistance-monitoring programs in Indonesia. The assessment of the baseline susceptibility would not only allow the assessment of the natural variation among field populations but can also be used to document shifts in susceptibility likely resulting from selection for resistance [30,31]. We hypothesize that there will be differences in the sensitivity of *O. furnacalis* populations. The goals of this study were to establish the baseline susceptibility of *O. furnacalis* to Cry1A.105 and Cry2Ab2 proteins in Indonesia and to estimate Cry1A.105 and Cry2Ab2 concentrations to be used as candidate diagnostic concentrations to monitor the development of resistance in *O. furnacalis* in Indonesia.

2. Results

2.1. Susceptibility of O. furnacalis to Cry1A.105

Populations of *O. furnacalis* collected from six provinces in 2013–2015 showed variation in susceptibility to Cry1A.105 (Table 1). The LC_{50} values for the field-collected populations ranged from 0.006 to 0.401 μg/mL, and for the laboratory population, these were 0.067 μg/mL. The LC_{95} of the laboratory population was 0.890 μg/mL, while those of the field-collected populations varied from 0.069 to 15.233 μg/mL. The resistance ratios (RRs) of the 24 field populations relative to the laboratory population were 0.1–6.0 based on the LC_{50} values (Table 1). Some field populations were more susceptible (RR < 1) to the protein than the laboratory population, while other field populations were more resistant (RR > 1) than the laboratory population. The difference between LC_{50} values of any two populations from Cry1A.105 assays was considered significant if their 95% fiducial limits (FLs) did not overlap [32,33].

Table 1. Susceptibility to Cry1A.105 of field-collected populations of *Ostrinia furnacalis* from major corn-growing areas in Indonesia.

Population			N	Slope (±SE)	LC_{50} (95% FL) (µg/mL)	LC_{95} (95% FL) (µg/mL)	RR [a]	χ^2 [c]
Province	**District**	**Village**						
North Sumatra	Langkat	Pasar VI Kwala Mencirim	440	2.05 ± 0.32	0.035 (0.012–0.066)	0.224 (0.112–1.173)	0.5	5.90
		Purwo Binangun	330	1.44 ± 0.32	0.039 (0.001–108)	0.535 (0.176–355.074)	0.6	6.38
	Deli Serdang	SM Diski	330	1.96 ± 0.42	0.010 (0.001–0.020)	0.069 (0.031–3.746)	0.1	1.31
		Suka Dame	330	1.27 ± 0.36	0.013 (N/A–N/A) *	0.250 (N/A–N/A) *	0.2	8.70 **
	Simalungun	Silenduk	550	0.77 ± 0.07	0.058 (0.031–0.105)	7.872 (2.647–45.801)	0.9	7.74
		Tangga Batu	550	0.93 ± 0.10	0.047 (0.023–0.081)	2.746 (1.311–8.273)	0.7	5.27
Lampung	South Lampung	Lematang	330	1.57 ± 0.42	0.045 (0.008–0.084)	0.495 (0.254–3.374)	0.7	3.57
		Kelau	550	1.33 ± 0.13	0.080 (0.044–0.132)	1.389 (0.701–4.166)	1.2	9.12
	Central Lampung	Trimurjo	330	0.86 ± 0.15	0.185 (0.066–0.357)	15.233 (5.345–127.917)	2.8	2.49
	Metro	Mulyojati	550	1.46 ± 0.20	0.032 (0.015–0.055)	0.429 (0.214–1.606)	0.5	4.17
	East Lampung	Gondang Rejo	550	1.70 ± 0.21	0.137 (0.102–0.176)	1.274 (0.906–2.003)	2.0	1.68
Gorontalo	Pohuwato	Manawa	550	1.65 ± 0.17	0.313 (0.238–0.403)	3.085 (2.062–5.408)	4.7	1.80
	Gorontalo	Tenilo	550	1.37 ± 0.20	0.273 (0.115–0.451)	4.316 (2.237–16.451)	4.1	2.18
Central Java	Purworejo	Ketawangrejo	550	1.38 ± 0.18	0.166 (0.071–0.292)	2.597 (1.148–16.906)	2.5	3.59
		Wonosari	550	1.06 ± 0.11	0.401 (0.233–0.630)	14.521 (7.038–45.194)	6.0	4.76
Yogyakarta	Sleman	Purwomartani	550	0.96 ± 0.11	0.051 (0.020–0.104)	2.604 (0.790–38.314)	0.8	6.87
		Widodomartani	550	1.02 ± 0.13	0.081 (0.022–0.197)	3.336 (0.932–98.888)	1.2	8.05
	Kulon Progo	Kedungsari	550	0.95 ± 0.10	0.053 (0.028–0.090)	2.907 (1.199–12.762)	0.8	5.77
		Kedundang	550	0.81 ± 0.07	0.101 (0.059–0.173)	0.730 (3.970–49.262)	1.5	6.55

Table 1. *Cont.*

Population			N	Slope (±SE)	LC_{50} (95% FL) (μg/mL)	LC_{95} (95% FL) (μg/mL)	RR [a]	$\chi^{2\,c}$
Province	District	Village						
East Java	Kediri	Papar	550	1.08 ± 0.10	0.043 (0.029–0.063)	1.404 (0.716–3.727)	0.6	3.90
		Mekikis	550	0.88 ± 0.11	0.006 (0.002–0.010)	0.418 (0.170–2.088)	0.1	3.16
	Nganjuk	Watu Dandang	550	0.77 ± 0.09	0.014 (0.006–0.025)	1.905 (0.659–12.081)	0.2	5.24
		Banjar Sari	550	0.83 ± 0.11	0.009 (0.003–0.020)	0.858 (0.231–20.417)	0.1	6.38
		Malangsari	550	0.91 ± 0.11	0.010 (0.004–0.021)	0.664 (0.225–6.057)	0.2	4.84
Laboratory [b]			330	1.47 ± 0.30	0.067 (0.004–0.164)	0.890 (0.336–49.547)	1.0	6.11

N = number of larvae. SE = standard error. 95% FL = 95% fiducial limits. [a] Resistance ratios (RRs) were calculated by dividing the values of LC_{50} of field populations by that of the laboratory population. [b] Collected from Yogyakarta in 2009 and maintained in the laboratory without *Bt* or insecticide selection. * The program did not give the 95% FL values. [c] Double asterisk (**) indicates significant differences based on the *p*-value (α = 0.05).

2.2. Susceptibility of *O. furnacalis* to Cry2Ab2

Populations of *O. furnacalis* collected from six provinces in 2013–2015 also showed variation in susceptibility to Cry2Ab2, as evidenced by the 0.1–2.0 RR values based on the LC_{50} (Table 2). The LC_{50} values of Cry2Ab2 against 24 field-collected populations ranged from 0.044 μg/mL to 4.490 μg/mL, whereas that of the laboratory population was 2.276 μg/mL. The LC_{95} value of the laboratory population was 22.984 μg/mL, and for the 24 field populations, it ranged from 3.320 to 277.584 μg/mL. Similarly to responses observed in the Cry1A.105 assays, some field populations were more susceptible to Cry2Ab2 than the laboratory population. The difference between LC_{50} values of any two populations from Cry2Ab2 assays was considered significant if their 95% fiducial limits (FLs) did not overlap [32,33].

Table 2. Susceptibility to Cry2Ab2 of field-collected populations of *Ostrinia furnacalis* from major corn-growing areas in Indonesia.

Population			N	Slope (±SE)	LC_{50} (95% FL) (μg/mL)	LC_{95} (95% FL) (μg/mL)	RR [a]	$\chi^{2\,c}$
Province	District	Village						
North Sumatra	Langkat	Pasar VI Kwala Mencirim	440	1.18 ± 0.16	1.527 (0.657–3.478)	37.665 (11.088–1150.551)	0.7	4.12
		Purwo Binangun	440	0.91 ± 0.19	4.361 (1.886–10.182)	277.584 (62.854–14,840.756)	1.9	2.70
	Deli Serdang	SM Diski	330	0.96 ± 0.18	3.416 (1.311–17.291)	180.121 (27.658–426,921.2)	1.5	3.44
		Suka Dame	440	2.23 ± 0.50	2.054 (0.490–3.540)	11.185 (5.908–130.605)	0.9	4.57
	Simalungun	Silenduk	550	0.83 ± 0.10	1.062 (0.673–1.684)	99.658 (36.621–502.575)	0.5	2.18
		Tangga Batu	440	1.58 ± 0.39	4.490 (2.313–7.002)	49.531 (22.220–511.071)	2.0	2.24

Table 2. *Cont.*

Population			N	Slope (±SE)	LC_{50} (95% FL) (μg/mL)	LC_{95} (95% FL) (μg/mL)	RR [a]	χ^{2} [c]
Province	District	Village						
Lampung	South Lampung	Lematang	550	1.19 ± 0.23	0.667 (0.268–1.181)	16.275 (6.610–132.775)	0.3	4.03
		Kelau	330	2.17 ± 0.56	4.018 (1.242–6.506)	23.015 (11.971–370.390)	1.8	5.09
	Central Lampung	Trimurjo	550	1.09 ± 0.19	2.074 (0.396–5.564)	67.414 (17.091–12,346.001)	0.9	3.89
	Metro	Mulyojati	550	0.99 ± 0.20	4.034 (0.966–12.182)	183.278 (36.318–284,609.920)	1.8	5.50
	East Lampung	Gondang Rejo	330	1.22 ± 0.27	1.662 (0.720–2.884)	37.164 (16.372–205.583)	0.7	1.71
Gorontalo	Pohuwato	Manawa	330	1.17 ± 0.28	1.590 (0.040–4.273)	40.540 (10.897–459,863.240)	0.7	5.71
	Gorontalo	Tenilo	550	1.01 ± 0.11	0.525 (0.212–1.110)	22.666 (6.744–387.286)	0.2	7.83
Central Java	Purworejo	Ketawangrejo	550	1.40 ± 0.16	0.989 (0.481–1.824)	14.861 (5.979–122.810)	0.4	4.39
		Wonosari	550	1.08 ± 0.11	0.254 (0.177–0.351)	8.371 (4.952–16.923)	0.1	2.20
Yogyakarta	Sleman	Purwomartani	550	1.09 ± 0.11	0.322 (0.175–0.550)	10.463 (4.533–41.801)	0.1	6.49
		Widodomartani	550	1.26 ± 0.16	0.163 (0.077–0.272)	3.320 (1.520–16.269)	0.1	2.57
	Kulon Progo	Kedungsari	550	1.20 ± 0.12	0.741 (0.401–1.340)	17.198 (6.732–104.303)	0.3	6.27
		Kedundang	550	1.01 ± 0.12	0.276 (0.134–0.477)	11.710 (4.766–61.882)	0.1	4.18
East Java	Kediri	Papar	330	0.66 ± 0.07	0.101 (0.058–0.176)	30.870 (10.553–144.161)	0.1	5.41
	Nganjuk	Mekikis	330	0.62 ± 0.07	0.095 (0.039–0.229)	41.121 (8.410–667.267)	0.1	11.65
		Watu Dandang	550	1.01 ± 0.10	0.585 (0.315–1.102)	24.771 (8.645–164.730)	0.3	8.48
		Banjar Sari	550	0.61 ± 0.06	0.044 (0.019–0.099)	22.509 (4.654–375.668)	0.1	12.86
		Malangsari	550	0.97 ± 0.08	0.522 (0.309–0.907)	25.653 (9.980–113.189)	0.2	8.43
Laboratory [b]			440	1.64 ± 0.48	2.276 (0.916–3.803)	22.984 (12.650–77.799)	1.0	0.62

N = number of larvae. SE = standard error. 95% FL = 95% fiducial limits. [a] Resistance ratios (RRs) were calculated by dividing the values of LC_{50} of field populations by that of the laboratory population. [b] Collected from Yogyakarta in 2009 and maintained in the laboratory without *Bt* or insecticide selection. [c] No significant differences based on the *p*-value (α = 0.05).

2.3. Candidate Diagnostic Concentrations

The *O. furnacalis* LC_{99} value for Cry1A.105 was significantly higher than its LC_{95} value as indicated by non-overlapping lower and upper limits of the 95% FL (Table 3). However, there was no significant difference observed between LC_{99} and LC_{95} values for Cry2Ab2. We propose the LC_{99} and its upper and lower fiducial limits (95% FLs) as the candidate diagnostic concentrations: 13.240 (6.716–33.831) μg/mL for Cry1A.105 and 127.320 (46.616–676.792) μg/mL for Cry2Ab2. These concentrations need to be tested with several field populations for further validation, followed by the selection of one diagnostic concentration for each protein to be used in monitoring programs.

Table 3. Candidate diagnostic concentrations of Cry1A.105 and Cry2Ab2 estimated using baseline susceptibility data of 24 field-collected populations of *Ostrinia furnacalis*.

Protein	Slope	No. Larvae	LC_{95} (95% FL) (μg/mL)	LC_{99} (95% FL) (μg/mL)
Cry1A.105	0.99 ± 0.03	11,200	2.720 (1.680–5.158) a	13.240 (6.716–33.831) b
Cry2Ab2	1.04 ± 0.03	11,440	28.050 (13.600–89.795) a	127.320 (46.616–676.792) a

SE = standard error. 95% FL = 95% fiducial limits. For each protein, LC_{95} and LC_{99} values followed by different letters were significantly different based on non-overlapping 95% fiducial limits.

3. Discussion

There is a need to develop a robust insect resistance-management (IRM) strategy for MON 89034 in Indonesia to prolong its durability in the field and, thus, delay the development of practical resistance once the event is approved for cultivation [27,34,35]. Baseline susceptibility data are essential for resistance-monitoring purposes, particularly to provide information on the level of susceptibility of *O. furnacalis* to Cry1A.105 and Cry2Ab2 before the introduction of MON 89034 corn and to provide benchmark data for future comparison to detect susceptibility shifts. In this study, the baseline susceptibility of *O. furnacalis* to Cry1A.105 and Cry2Ab2 was established based on the populations collected from major corn-producing provinces in Indonesia.

Differences in susceptibility to *Bt* proteins before the onset of resistance have been reported among geographically distinct populations in many different species of insects attacking corn [20,36–38]. For *O. furnacalis*, several studies have been conducted using five proteins, i.e., Cry1Ab, Cry1Ac, Cry1F, Cry1A.105, and Cry2Ab2. Field-collected populations of *O. furnacalis* in Vietnam differed in susceptibility to Cry1Ab by up to three-fold, which was reflective of natural variability among the 11 populations used in the study [39]. In China, the differences in susceptibility for this species were up to eight-fold for Cry1Ab [40], and in different study, up to two-fold for Cry1Ab, Cry1Ac, and Cry1F [41]. Contrastingly, in yet another study that assayed the bioactivity of Cry1Ab with 25 field populations of *O. furnacalis* in China [33], 80-fold and 309-fold variations at LC_{50} and LC_{95}, respectively, were demonstrated. Furthermore, Alcantara et al. [20] reported that the differences in the Philippines were six- and seven-fold to Cry1A.105 and Cry2Ab2, respectively.

We reported higher levels of variation in the susceptibility of *O. furnacalis* field populations to Cry1A.105 and Cry2Ab2 than were reported in the Philippines. However, direct comparison between these two studies (diet dipping vs. diet overlay) was not possible since different bioassay procedures were employed. Differences in the susceptibility of *O. furnacalis* to these two proteins may represent natural variation among populations because transgenic corn has not yet been planted commercially in Indonesia, and commercial *Bt* formulations for controlling *O. furnacalis* are not commonly utilized by Indonesian farmers [4]. A high difference in the susceptibility of field-collected populations was previously reported with *O. furnacalis* to Cry1Ab [33] and in *Helicoverpa armigera* to Cry1Ac (67-fold) in India [42]. In addition, the innate heterogeneity within insect populations tested in routine bioassays using the same methodology may account for three- to six-fold, or even twelve-fold, variation in laboratory-reared population comparisons [43,44]. Therefore, the heterogeneity may be even greater across field-derived populations, and the method chosen for bioassay needs careful consideration in the context of resistance monitoring.

Cry1A.105 was more toxic than Cry2Ab2 against *O. furnacalis* in Indonesia (Tables 1 and 2) and in the Philippines [20]. Similar results were reported for one species of corn stem borer, *Chilo partellus*, in India [38]. In contrast, Cry2Ab2 was more toxic than Cry1A.105 against the other species of corn borer in India, *Sesamia inferens* [38]. The same researchers also reported that Cry1A.105 was more toxic than Cry2Ab2 against *H. armigera*. Hernandez-Rodriguez et al. [45] also reported that Cry1A.105 was more toxic than Cry2Ab2 to *Ostrinia nubilalis* and *S. frugiperda*. These studies provide evidence that different species of corn borers may have different levels of susceptibility to different proteins and that their susceptibility may differ among geographically distinct populations.

A laboratory concentration of Cry1A.105 and Cry2Ab2 that reliably causes 99% mortality provides a candidate for use as a discriminating concentration [46]. The proposed diagnostic concentrations of 6.7, 13.2, and 33.8 μg/mL for Cry1A.105 and of 46.6, 127.3, and 676.8 μg/mL for Cry2Ab2 need to be validated against field-collected populations. Based on these tests, the determination of diagnostic concentrations for each protein is planned and these should be available before the commercialization of MON 89304.

The expression of Cry1A.105 and Cry2Ab2 in MON 89034 corn varies depending on the tissues with the highest expression occurred in the young leaves of V2–V4 with the average of 520 and 180 μg/g dry weight tissue [47]. This information in combination with baseline data may be used in developing IRM strategies in Indonesia.

4. Conclusions

This study reports the baseline susceptibility of distinct geographical *O. furnacalis* populations of Indonesia to Cry1A.105 and Cry2Ab2 proteins. Our findings demonstrate that *O. furnacalis* populations are highly sensitive to Cry1A.105 and Cry2Ab2 proteins, although there are differences in sensitivity among populations, which is a natural variation. The baseline susceptibility data are used to establish candidate diagnostic concentrations for further validations. The data on baseline susceptibility and the diagnostic concentrations provide valuable tools for future resistance-monitoring programs to detect early shifts in sensitivity among the field populations of *O. furnacalis* once transgenic corn expressing these two proteins is commercialized in Indonesia.

5. Materials and Methods

5.1. Field-Collected Populations

Egg masses, larvae, and pupae of *O. furnacalis* were collected from six provinces in Indonesia: three on Java Island (East Java, Central Java, and Yogyakarta Special Region), two on Sumatra Island (North Sumatra and Lampung), and the province of Gorontalo on Sulawesi (Figure 1). With the exception of Yogyakarta, these provinces are the major corn production sites in Indonesia, and insect pressure in these provinces were prominent based on the previous field observations. Field collections were made from non-*Bt* corn plants between 55 and 70 days after sowing with the intention of collecting insects from the second generation of *O. furnacalis*. Twenty-four populations were collected from 14 districts distributed in the six selected provinces between November 2013 and May 2015. The number of collected *O. furnacalis* from each location varied, and 83% of the collected populations were between 65 and 113 larvae and pupae. Only three of the collections were egg masses, each obtained from three different sites. The lowest number of *O. furnacalis* collected from one population was 12 larvae + 12 pupae + 1 egg mass. Each population was collected from at least two corn farms in one village to capture the genetic variability among individuals within a population and also because of the low infestations of *O. furnacalis* during the period of collection.

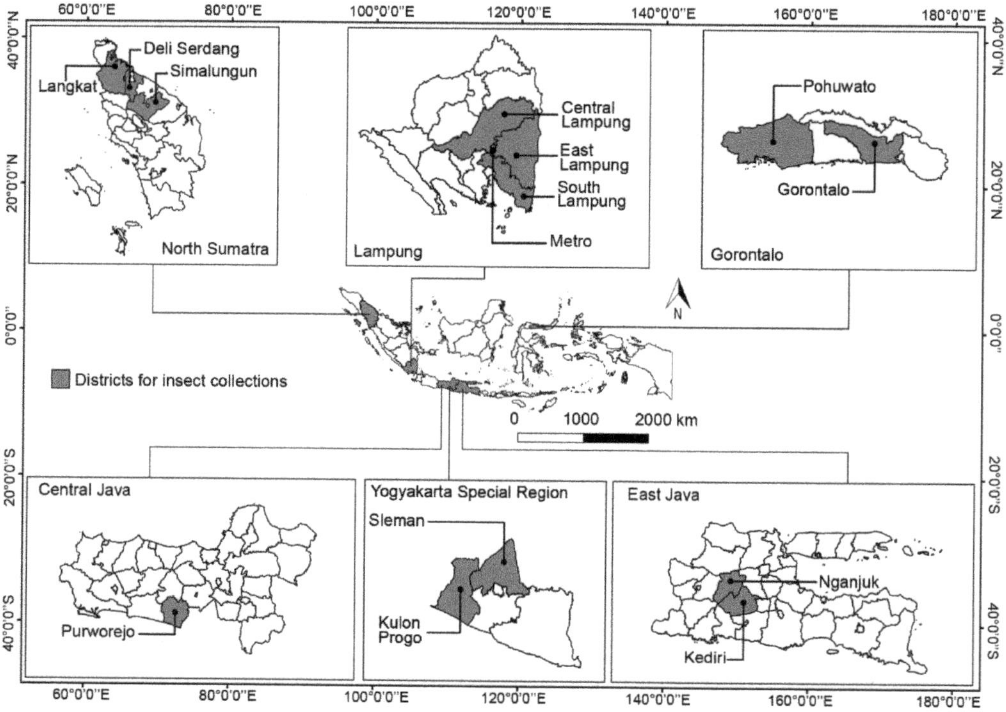

Figure 1. The six provinces in Indonesia selected for collecting field populations of *Ostrinia furnacalis* in 2013–2015 during the peak corn season in each province. With the exception of the Yogyakarta Special Region, these provinces were the top corn-producing areas in the country.

O. furnacalis at different life stages collected from the fields were handled using different methods. *O. furnacalis* egg masses were transferred into jars (18 cm tall, 7 cm diameter) layered with wetted filter paper to maintain the freshness of the eggs. Newly hatched larvae and bigger larvae were transferred individually into clear plastic containers with a screw cap (4.3 cm tall, 3.3 cm diameter) containing an artificial diet (ca. 5 g) [48] to minimize field-derived diseases. Collected pupae were put together in a container cup (the same size as for the larvae layered with filter paper) with a maximum of 10 pupae in each container. All collected larvae were healthy and pupated. Pupae produced from collected larvae, as well as those collected directly from the field, were placed into a Petri dish (9 cm diameter) layered with filter paper, and the Petri dish with the pupae was placed in a wire mesh cage (20 cm in diameter and 20 cm in height) upon arrival in the laboratory. Moisture was maintained by adding water to the filter paper in the plastic cups containing egg masses. *Ostrinia furnacalis* populations collected from different locations were reared in separate cups, trays, and cages to avoid contamination between populations. Additional details are provided below in "Insect-Rearing Procedure".

5.2. Insect-Rearing Procedure

The field-collected and laboratory *O. furnacalis* populations were reared using a similar artificial diet and standardized laboratory procedures [48]. Larvae were fed on a red-bean-based artificial diet (2–3 larvae per ~5 g of diet) in clear plastic cups (3.3 cm in diameter and 4.3 cm height) until pupation. A maximum of fifty cups were placed in a plastic tray. The trays were labelled with the locations of the populations. Pupae were collected daily and approximately 200 pupae were placed in a Petri dish (9 cm in diameter). The Petri dish containing the pupae were placed in the middle of a mating cage made of wire (21 cm in diameter and 21 cm height) covered with white paper on the top for oviposition. Emerging adults were fed with a 10% honey solution, and wet cotton was placed in the cage for maintaining high relative humidity during the day. The paper containing egg masses was removed every other day or daily as needed to collect larvae of similar ages. Egg masses were incubated in glass jars (6.5 cm in diameter and 15.5 cm height) containing moistened filter paper until hatching. A portion of the newly hatched larvae (ca. 300 larvae) was transferred individually into plastic cups as described above for the next generations, and the other portion was used for bioassays. All insect life stages were incubated at room temperature (24–28 °C) with relative humidity ranges of 60–85%. All equipment for making the artificial diet as well as for rearing was semi-sterilized by dipping in 10% sodium hypochlorite (SC Johnson, Indonesia). If mortality occurred during mass rearing, dead larvae were removed immediately from the colony to prevent contamination during mass rearing.

5.3. Susceptibility of *O. furnacalis* to Cry1A.105 and Cry2Ab2

The proteins (Cry1A.105 and Cry2Ab2) were provided by Bayer CropScience, St. Louis, MO, USA. Bioassays were carried out using mostly F_2 generation *O. furnacalis* neonates and a few using the F_1 or F_3 generation. The F_3 generation was used when neonates of F_2 was not sufficient to do the whole bioassays. The larvae of F_2 and F_3 were used if the F_1 neonates in uniform age were inadequate for the conduction of the bioassay. Bioassays were carried out by following the diet-dipping procedure described by Trisyono et al. [49] by dipping the diet in protein solutions or distilled water. A cube of diet (1 cm × 1 cm × 1 cm) was dipped for 10 s in a treated or control solution and then air dried for 20 min. After drying, the diet was transferred into a plastic cup of similar size to the ones used for rearing. For each replication, ten newly hatched *O. furnacalis* larvae were transferred individually into two plastic cups containing the treated or control diet (5 larvae per cup). Each treatment was replicated 3–5 times. In the Cry1A.105 bioassays, 10 concentrations from 0.002 to 48 µg/mL Cry1A.105 (three-fold dilutions) were tested to determine LC_{50} and LC_{95} of 24 field-collected *O. furnacalis* populations and the laboratory population. For Cry2Ab2, 10 concentrations ranging from 0.0007 to 15.5 µg/mL (three-fold dilutions) and the control were used for all bioassays to determine LC_{50} and LC_{95} against the same F_1–F_3 of field-collected and laboratory populations. Based on the results from the preliminary bioassays, the concentrations used for Cry1A.105 and Cry2Ab2 were different due to varying response of *O. furnacalis* to each protein. The selected concentrations based on the preliminary test were expected to result in larval mortality ranging from 2% to 98%. Prior to the actual assays, preliminary tests of both proteins were carried out using 10 newly hatched larvae per concentration in three replicates. Larvae were exposed to the treated or control diet continuously, and observed mortality was recorded 7 days after they were placed in the plastic cups containing the artificial diet.

5.4. Data Analysis

Probit analysis [50] was carried out using the PoloJR program within PoloSuite, Version 2.1 (LeOra Software 2016) to determine LC_{50} and LC_{95} values and their 95% fiducial limits (FLs) for each protein and population. The relative resistance ratios (RR) were calculated by dividing the values of LC_{50} of field populations by that of the laboratory population [51]. The LC_{95} and LC_{99} and their lower and upper limits for the Cry1A.105 and Cry2Ab2 proteins were determined from pooled baseline concentration–mortality data (24 field-collected populations) using the program PoloJr, and the estimated LC_{99} and its lower and upper limits of 95% were proposed as candidate diagnostic concentrations for each protein, as they killed 99% of the susceptible population [52,53].

Author Contributions: Conceptualization, Y.A.T.; methodology, Y.A.T.; formal analysis, V.E.F.A. and T.R.; investigation, V.E.F.A. and T.R.; writing—original draft, Y.A.T.; writing—review and editing, S.M., G.P.H., S.P. and L.R.C.; funding acquisition, Y.A.T. All authors have read and agreed to the published version of the manuscript.

Funding: This research was funded by Universitas Gadjah Mada and PT Branita Sandhini Indonesia (Bayer Indonesia) grant number [032/RA Dept/PTBS/2012] or [2137/PN/TU/VII/2012] and The APC was funded by Universitas Gadjah Mada and Bayer CropScience Indonesia.

Institutional Review Board Statement: Not applicable.

Informed Consent Statement: Not applicable.

Data Availability Statement: Not applicable.

Acknowledgments: We thank Muhamad Lihawa, Sriyanto Harjanto, many students, and field partners for assistance in collecting and rearing the insects.

Conflicts of Interest: The authors from UGM (Y.A.T., V.E.F.A., T.R.) worked on the project as study collaborators, and some of the authors are employed by Bayer CropScience (S.M., G.P.H., S.P., L.R.C.). The authors of the manuscript declare that no conflicts of interest could have appeared to influence the work reported here.

References

1. Gerpacio, R.V.; Pingali, P.L. *Tropical and Subtropical Maize in Asia: Production Systems, Constraints, and Research Priorities*; CIMMYT: Méx, Mexico, 2007; ISBN 978-970-648-155-9.
2. Areekul, S.; Skulpanich, U.; Teeravate, P. Some studies on the control of corn borer in Thailand. *Agric. Nat. Resour.* **1964**, *4*, 110–119.
3. Camarao, G.C. Population dynamics of the cornborer, *Ostrinia furnacalis* (Guenee), I. Life cycle, behavior, and generation cycles. *Philipp. Entomol.* **1976**, *3*, 179–200.
4. Nafus, D.M.; Schreiner, I.H. Location of *Ostrinia furnacalis* (Lepidoptera: Pyralidae) eggs and larvae on sweet corn in relation to plant growth stage. *J. Econ. Entomol.* **1987**, *80*, 411–416. [CrossRef]
5. Nafus, D.M.; Schreiner, I.H. Review of the biology and control of the Asian corn borer, *Ostrinia furnacalis* (Lep: Pyralidae). *Trop. Pest Manag.* **1991**, *37*, 41–56. [CrossRef]
6. Da-Lopez, Y.F.; Trisyono, Y.A.; Witjaksono, W.; Subiadi, S. Distribution pattern of *Ostrinia furnacalis* Guenée (Lepidoptera Crambidae) egg-mass on maize-field. *J. Entomol. Indones.* **2014**, *11*, 81–92. [CrossRef]
7. Trisyono, Y.A.; Aryuwandari, V.E.F.; Andika, I.P.; Sinulingga, N.G.H. *Assessment on the Economic Importance of Corn Borers in Indonesia*; Technical Report Submitted to Croplife Indonesia; Universitas Gadjah Mada: Yogyakarta, Indonesia, 2020; p. 41.
8. Morallo-Rejesus, B.; Buctuanon, E.M.; Rejesus, R.S. Defining the economic threshold determinants for the Asian corn borer, *Ostrinia furnacalis* (Guenee) in the Philippines. *Int. J. Pest Manag.* **1990**, *36*, 114–121. [CrossRef]
9. Subiadi, S.; Trisyono, Y.A.; Martono, E. Economic injury level (EIL) of *Ostrinia furnacalis* (Lepidoptera: Crambidae) larvae on three growth stages of corn. *J. Entomol. Indones.* **2014**, *11*, 19–26. [CrossRef]
10. CABI. *Ostrinia furnacalis* (Asian Corn Borer). In: Invasive Species Compendium. Available online: https://www.cabi.org/isc/datasheet/38026 (accessed on 2 October 2021).
11. Plantwise Knowledge Bank. Asian Corn Borer (*Ostrinia furnacalis*). Available online: https://www.plantwise.org/knowledgebank/datasheet/38026 (accessed on 2 October 2021).
12. Hussein, M.Y.; Kameldeer, A.K. A field study on the oviposition of *Ostrinia furnacalis* Guenee (Lepidoptera: Pyralidae) on maize in Selangor, Malaysia. *Int. J. Pest Manag.* **1988**, *34*, 44–47. [CrossRef]
13. Brookes, G.; Dinh, T.X. The impact of using Genetically Modified (GM) corn/maize in Vietnam: Results of the first farm-level survey. *GM Crops Food* **2021**, *12*, 71–83. [CrossRef]
14. International Service for the Acquisition of Agri-Biotech Applications (ISAAA). *Global Status of Commercialized Biotech/GM Crops in 2019: Biotech Crops Drive Socio-Economic Development and Sustainable Environment in the New Frontier*; ISAAA Brief No. 55; ISAAA: Ithaca, NY, USA, 2019.
15. ISAAA. ISAAA's GM Approval Database. Available online: http://www.isaaa.org/gmapprovaldatabase/ (accessed on 2 October 2021).
16. Biosafety Clearing-House (BCH). Country's Decisions and Other Communications. Available online: https://bch.cbd.int/en/registries/living-modified-organisms (accessed on 17 August 2023).
17. Indonesia Biosafety Clearing House. Keputusan Aman-Pangan. Available online: https://indonesiabch.menlhk.go.id/surat-keputusan/ (accessed on 17 August 2023).
18. Erasmus, A.; Marais, J.; Van den Berg, J. Movement and survival of Busseola fusca (Lepidoptera: Noctuidae) larvae within maize plantings with different ratios of non-Bt and Bt seed. *Pest. Manag. Sci.* **2016**, *72*, 2287–2294. [CrossRef]
19. Botha, A.S.; Erasmus, A.; du Plessis, H.; Van den Berg, J. Efficacy of Bt maize for control of Spodoptera frugiperda (Lepidoptera: Noctuidae) in South Africa. *J. Econ. Entomol.* **2019**, *112*, 1260–1266. [CrossRef] [PubMed]
20. Alcantara, E.; Atienza, M.M.; Camacho, L.; Parimi, S. Baseline susceptibility of Philippine *Ostrinia furnacalis* (Lepidoptera: Crambidae) populations to insecticidal Cry1A. *105 and Cry2Ab2 proteins and validation of candidate diagnostic concentration for monitoring resistance. Biodiversitass* **2021**, *22*, d220251. [CrossRef]

21. James, C. *Global Status of Commercialized Biotech/GM Crops: 2011*; ISAAA Briefs no 43; ISAAA: Ithaca, NY, USA, 2011.
22. International Service for the Acquisition of Agri-Biotech Applications (ISAAA). *Global Status of Commercialized Biotech/GM Crops: 2016*; ISAAA Brief No. 52; ISAAA: Ithaca, NY, USA, 2016.
23. Department of Agriculture. List of Registered Plant-Incorporated Protectants Derived from Modern Biotechnology 2021. Department of Agriculture Quezon City, Philippines. Available online: https://fpa.da.gov.ph/NW/images/FPAfiles/DATA/Regulation/Pesticide/Files-2021/ListofRegisteredPIP12312021.pdf (accessed on 5 January 2022).
24. Gould, F. Sustainability of transgenic insecticidal cultivars: Integrating pest genetics and ecology. *Annu. Rev. Entomol.* **1998**, *43*, 701–726. [CrossRef] [PubMed]
25. Tabashnik, B.E. Delaying insect resistance to transgenic crops. *Proc. Natl. Acad. Sci. USA* **2008**, *105*, 19029–19030. [CrossRef] [PubMed]
26. Tabashnik, B.E.; Carrière, Y. Evaluating cross-resistance between Vip and Cry toxins of Bacillus thuringiensis. *J. Econ. Entomol.* **2020**, *113*, 553–561. [CrossRef]
27. Roush, R.T. Two–toxin strategies for management of insecticidal transgenic crops: Can pyramiding succeed where pesticide mixtures have not? *Phil. Trans. R. Soc. Lond. B* **1998**, *353*, 1777–1786. [CrossRef]
28. Zhao, J.-Z.; Cao, J.; Li, Y.; Collins, H.L.; Roush, R.T.; Earle, E.D.; Shelton, A.M. Transgenic plants expressing two Bacillus thuringiensis toxins delay insect resistance evolution. *Nat. Biotechnol.* **2003**, *21*, 1493–1497. [CrossRef]
29. Carrière, Y.; Crickmore, N.; Tabashnik, B.E. Optimizing pyramided transgenic Bt crops for sustainable pest management. *Nat. Biotechnol.* **2015**, *33*, 161–168. [CrossRef]
30. Blanco, C.A.; Storer, N.P.; Abel, C.A.; Jackson, R.; Leonard, R.; Lopez, J.D.; Payne, G.; Siegfried, B.D.; Spencer, T.; N-Vargas, A.P.T. Baseline susceptibility of tobacco budworm (Lepidoptera: Noctuidae) to Cry1F toxin from Bacillus thuringiensis. *J. Econ. Entomol.* **2008**, *101*, 6. [CrossRef]
31. Leite, N.A.; Pereira, R.M.; Durigan, M.R.; Amado, D.; Fatoretto, J.; Medeiros, F.C.L.; Omoto, C. Susceptibility of Brazilian populations of Helicoverpa armigera and Helicoverpa zea (Lepidoptera: Noctuidae) to Vip3Aa20. *J. Econ. Entomol.* **2018**, *111*, 399–404. [CrossRef]
32. Robertson, J.L.; Preisler, H.K. *Pesticide Bioassays with Arthropods*; CRC Press: Boca Raton, FL, USA, 1992; ISBN 0-8493-6463-9.
33. Liu, X.; Liu, S.; Long, Y.; Wang, Y.; Zhao, W.; Shwe, S.M.; Wang, Z.; He, K.; Bai, S. Baseline susceptibility and resistance allele frequency in *Ostrinia furnacalis* in relation to Cry1Ab toxins in China. *Toxins* **2022**, *14*, 255. [CrossRef] [PubMed]
34. Tabashnik, B.E.; Re, Y.C.; Dennehy, T.J.; Morin, S.; Sisterson, M.S.; Roush, R.T.; Shelton, A.M.; Zhao, J.-Z. Insect resistance to transgenic Bt crops: Lessons from the laboratory and field. *J. Econ. Entomol.* **2003**, *96*, 1031–1038. [CrossRef] [PubMed]
35. Huang, F.; Andow, D.A.; Buschman, L.L. Success of the high-dose/refuge resistance management strategy after 15 years of Bt crop use in North America: Bt crops and resistance management. *Entomol. Exp. Appl.* **2011**, *140*, 1–16. [CrossRef]
36. Alcantara, E.; Estrada, A.; Alpuerto, V.; Head, G. Monitoring Cry1Ab Susceptibility in Asian corn borer (Lepidoptera: Crambidae) on Bt corn in the Philippines. *Crop Prot.* **2011**, *30*, 554–559. [CrossRef]
37. Marçon, P.C.R.G.; Young, L.J.; Steffey, K.L.; Siegfried, B.D. Baseline susceptibility of European corn borer (Lepidoptera: Crambidae) to Bacillus thuringiensis toxins. *J. Econ. Entomol.* **1999**, *92*, 279–285. [CrossRef]
38. Jalali, S.K.; Yadavalli, L.; Ojha, R.; Kumar, P.; Sulaikhabeevi, S.B.; Sharma, R.; Nair, R.; Kadanur, R.C.; Kamath, S.P.; Komarlingam, M.S. Baseline sensitivity of maize borers in India to the Bacillus thuringiensis insecticidal proteins Cry1A.105 and Cry2Ab2: Bt baseline sensitivity of Indian maize Lepidopteran Pests. *Pest. Manag. Sci.* **2015**, *71*, 1082–1090. [CrossRef]
39. Le, D.K.; Le, Q.K.; Tran, T.T.H.; Nguyen, D.V.; Dao, T.H.; Nguyen, T.T.; Truong, X.L.; Nguyen, Q.C.; Pham, H.P.; Phan, T.T.T.; et al. Baseline susceptibility of Asian corn borer (*Ostrinia furnacalis* (Guenée)) populations in Vietnam to Cry1Ab insecticidal protein. *J. Asia-Pac. Entomol.* **2019**, *22*, 493–498. [CrossRef]
40. He, K.; Wang, Z.; Wen, L.; Bai, S.; Ma, X.; Yao, Z. Determination of baseline susceptibility to Cry1Ab protein for Asian corn borer (Lep., Crambidae). *J. Appl. Entomol.* **2005**, *129*, 407–412. [CrossRef]
41. Li, G.; Huang, J.; Ji, T.; Tian, C.; Zhao, X.; Feng, H. Baseline susceptibility and resistance allele frequency in *Ostrinia furnacalis* related to Cry1 toxins in the Huanghuaihai summer corn region of China. *Pest Manag. Sci.* **2020**, *76*, 4311–4317. [CrossRef]
42. Kranthi, K.R.; Kranthi, S.; Wanjari, R.R. Baseline toxicity of Cry1A toxins to Helicoverpa armigera (Hubner) (Lepidoptera: Noctuidae) in India. *Int. J. Pest Manag.* **2001**, *47*, 141–145. [CrossRef]
43. Bird, L.J.; Akhurst, R.J. Variation in susceptibility of Helicoverpa armigera (Hübner) and Helicoverpa punctigera (Wallengren) (Lepidoptera: Noctuidae) in Australia to two Bacillus thuringiensis Toxins. *J. Invertebr. Pathol.* **2007**, *94*, 84–94. [CrossRef] [PubMed]
44. Siegfried, B.D.; Spencer, T.; Crespo, A.L.; Storer, N.P.; Head, G.P.; Owens, E.D.; Guyer, D. Ten years of Bt resistance monitoring in the European corn borer: What we know, what we don't know, and what we can do better. *Am. Entomol.* **2007**, *3*, 208–214. [CrossRef]
45. Hernández-Rodríguez, C.S.; Hernández-Martínez, P.; Van Rie, J.; Escriche, B.; Ferré, J. Shared midgut binding sites for Cry1A.105, Cry1Aa, Cry1Ab, Cry1Ac and Cry1Fa proteins from Bacillus thuringiensis in two important corn pests, Ostrinia nubilalis and Spodoptera frugiperda. *PLoS ONE* **2013**, *8*, e68164. [CrossRef]
46. Marçon, P.C.R.G.; Siegfried, B.D.; Spencer, T.; Hutchison, W.D. Development of diagnostic concentrations for monitoring Bacillus thuringiensis resistance in European corn borer (Lepidoptera: Crambidae). *J. Econ. Entomol.* **2000**, *93*, 925–930. [CrossRef] [PubMed]

47. Monsanto Company. Petition for the Determination of Non-Regulated Status for MON 89034. Available online: https://www.aphis.usda.gov/brs/aphisdocs/06_29801p.pdf (accessed on 17 August 2023).
48. Rahayu, T.; Trisyono, Y.A.; Witjaksono. Fitness of Asian corn borer, *Ostrinia furnacalis* (Lepidoptera: Crambidae) reared in an artificial diet. *J. Asia-Pac. Entomol.* **2018**, *21*, 823–828. [CrossRef]
49. Trisyono, Y.A.; Rahayu, S.T.S.; Margino, S. Bioactivity of a Bacillus thuringiensis Cry1Ac toxin to Spodoptera litura. *J. Perlindungan Tanam. Indones.* **2004**, *10*, 53–62.
50. Finney, D.J. *Probit Analysis*; Cambridge University Press: Cambridge, UK, 1971.
51. Oppenoorth, F.J.; Welling, W. Biochemistry and Physiology of Resistance. In *Insecticide Biochemistry and Physiology*; Wilkinson, C.F., Ed.; Springer: Boston, MA, USA, 1976; pp. 507–551. ISBN 978-1-4899-2214-4.
52. Roush, R.T.; Miller, G.L. Considerations for design of insecticide resistance monitoring programs. *J. Econ. Entomol.* **1986**, *79*, 293–298. [CrossRef]
53. Menger, J.; Beauzay, P.; Chirumamilla, A.; Dierks, C.; Gavloski, J.; Glogoza, P.; Hamilton, K.; Hodgson, E.W.; Knodel, J.J.; MacRae, I.V.; et al. Implementation of a diagnostic-concentration bioassay for detection of susceptibility to pyrethroids in soybean aphid (Hemiptera: Aphididae). *J. Econ. Entomol.* **2020**, *113*, 932–939. [CrossRef]

toxins

MDPI

Article

Analysis of Synergism between Extracellular Polysaccharide from *Bacillus thuringensis* subsp. *kurstaki* HD270 and Insecticidal Proteins

Bai Xue [1,2], Meiling Wang [2], Zeyu Wang [2], Changlong Shu [2], Lili Geng [2,*] and Jie Zhang [1,2,*]

1 College of Life Sciences, Northeast Agricultural University, Harbin 150030, China
2 State Key Laboratory for Biology of Plant Diseases and Insect Pests, Institute of Plant Protection, Chinese Academy of Agricultural Sciences, Beijing 100193, China
* Correspondence: genglili@caas.cn (L.G.); zhangjie05@caas.cn (J.Z.)

Abstract: *Bacillus thuringiensis* (Bt) is the most widely used biopesticide worldwide and can produce several insecticidal crystal proteins and vegetative insecticidal proteins (Vips) at different growth stages. In our previous study, extracellular polysaccharides (EPSs) of Bt strain HD270 were found to enhance the insecticidal activity of Cry1Ac protoxin against *Plutella xylostella* (L.) and promote the binding of Cry1Ac to the intestinal brush border membrane vesicles (BBMVs). Whether the synergistic activity of Bt EPSs is common to other Cry1-type or Vip proteins is unclear, as is the potential synergistic mechanism. In this study, crude EPS-HD270 was found to increase the toxicity of Cry1-type toxins and Vip3Aa11 against different lepidopteran pests by approximately 2-fold. The purified EPS-HD270 also possessed synergistic activity against the toxicity of Cry1Ac and Vip3Aa11 against *Spodoptera frugiperda* (J.E. Smith) and *Helicoverpa armigera* (Hübner). Furthermore, we found that EPS-HD270 had a strong binding ability with Vip3Aa11 and promoted the binding of Vip3Aa11 to the BBMVs of *H. armigera* and *S. frugiperda*. Bt EPS-HD270 also protected Vip3Aa11 from proteolytic processing in larval midgut juice. Bt EPSs had universal synergistic effects on Cry1-type or Vip toxins against *S. frugiperda* and *H. armigera*. Bt EPS-HD270 exhibited synergistic activity with Vip3Aa through promotion of binding to BBMVs and protection from digestion by midgut protease. The results indicated that synergistic activity with Bt toxins was an important function of Bt EPSs, which was very different from other *Bacillus* spp.

Keywords: *Bacillus thuringiensis*; microbial pesticides; extracellular polysaccharides; synergistic effects; BBMVs; midgut juice

Key Contribution: The EPS-HD270 could commonly increase the toxicity of both Cry and Vip proteins against different target insect pests. The results reveal multi-biological functions of Bt EPSs and lay a theoretical foundation for the application of EPSs in Bt formulation.

Citation: Xue, B.; Wang, M.; Wang, Z.; Shu, C.; Geng, L.; Zhang, J. Analysis of Synergism between Extracellular Polysaccharide from *Bacillus thuringensis* subsp. *kurstaki* HD270 and Insecticidal Proteins. *Toxins* **2023**, *15*, 590. https://doi.org/10.3390/toxins15100590

Received: 24 August 2023
Revised: 22 September 2023
Accepted: 26 September 2023
Published: 28 September 2023

1. Introduction

Bacterial extracellular polysaccharides (EPSs) are synthesized inside the cells and secreted into the extracellular environment or synthesized outside the cells by cell wall-anchored enzymes during bacterial growth and metabolism [1,2]. Exopolysaccharides are macromolecules composed of monosaccharide units that form complex structures and exhibit multiple functions. Bacteria produce extracellular polysaccharides to protect themselves from various environmental stresses, including desiccation, ionic stress, and other biotic stresses, such as predation by amoebae [3], facilitating their adaptation in different habitats by forming biofilms. *Bacillus* bacteria have also been reported to produce extracellular polysaccharides with flocculation activity [4], and the abilities to absorb lead [5] and induce systemic resistance to plant pathogenic bacteria [6], in addition to their well-known function in the biofilm matrix of *Bacillus subtilis* [7]. In our recent study, we found that most *Bacillus thuringiensis* (Bt) strains could produce extracellular polysaccharides. EPS produced by the Bt HD270 strain increased the activity of insecticidal crystal proteins [8].

Bt is an aerobic, Gram-positive entomopathogenic bacterium [9] that produces insecticidal proteins at different growth periods [10]. These proteins are commonly known as crystal (Cry) proteins when produced during the spore-forming stage and vegetative insecticidal proteins (Vips) when produced at the vegetative growth stage. Due to their specific insecticidal activity against a large number of lepidopteran and coleopteran pests in agriculture, they are widely used in biopesticides and transgenic crops [11]. The two types of proteins have different structures and different insecticidal mechanisms, but both need to bind to receptors in the intestinal brush border membrane vesicles (BBMVs) to exert insecticidal activity.

The exploration of Bt strains and insecticidal proteins has entered a bottleneck period, making it difficult to explore new insecticidal resources. Therefore, there is growing interest in synergistic substances that could improve the larvicidal activity of Bt proteins. In our previous study, extracellular polysaccharides produced by the Bt strain HD270 (EPS-HD270) were observed to increase the toxicity of the total toxins isolated from HD270 and Cry1Ac proteins against *Plutella xylostella* (L.) [8]. The molecular weight of EPS-HD270 was 58.0 kDa, and EPS-HD270 was composed mainly of mannose (44.2%) and GlcN (35.5%) [8]. However, it is not clear whether Bt EPSs would commonly increase the toxicity of both Cry and Vip proteins against different target insect pests. Clarification of this issue will reveal multi-biological functions of Bt EPSs and lay a theoretical foundation for using EPSs to improve the control effect of Bt formulations.

In this study, the synergistic effects of EPS with Cry and Vip proteins were analyzed. EPS-HD270 not only enhanced the toxicity of the Cry1Ac, Cry1Ab, and Cry1Ah proteins but also increased that of the Vip3Aa11 protein against *Helicoverpa armigera* (Hübner) and *Spodoptera frugiperda* (J.E. Smith). The potential mechanism involved in the synergism of EPSs and Vip3Aa11 toxin was the protection of Vip3Aa11 from proteolytic processing in larval midgut juice and the increased binding affinity of Vip3Aa11 to BBMVs of *H. armigera* and *S. frugiperda* caused by adding EPSs.

2. Results

2.1. Universal Synergistic Activity of Bt EPSs on Insecticidal Proteins

EPS-HD270 was found to enhance the virulence of the Cry1Ac protoxin against *P. xylostella* in our previous work [8], but the universality of the synergistic effect is unclear. In this study, the synergistic activities of EPS-HD270 with Cry1Ac protein against larvae of other lepidopteran insect pests, including *S. frugiperda* and *H. armigera*, were further analyzed. The LC_{50} values of Cry1Ac protoxin against first instar larvae of *S. frugiperda* and *H. armigera* were 68.26 μg/g and 12.40 μg/g, respectively, while the LC_{50} values of Cry1Ac protoxin supplemented with 2.0 mg/g crude EPS were 33.64 μg/g and 5.92 μg/g (Table 1), respectively, indicating a 2.03-fold and 2.09-fold increase in insecticidal activity. Crude EPS was treated with proteinase K and further purified with an anion-exchange chromatography column and a gel filtration chromatography column, as in our previous study [8], to obtain a single component (shown in Figure S1). Moreover, the addition of 0.5 mg/g purified EPS significantly increased the 7-day mortality of 50 and 75 μg/g Cry1Ac protoxin against *S. frugiperda* to 63.9% and 66.7% ($p < 0.05$, Figure 1A); 7-day mortality of 12 and 16 μg/g Cry1Ac protoxin against *H. armigera* increased to 69.4% and 87.5% ($p < 0.05$, Figure 1B). However, crude and purified EPS showed no toxicity against *S. frugiperda* and *H. armigera* larvae.

In addition, the synergistic activity of EPS-HD270 with Cry1Ab (which shares 86% of the amino acid sequence identity of Cry1Ac) and Cry1Ah protein (which shares 88% of the amino acid sequence identity of Cry1Ac) was investigated (Figure 2A). The bioassay results showed that the 3-day mortality of 20 μg/mL Cry1Ab against *P. xylostella* was increased to 92.9% ($p < 0.01$) and 100.0% ($p < 0.001$) by adding 5 and 50 mg/mL crude EPS, respectively (Figure 2B). The 3-day mortality of 0.8 μg/mL Cry1Ah against *P. xylostella* was increased to 76.2% ($p < 0.01$) and 95.2% ($p < 0.001$) by adding 5 and 50 mg/mL crude EPS, respectively, and the mortality of 1.6 μg/mL Cry1Ah was also significantly increased to 100.0% ($p < 0.01$) by adding 50 mg/mL crude EPS (Figure 2C). Overall, Bt EPS-HD270 enhanced the virulence of the Cry1Ac protoxin against lepidopteran insect pests, including *P. xylostella*, *S. frugiperda* and *H. armigera*, and synergistic activities of EPS-HD270 with other Cry1-type proteins, including Cry1Ab and Cry1Ah, were observed.

Table 1. The LC_{50} value of Cry1Ac and crude EPS against *S. frugiperda* and *H. armigera*.

Insect Pests	Treatment	LC_{50} (μg/g) (95% Fiducial Limits)	Slope ± SE	χ^2	Fold Change
S. frugiperda	Cry1Ac	68.26 (46.95–97.97)	2.67 ± 0.30	7.25	2.03
	Cry1Ac + EPS	33.64 (29.76–37.60)	6.88 ± 0.83	5.31	
H. armigera	Cry1Ac	12.40 (9.57–16.49)	1.22 ± 0.17	2.30	2.09
	Cry1Ac + EPS	5.92 (4.28–8.47)	1.72 ± 0.19	3.17	

The concentration of EPS in artificial diets was 2 mg/g. Three replications of each sample.

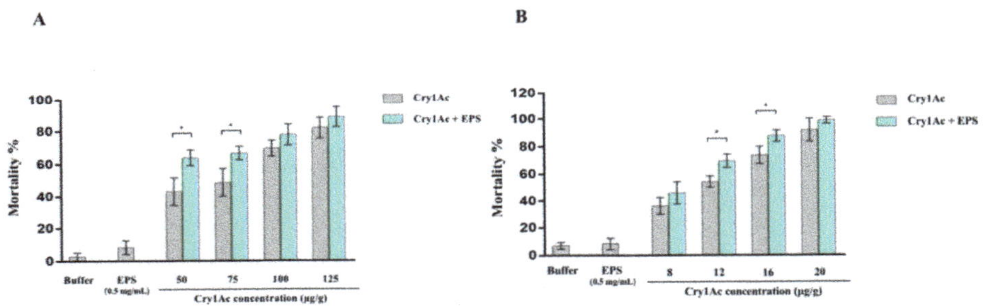

Figure 1. Insecticidal activity of Cry1Ac protein by adding purified EPS of Bt HD270 against *S. frugiperda* (**A**) and *H. armigera* (**B**). Data are the average ± standard deviation (SD) from three independent experiments. (* $p < 0.05$).

In addition to Cry-1 type proteins with a typical three-domain structure, the effects of Bt EPSs on the toxicity of Vip3Aa were analyzed. The LC_{50} of Vip3Aa protein against first instar larvae of *S. frugiperda* and *H. armigera* were 1.22 μg/g and 12.57 μg/g, respectively, while the LC_{50} of Vip3Aa protein added with 2.0 mg/g crude EPS were 0.49 μg/g and 5.51 μg/g (Table 2), indicating that the virulence was enhanced 2.49-fold and 2.28-fold, respectively. Furthermore, the addition of 0.5 mg/g purified EPSs significantly increased the 7-day mortality of 0.8 and 1.2 μg/g Vip3Aa protein against *S. frugiperda* to 55.6% ($p < 0.05$) and 77.7% ($p < 0.01$, Figure 3A). The 7-day mortality of 9 and 16 μg/g Vip3Aa protein against *H. armigera* was also enhanced to 50.0% and 72.2%, respectively ($p < 0.05$, Figure 3B). In summation, Bt EPS-HD270 exhibited synergistic activities with Vip3Aa against *S. frugiperda* and *H. armigera* as well.

Table 2. The LC_{50} values of Vip3Aa and crude EPS against *S. frugiperda* and *H. armigera*.

Insect Pests	Samples	LC_{50} (μg/g) (95% Fiducial Limits)	Slope ± SE	χ^2	Fold Change
S. frugiperda	Vip3Aa	1.22 (0.96–1.58)	1.34 ± 0.17	1.30	2.49
	Vip3Aa + EPS	0.49 (0.34–0.70)	0.92 ± 0.16	0.74	
H. armigera	Vip3Aa	12.57 (9.62–16.98)	1.17 ± 0.17	1.58	2.28
	Vip3Aa + EPS	5.51 (4.28–7.20)	1.25 ± 0.17	2.63	

The concentration of EPS in the artificial diet was 2 mg/g. Three replications of each sample.

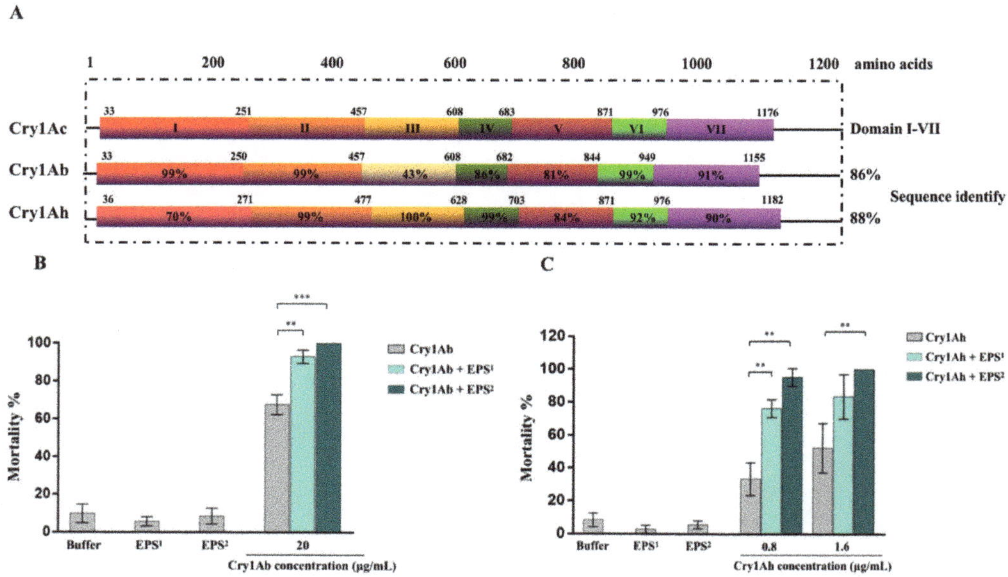

Figure 2. Insecticidal activity of crude EPS–HD270 and Cry1A protein against *P. xylostella*. (**A**) Analysis of the amino acid sequence identity of Cry1Ab and Cry1Ah with Cry1Ac protoxins. Percentage represents amino acid sequence identity with domain I to VII of Cry1Ac protein. (**B**) Cry1Ab protein and EPS. [1] means the concentration of crude EPS was 5 mg/mL. [2] means the concentration of crude EPS was 50 mg/mL. (**C**) Cry1Ah protein and EPS. Data are the mean value ± standard deviation from three independent experiments. (** $p < 0.01$, *** $p < 0.001$).

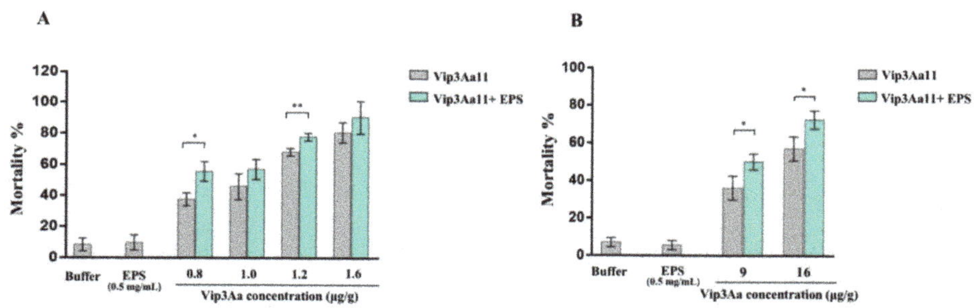

Figure 3. Insecticidal activity of Vip3Aa protein and purified EPS-HD270 EPS against *S. frugiperda* (**A**) and *H. armigera* (**B**). Data are the mean value ± standard deviation from three independent experiments. (* $p < 0.05$, ** $p < 0.01$).

2.2. EPS-HD270 Formed Specific Binding with Vip3Aa11 Protein

The Vip3A protoxin contains five distinct structural domains (domains I–V), and the carbohydrate-binding module domains were found in domains IV and V, which were exposed to the solvent [12,13]. We hypothesized that EPS might bind to the Vip3A protoxin, which in turn affects the toxicity of the Vip3A protein against *S. frugiperda* and *H. armigera*. The interaction between EPS-HD270 and the Vip3Aa protein was determined by ELISA using an anti-Vip3A antibody. The results showed that EPS-HD270 specifically bound to the Vip3Aa11 protoxin with a K_d value of 6.43 ± 1.31 nM (Figure 4).

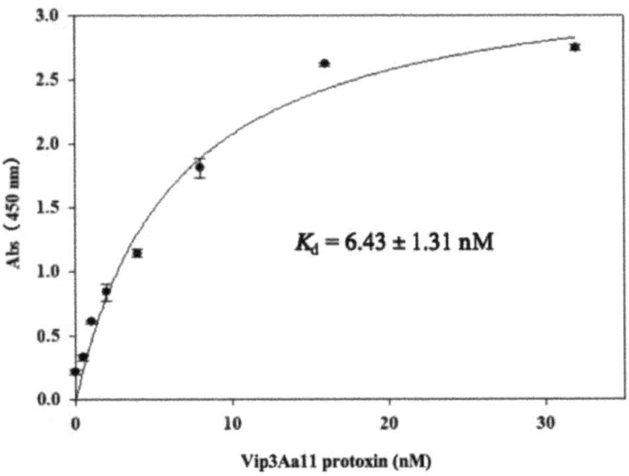

Figure 4. Analysis of the interaction between Vip3Aa11 and purified EPS. Error bars represent SD. Abs represents absorbance.

2.3. EPS-HD270 Delayed Proteolytic Processing of Cry1Ac and Vip3Aa Proteins in Midgut Juice

Purified EPS-HD270 specifically bound with Cry1Ac [8] and Vip3Aa, which might protect proteins from being digested by proteases in the midgut. The effects of EPS on the proteolytic processing of Bt proteins in midgut juice were analyzed. Cry1Ac and Vip3Aa proteins were mixed with purified EPS-HD270 in the proportions used in the bioassay. The intensity of the protein bands was determined using ImageJ software, and untreated Cry1Ac or Vip3Aa was used as a 100% reference. After 40 µg of Cry1Ac processing by *S. frugiperda* midgut juice ratios at 0.003% and 0.001%, the density of protein with a molecular mass higher than 70 kDa was approximately 39% and 72%, while with the addition of 400 µg EPS-HD270, the density of protein greater than 70 kDa was approximately 67% and 91% (Figure 5A). The same phenomenon was observed when 10 µg of Cry1Ac was mixed with 400 µg of EPS-HD270 processed with *H. armigera* midgut juice at ratios of 0.003% and 0.001% (Figure 5A). EPS-HD270 delayed proteolytic processing of Cry1Ac protoxin in the midgut juice of *S. frugiperda* and *H. armigera*.

After 6 µg of Vip3Aa protoxin processing with *S. frugiperda* midgut juice ratios at 0.03% and 0.01%, the density of protoxin was approximately 29% and 66%, while with the addition of 400 µg EPS-HD270, the density of protoxin was increased to 60% and 91%, respectively (Figure 5B). Similar results were obtained when the density of protoxin was increased to 68% and 93% with the addition of 400 µg of EPS-HD270 after processing with *H. armigera* midgut juice ratios at 0.01% and 0.003% (Figure 5B). Overall, EPS-HD270 delayed proteolytic processing of the Cry1Ac and Vip3A protoxins in the midgut juice of *H. armigera* and *S. frugiperda*.

2.4. EPS-HD270 Enhances the Binding Affinity of Vip3Aa11 to BBMVs of H. armigera and S. frugiperda

Cry1-type proteins and Vip3A proteins are pore-forming toxins, and the ability to bind to receptors on BBMVs is a primary factor affecting their virulence [14]. Our previous work found that EPS-HD270 promotes the binding ability of Cry1Ac protoxin to BBMVs of *P. xylostella*. Therefore, the effects of EPS on the binding of Vip3Aa11 protoxin to the BBMVs of *S. frugiperda* and *H. armigera* were further analyzed in the present study. Saturation binding assays were conducted by using increasing concentrations of Vip3Aa proteins to bind to the BBMVs of *H. armigera* and *S. frugiperda*. Western blot results showed that binding of Vip3Aa11 protein to BBMVs of these two insects was saturable (Figure 6A,B). Subsequently, 80 nmol/L of Vip3Aa11 at unsaturated binding was used, and the addition of EPS-HD270 at mass ratio of 1:10 and 1:50 enhanced the signal of Vip3Aa11 protoxin binding to BBMVs of *S. frugiperda* 2.41- and 3.16-fold (Figure 6C). In addition, for BBMVs of *H. armigera*, 160 nmol/L of Vip3Aa11 at unsaturated binding was used, and the addition of EPS-HD270 at mass ratio of 1:50 and 1:100 enhanced the signal of Vip3Aa11 protoxin binding to BBMVs 4.07- and 6.18-fold (Figure 6D). Thus, the EPS-HD270 interacted with Vip3Aa11 protoxin directly to enhance binding to BBMVs, which was correlated with increased toxicity against *S. frugiperda* and *H. armigera*.

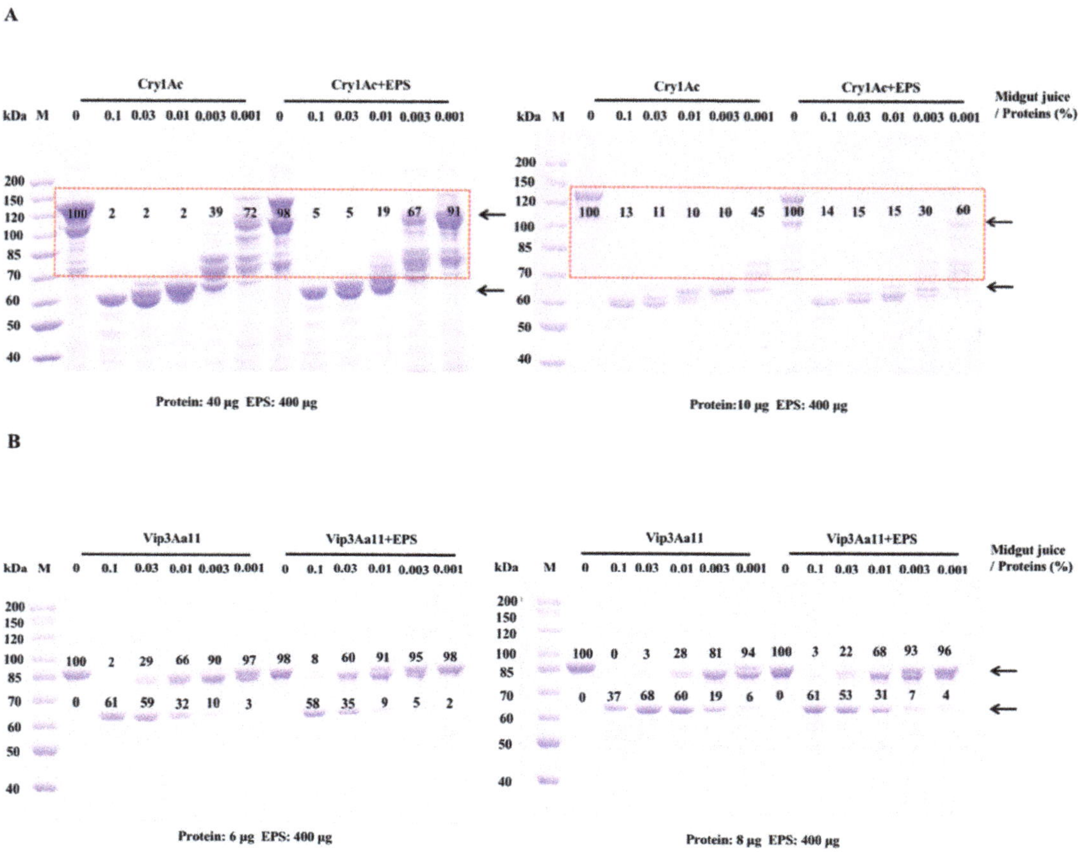

Figure 5. Proteolytic processing of Cry1Ac and Vip3Aa11 in midgut juice. (**A**) SDS-PAGE analysis of Cry1Ac digested by S. frugiperda (left) and H. armigera (right) midgut juice. The arrows indicate Cry1Ac protoxin (~130 kDa) and activated toxin (~65 kDa). (**B**) SDS-PAGE analysis of Vip3Aa11 digested by S. frugiperda (left) and H. armigera (right) midgut juice. The arrow indicates Vip3Aa11 protoxin (~88 kDa) and activated toxin (~66 kDa). M, molecular weight marker.

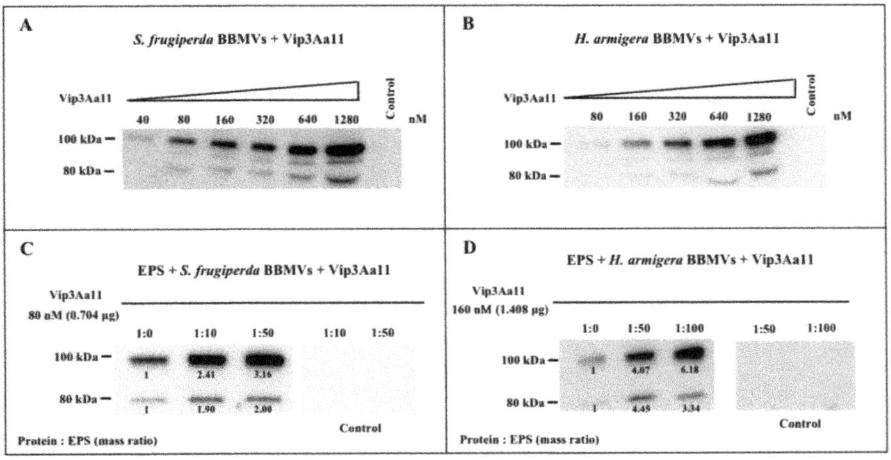

Figure 6. Binding assay of Vip3Aa11 protoxins and purified EPS to BBMVs of *S. frugiperda* and *H. armigera*. Saturation binding assays of Vip3Aa11 to *S. frugiperda* (**A**) and *H. armigera* BBMVs (**B**). The control represents the highest concentration of Vip3Aa11 without BBMVs, indicating that the toxins did not precipitate in the absence of BBMVs. (**C**) The effects of EPS on the binding ability of Vip3Aa11 (80 nM) to *S. frugiperda* BBMVs. (**D**) The effects of EPS on the binding ability of Vip3Aa11 (160 nM) to *H. armigera* BBMVs. Control represents the absence of BBMVs, and the result indicates that toxins mixed with EPS did not precipitate in the absence of BBMVs.

3. Discussion

Extracellular polysaccharides of Bt exist in fermentation broth during production and are discarded as byproducts. In addition, due to the requirements of environmental protection, fermentation wastewater can only be discharged after strict treatment, which undoubtedly increases the production costs of Bt products, increases the economic burden on manufacturers, and increases costs for farmers using biological pesticides.

In this study, we found that exopolysaccharides produced by Bt increased the toxicity of Cry1-type and Vip3 proteins against Lepidopteran insect pests. In other words, EPS produced by excellent strains such as HD270 can improve the insecticidal activity of a variety of insecticidal proteins against a variety of insect pests. If these polysaccharides are effectively recycled and added to Bt products, it will reduce the environmental pollution caused by emissions; more importantly, it will improve the insecticidal effect of Bt products, save production costs, and be an innovative move that kills two birds with one stone. To our knowledge, Bt EPS were the first to exhibit synergistic activities with self-producing active substances of bacteria. This study lays a theoretical foundation for the production and application of Bt extracellular polysaccharides and provides important references for the creation of other biological pesticides.

Chakroun et al. reported that trypsin-activated Vip3Aa protein showed higher toxicity than nickel-purified Vip3Aa protoxin against *S. frugiperda* [15]. In this study, only nickel-purified Vip3Aa11 protoxin was used. Further study is needed to investigate the effects of EPS-HD270 on the activated Vip3Aa protein. In addition, some Bt toxins were found to evolve resistance to lepidopteran insect pests, including *P. xylostella* with resistance against Cry1Ac [16,17], *H. armigera* with resistance against Vip3Aa [18], and *S. frugiperda* with resistance against Cry1F and Cry1Ab [19–21]. Whether EPS produced by Bt can help Bt toxins deal with the resistance is a question that needs to be analyzed.

The synthesis of extracellular polysaccharides is an energy-consuming process [22]. Our previous work found that 96.5% of 170 Bt strains cultured in LB medium produced exopolysaccharides [8]. Exopolysaccharides produced by *Bacillus* can not only help *Bacillus* spp. deal with external pressures [7], but also improve their virulence to the hosts, which is very important for the infection and epidemic of Bt strains [8]. These important ecological functions may be the reasons why most Bt strains need to synthesize extracellular polysaccharides at the expense of energy.

Whether Cry1-type or Vip3 proteins are activated by proteases in the midgut is the key factor in exerting their insecticidal activities [23,24]. The difference is that domains IV and V of Cry-1-type proteins are released, while domains I and II of Vip3 proteins are removed after proteolytic processing [23–25]. Over-digestion or insufficient processing of Cry protoxins has been reported to affect their virulence [26]. In addition, Vip3Aa mutants with greater stability have been shown to exhibit higher toxicity against *S. frugiperda* and *H. armigera* [27]. In the present study, the addition of EPS delayed proteolytic processing of Cry1Ac and Vip3Aa proteins in the gut juice of *H. armigera* and *S. frugiperda* (Figure 5), which was related to their enhanced toxicity. Because EPS formed specific bonds with Cry1Ac [8] and Vip3A proteins (Figure 4), we speculated that EPS might cover the proteolytic processing sites and make them inaccessible. This is consistent with the extracellular polysaccharide alginate of *Pseudomonas aeruginosa* protecting lipase LipA from degradation by the extracellular protease elastase covering cleavage sites [28].

In our previous study, EPS-HD270 was found to specifically bind to Cry1Ac protoxin (K_d values of 113.0 ± 35.1 nmol/L) [8] and promote the binding of Cry1Ac to BBMVs of *P. xylostella*. Vip3Aa protoxin showed a higher affinity for EPS (K_d values of 6.43 ± 1.31 nM) than Cry1Ac (Figure 4), and binding to BBMVs of *H. armigera* and *S. frugiperda* was promoted by interaction with EPS-HD270 (Figure 6). The insecticidal activity of the Vip3Aa protein increased with the promotion of its ability to bind to BBMVs [27].

Domains IV and V of Vip3Aa are predicted to contain a carbohydrate-binding motif [13]. In this study, EPS-HD270 was found to show a higher affinity for Vip3Aa than Cry1Ac (Figure 4). Moreover, it promoted the binding of the Vip3Aa protoxin to BBMVs of *S. frugiperda* and *H. armigera*. (Figure 6). Since the insecticidal activity of the Vip3Aa protein increased with the promotion of its ability to bind to BBMVs [27], EPS-HD270 was found to specifically bind to the Cry1Ac protoxin and promote the binding of Cry1Ac to the BBMV of *P. xylostella* [8]. We speculate that EPS-HD270 increased the virulence of Vip3Aa by enhancing its binding to BBMVs.

4. Conclusions

In this study, we found that the extracellular polysaccharide from Bt HD270, a strain belonging to subsp. *kurstaki*, can not only enhance the virulence of insecticidal proteins with different structures commonly contained in subsp. *Kurstaki*, but it also has a synergistic effect against a variety of lepidopteran insect pests. The synergistic activities of EPS-HD270 with Cry-1 type and Vip3Aa proteins were mainly related to the delay in proteolytic processing and promotion of binding to BBMVs.

5. Materials and Methods

5.1. Strains

Bt serovar *kurstaki* HD270 was used for EPS production. The recombinant Bt strain HD73$^-$-*cry1Ab* (Bt HD73-strain containing the *cry1Ab* gene) and the HD73$^-$ strain containing the *cry1Ah* gene were used for Cry1Ab and Cry1Ah protein extraction. The Vip3Aa11 protein was derived from an *Escherichia coli* expression system, and the plasmid pET28a carried the *vip3Aa11* gene. All of the strains mentioned here were stored in our laboratory. Luria–Bertani (LB) media were used for all strains to grow at 30 °C.

5.2. Isolation and Purification of Insecticidal Proteins

Bt strains HD73-*cry1Ab* and HD73-*cry1Ah* were grown in 1/2 liquid Luria–Bertani (LB) media until 50–60% of insecticidal crystals were released. Cry1Ab and Cry1Ah proteins were extracted according to the continuous crystal solubilization method mentioned by Zhou et al. [29], purified using an ÄKTA avant 150 system (GE Healthcare Life Sciences; Piscataway, NJ, USA) and affinity chromatography (HiTrap Q Sepharose High Performance

5 mL, GE Healthcare, Uppsala, Sweden). Vip3Aa11 protein was extracted and purified according to a previous publication by Wang et al. [30]. Vip3Aa protein carrying a His-tag was purified by Ni^{2+}-affinity chromatography and finally desalted. The concentrations of Cry1Ab/h and Vip3Aa11 purified proteins were measured via sodium dodecyl sulfate-polyacrylamide gel electrophoresis (SDS-PAGE) and quantified using ImageJ software (Version 1.42I).

5.3. Culture of Strain and Preparation of EPS-HD270

The extraction and purification of EPS-HD270 were performed with reference to the description published by Wang et al. [8]. The Bt serovar *kurstaki* HD270 strain, kept in our laboratory, was cultured at 30 °C using liquid LB medium. The bacterial suspension was collected via centrifugation at 12,000× *g* for 30 min before all spores were released. The supernatant obtained after centrifugation was mixed with 95% ethanol at a volume ratio of 1:3 and placed at 4 °C overnight to precipitate EPS. After centrifugation, the precipitate was resuspended in ultra-pure water, and proteinase K was added at one-tenth of the protein mass in the mixture and digested at 50 °C for 2 h. Proteinase K was then inactivated for 20 min at 100 °C to obtain a crude EPS solution.

The proteinase K-treated solution was purified on an ÄKTA avant 150 system. After small molecule impurities had been removed using a desalting column, the targets were purified using an anion-exchange chromatography column (Q HP column). All fractions were collected and identified via the phenol–sulfuric acid method [31]. The target samples were further separated using a gel filtration chromatography column (HiLoad 26/600 Superdex 200, GE Healthcare Life Sciences). The main components were recovered, and NaCl was removed using a desalting column. EPS-HD270 concentrations were tested using the phenol–sulfuric acid method.

5.4. Bioassay

The first instar larvae of *H. armigera* were provided by the Jilin Academy of Agricultural Sciences. The first instar larvae of *S. frugiperda* and the second instar larvae of *P. xylostella* were maintained in our laboratory. Bioassays were performed using proteins, EPS and mixtures of proteins with EPS.

The bioassay method for *P. xylostella* was conducted as mentioned previously [8]. Briefly, fresh cabbage leaves with a diameter of 6 cm were immersed in a gradient of different concentrations of proteins and EPS. The leaf surfaces were dried, and each leaf was placed in a plastic Petri dish (9 cm diameter) with 30 larvae. Cry1A buffer (20 mmol/L Na_2CO_3-$NaHCO_3$, pH 9.8) and EPS buffer (ultra-pure water) were used as controls. Mortality was calculated 3 days later. Each treatment was performed in three replicates. Thirty larvae were used for each replicate.

The bioassays of *S. frugiperda* and *H. armigera* were performed using first instar larvae [27]. Artificial diets containing soy flour, wheat bran, yeast, and vitamins were prepared as described by Liang et al. [32]. Fifteen grams of artificial diets were weighed separately in each Petri dish (9 cm diameter). The protein–exopolysaccharide mixture (3 mL) was added to the diet and equally distributed into 24-well culture plates after moderate moisture was evaporated. The larvae were placed in the treated diet and covered with tissue. Each treatment was conducted in three replicates and 24 larvae were used for each replicate. The same ratio of Cry1Ac/Vip3Aa11 buffer (20 mmol/L Na_2CO_3-$NaHCO_3$/20 mmol/L Tris-HCl) and EPS buffer (ultra-pure water) was used as the control. Seven-day mortality was counted.

In all insect bioassay experiments, larvae that did not respond to being poked were determined to be dead larvae.

5.5. Preparation of BBMVs

The insects were maintained until the third instar, and the midguts were obtained after dissection. After the preparation of BBMVs via magnesium precipitation [33], the enzyme activity of aminopeptidase N (APN) was calculated to determine the purity of the BBMVs [34].

5.6. ELISA Analysis of the Binding of Vip3Aa11 Protoxin to EPS-HD270

EPS-HD270 (2.0 mg/mL, 100 μL) was loaded into 96-well ELISA plates (Nunc Maxisorb, Thermo) and immobilized at 4 °C overnight. The ELISA plates were washed three times with Tris-buffered saline (TBS) and blocked with TBST (TBS containing 0.1% Tween-20) containing 2.0% BSA (200 μL) at 37 °C for 2 h. The plates were washed with TBST three times and incubated with different concentrations of Vip3Aa11 (0, 0.5, 1, 2, 4, 8, 16, 32 nmol/L) (100 μL) for 1 h at 37 °C. After washing with TBST buffer 3 times, anti-Vip3A antibody (TBST 1:5000 dilution, 100 μL) was added. TBST (100 μL) containing 1/10,000 HRP-conjugated goat anti-mouse IgG (Solarbio Life Sciences, Beijing, China) was incubated at 37 °C for 1 h. For each treatment, three replicates were performed. After washing, the reaction was tested with 3,3′,5,5′-tetramethylbenzidine (TMB) solution (Solarbio Life Sciences, Beijing, China) (100 μL) for 15 min in the dark at 37 °C. The reaction was terminated with HCl (2.0 mol/L, 100 μL), and the absorbance was immediately read at 450 nm using a microplate reader. The equilibrium dissociation constant (K_d) was analyzed using Sigma-plot Software (Version 12.0).

5.7. Western Blot Analysis of the Binding of Vip3Aa11 Protoxin with EPS-HD270 to BBMVs of S. frugiperda and H. armigera

The EPS-HD270 was incubated with Vip3Aa11 protoxin at 4 °C overnight after mixing with BBMVs (20 μg). The control was EPS buffer (ultrapure water) incubated with BBMVs (20 μg) and Vip3Aa11 protoxin. After centrifugation at 18,000× *g* for 10 min at 4 °C, proteins were separated by SDS-PAGE and electrotransferred to PVDF membranes. After blocking with TBST containing 5% skimmed milk powder on a low-speed spinner for 1 h, the PVDF membranes were incubated with Vip3A antibody at a 1:5000 dilution for 1 h and washed with TBST. HRP-conjugated goat anti-mouse IgG at a 1:10,000 dilution in TBST was added and incubated for 1 h. After they had been washed three times, the membranes were detected with SuperSignal™ West Pico PLUS Chemiluminescent Substrate chemiluminescent substrate (Thermo Scientific, Waltham, MA, USA) using a LAS-4000 mini-imaging system (GE Healthcare, Chicago, IL, USA).

5.8. Proteolytic Processing Analysis of Insecticidal Protoxin

The fresh intact midguts of *S. frugiperda* and *H. armigera* third instar larvae were centrifuged at 14,000× *g* for 20 min at 4 °C to obtain the supernatant. The concentration gradient of protein to midgut juice was set at a volume ratio, and the final volume of protein was 100 μL. EPS was added at the mass ratio (EPS: protoxin) used in the bioassay, and the final volume of EPS was 100 μL. The digestion of proteins by the midgut fluid was analyzed via SDS-PAGE after 1 h of incubation in a water bath at 30 °C.

5.9. Data Analysis

For each treatment, mortality data of five different concentrations were used to calculate LC_{50} by SPSS 20 (IBM, Armonk, NY, USA) using probit analysis. The statistical analyses in the figure were performed using GraphPad Prism (version 7.0). Before the analysis was performed, percentages were arcsine square root transformed, and pairwise comparisons were performed via *t*-test.

Supplementary Materials: The following supporting information can be downloaded at: https://www.mdpi.com/article/10.3390/toxins15100590/s1, Figure S1. The preparation flow diagram of purified EPS-HD270.

Author Contributions: Conceptualization, L.G. and J.Z.; methodology, B.X.; software, B.X. and M.W.; validation, Z.W. and C.S.; formal analysis, B.X.; investigation, B.X.; resources, M.W.; data curation, Z.W.; writing—original draft preparation, L.G. and B.X.; writing—review and editing, L.G. and J.Z.;

supervision, L.G. and J.Z.; project administration, L.G. and J.Z.; funding acquisition, L.G. and J.Z. All authors have read and agreed to the published version of the manuscript.

Funding: This study was funded by the National Key Research and Development Program of China (2022YFE0116500), and the Agricultural Science and Technology Innovation Program of the Chinese Academy of Agricultural Sciences (CAAS-ZDRW202108).

Institutional Review Board Statement: Not applicable.

Informed Consent Statement: Not applicable.

Data Availability Statement: No new sequencing data were created or analyzed in this study.

Conflicts of Interest: The authors declare no conflict of interest.

References

1. Bazaka, K.; Crawford, R.J.; Nazarenko, E.L.; Ivanova, E.P. Bacterial extracellular polysaccharides. *Adv. Exp. Med. Biol.* **2011**, *715*, 213–226. [PubMed]
2. Nwodo, U.U.; Green, E.; Okoh, A.I. Bacterial exopolysaccharides: Functionality and prospects. *Int. J. Mol. Sci.* **2012**, *13*, 14002–14015. [CrossRef] [PubMed]
3. Limoli, D.H.; Jones, C.J.; Wozniak, D.J. Bacterial extracellular polysaccharides in biofilm formation and function. *Microbiol. Spectr.* **2015**, 3, 1–19. [CrossRef] [PubMed]
4. Kumar, C.G.; Joo, H.S.; Choi, J.W.; Koo, Y.M.; Chang, C.S. Purification and characterization of an extracellular polysaccharide from haloalkalophilic *Bacillus* sp. I-450. *Enzyme Microb. Technol.* **2004**, *34*, 673–681. [CrossRef]
5. Shameer, S. Biosorption of lead, copper and cadmium using the extracellular polysaccharides (EPS) of *Bacillus* sp. from solar salterns. *3 Biotech.* **2016**, *6*, 194. [CrossRef]
6. Jiang, C.H.; Fan, Z.H.; Xie, P.; Guo, J.H. *Bacillus cereus* AR156 extracellular polysaccharides served as a novel micro-associated molecular pattern to induced systemic immunity to Pst DC3000 in Arabidopsis. *Front. Microbiol.* **2016**, *7*, 664. [CrossRef]
7. Vlamakis, H.; Chai, Y.; Beauregard, P.; Losick, R.; Kolter, R. Sticking together: Building a biofilm the *Bacillus subtilis* way. *Nat. Rev. Microbiol.* **2013**, *11*, 157–168. [CrossRef]
8. Wang, M.L.; Geng, L.L.; Xue, B.; Wang, Z.Y.; Xu, W.Y.; Shu, C.L.; Zhang, J. Structure characteristics and function of a novel extracellular polysaccharide from *Bacillus thuringiensis* strain 4D19. *Int. J. Biol. Macromol.* **2021**, *189*, 956–964. [CrossRef]
9. Harris, M.K. *Bacillus thuringiensis* and pest control. *Science* **1991**, *253*, 1194. [CrossRef]
10. Palma, L.; Muñoz, D.; Berry, C.; Murillo, J.; Caballero, P. *Bacillus thuringiensis* toxins: An overview of their biocidal activity. *Toxins.* **2014**, *6*, 3296–3325. [CrossRef]
11. Sanahuja, G.; Banakar, R.; Twyman, R.M.; Capell, T.; Christou, P. *Bacillus thuringiensis*: A century of research, development and commercial applications. *Plant Biotechnol J.* **2021**, *9*, 283–300. [CrossRef] [PubMed]
12. Núñez-Ramírez, R.; Huesa, J.; Bel, Y.; Ferré, J.; Casino, P.; Arias-Palomo, E. Molecular architecture and activation of the insecticidal protein Vip3Aa from *Bacillus thuringiensis*. *Nat. Commun.* **2020**, *11*, 3974. [CrossRef] [PubMed]
13. Jiang, K.; Zhang, Y.; Chen, Z.; Wu, D.L.; Cai, J.; Gao, X. Structural and functional insights into the C-terminal fragment of insecticidal Vip3A toxin of *Bacillus thuringiensis*. *Toxins* **2020**, *12*, 438. [CrossRef] [PubMed]
14. Chakrabarty, S.; Jin, M.; Wu, C.; Chakraborty, P.; Xiao, Y. *Bacillus thuringiensis* vegetative insecticidal protein family Vip3A and mode of action against pest Lepidoptera. *Pest Manag. Sci.* **2020**, *76*, 1612–1617. [CrossRef]
15. Chakroun, M.; Bel, Y.; Caccia, S.; Abdelkefi-Mesrati, L.; Escriche, B.; Ferré, J. Susceptibility of *Spodoptera frugiperda* and *Spodoptera exigua* to *Bacillus thuringiensis* Vip3Aa insecticidal protein. *J. Invertebr. Pathol.* **2012**, *110*, 334–339. [CrossRef]
16. Baxter, S.W.; Badenes-Pérez, F.R.; Morrison, A.; Vogel, H.; Crickmore, N.; Kain, W.; Wang, P.; Heckel, D.G.; Jiggins, C.D. Parallel evolution of *Bacillus thuringiensis* toxin resistance in lepidoptera. *Genetics* **2011**, *189*, 675–679. [CrossRef]
17. Guo, Z.; Kang, S.; Chen, D.; Wu, Q.; Wang, S.; Xie, W.; Zhu, X.; Baxter, S.W.; Zhou, X.; Jurat-Fuentes, J.L.; et al. MAPK signaling pathway alters expression of midgut ALP and ABCC genes and causes resistance to *Bacillus thuringiensis* Cry1Ac toxin in diamondback moth. *PLoS Genet* **2015**, *11*, e1005124. [CrossRef]
18. Mahon, R.J.; Downes, S.J.; James, B. Vip3A resistance alleles exist at high levels in Australian targets before release of cotton expressing this toxin. *PLoS ONE* **2012**, *7*, e39192. [CrossRef]
19. Farias, J.R.; Andow, D.A.; Horikoshi, R.J.; Sorgatto, R.J.; Fresia, P.; Dos Santos, A.C.; Omoto, C. Field-evolved resistance to Cry1F maize by *Spodoptera frugiperda* (Lepidoptera: Noctuidae) in Brazil. *Crop Prot.* **2014**, *64*, 150–158. [CrossRef]
20. Huang, F.; Qureshi, J.A.; Meagher, R.L.; Jr Reisig, D.D.; Head, G.P.; Andow, D.A.; Ni, X.; Kerns, D.; Buntin, G.D.; Niu, Y.; et al. Cry1F resistance in fall armyworm *Spodoptera frugiperda*: Single gene versus pyramided Bt maize. *PLoS ONE* **2014**, *9*, e112958. [CrossRef]
21. Storer, N.P.; Babcock, J.M.; Schlenz, M.; Meade, T.; Thompson, G.D.; Bing, J.W.; Huckaba, R.M. Discovery and characterization of field resistance to Bt maize: *Spodoptera frugiperda* (Lepidoptera: Noctuidae) in Puerto Rico. *J. Econ. Entomol.* **2010**, *103*, 1031–1038. [CrossRef] [PubMed]

22. Madhuri, K.; Prabhakar, K. Microbial exopolysaccharides: Biosynthesis and potential applications. *Orient J. Chem.* **2014**, *30*, 1401–1410. [CrossRef]
23. Chakroun, M.; Banyuls, N.; Bel, Y.; Escriche, B.; Ferré, J. Bacterial vegetative insecticidal proteins (Vip) from entomopathogenic bacteria. *Microbiol. Mol. Biol. Rev.* **2016**, *80*, 329–350. [CrossRef]
24. Adang, M.J.; Crickmore, N.; Fuentes, J.L.J. Diversity of *Bacillus thuringiensis* crystal toxins and mechanism of action. *Adv. Insect. Physiol.* **2014**, *47*, 39–87.
25. Byrne, M.J.; Iadanza, M.G.; Perez, M.A.; Maskell, D.P.; George, R.M.; Hesketh, E.L.; Beales, P.A.; Zack, M.D.; Berry, C.; Thompson, R.F. Cryo-EM structures of an insecticidal Bt toxin reveal its mechanism of action on the membrane. *Nat. Commun.* **2021**, *12*, 2791. [CrossRef] [PubMed]
26. Lightwood, D.J.; Ellar, D.J.; Jarrett, P. Role of proteolysis in determining potency of *Bacillus thuringiensis* Cry1Ac δ-Endotoxin. *Appl. Environ. Microbiol.* **2000**, *66*, 5174–5181. [CrossRef]
27. Yang, X.X.; Wang, Z.Y.; Geng, L.L.; Chi, B.Y.; Liu, R.M.; Li, H.T.; Gao, J.G.; Zhang, J. Vip3Aa domain IV and V mutants confer higher insecticidal activity against *Spodoptera frugiperda* and *Helicoverpa armigera*. *Pest Manag. Sci.* **2022**, *78*, 2324–2331. [CrossRef]
28. Tielen, P.; Kuhn, H.; Rosenau, F.; Jaeger, K.E.; Wingender, J. Interaction between extracellular lipase LipA and the polysaccharide alginate of *Pseudomonas aeruginosa*. *BMC Microbiol.* **2013**, *13*, 159. [CrossRef]
29. Zhou, Z.S.; Yang, S.J.; Shu, C.L.; Song, F.P.; Zhou, X.P.; Zhang, J. Comparison and optimization of the method for Cry1Ac protoxin preparation in HD73 strain. *J. Integr. Agr.* **2015**, *14*, 1598–1603. [CrossRef]
30. Wang, Z.Y.; Fang, L.F.; Zhou, Z.S.; Pacheco, S.; Gómez, I.; Song, F.P.; Soberón, M.; Zhang, J.; Bravo, A. Specific binding between *Bacillus thuringiensis* Cry9Aa and Vip3Aa toxins synergizes their toxicity against Asiatic rice borer (*Chilo suppressalis*). *J. Biol. Chem.* **2018**, *293*, 11447–11458. [CrossRef]
31. Dubois, M.; Gilles, K.; Hamilton, J.K.; Rebers, P.A.; Smith, F. A colorimetric method for the determination of sugars. *Nature* **1951**, *168*, 167. [CrossRef] [PubMed]
32. Liang, G.M.; Tian, W.J.; Guo, Y.Y. An improvement in the technique of artificial rearing of the cotton bollworm. *Plant Protection.* **1999**, *25*, 15–17.
33. Wolfersberger, M.; Luethy, P.; Maurer, A.; Parenti, P.; Sacchi, F.V.; Giordana, B.; Hanozet, G.M. Preparation and partial characterization of amino acid transporting brush border membrane vesicles from the larval midgut of the cabbage butterfly (*Pieris brassicae*). *Comp. Biochem. Physiol. Part A Physiol.* **1987**, *86*, 301–308. [CrossRef]
34. Lorence, A.; Darszon, A.; Bravo, A. Aminopeptidase dependent pore formation of *Bacillus thuringiensis* Cry1Ac toxin on *Trichoplusia ni* membranes. *FEBS Lett.* **1997**, *414*, 303–307. [PubMed]

toxins

MDPI

Communication

Processing Properties and Potency of *Bacillus thuringiensis* Cry Toxins in the Rice Leaffolder *Cnaphalocrocis medinalis* (Guenée)

Yajun Yang [1], Zhihong Wu [1], Xiaochan He [2], Hongxing Xu [1] and Zhongxian Lu [1,*]

[1] State Key Laboratory for Managing Biotic and Chemical Threats to the Quality and Safety of Agro-Products, Institute of Plant Protection and Microbiology, Zhejiang Academy of Agricultural Sciences, Hangzhou 310021, China

[2] Jinhua Academy of Agricultural Sciences, Jinhua 321000, China

* Correspondence: luzxmh@163.com

Abstract: Different Cry toxins derived from *Bacillus thuringiensis* (Bt) possess different insecticidal spectra, whereas insects show variations in their susceptibilities to different Cry toxins. Degradation of Cry toxins by insect midgut extracts was involved in the action of toxins. In this study, we explored the processing patterns of different Cry toxins in *Cnaphalocrocis medinalis* (Lepidoptera: Crambidae) midgut extracts and evaluated the impact of Cry toxins degradation on their potency against *C. medinalis* to better understand the function of midgut extracts in the action of different Cry toxins. The results indicated that Cry1Ac, Cry1Aa, and Cry1C toxins could be degraded by *C. medinalis* midgut extracts, and degradation of Cry toxins by midgut extracts differed among time or concentration effects. Bioassays demonstrated that the toxicity of Cry1Ac, Cry1Aa, and Cry1C toxins decreased after digestion by midgut extracts of *C. medinalis*. Our findings in this study suggested that midgut extracts play an important role in the action of Cry toxins against *C. medinalis*, and the degradation of Cry toxins by *C. medinalis* midgut extracts could reduce their toxicities to *C. medinalis*. They will provide insights into the action of Cry toxins and the application of Cry toxins in *C. medinalis* management in paddy fields.

Keywords: Cry toxin; midgut extracts; *Cnaphalocrocis medinalis*; digestion; potency

Key Contribution: Midgut extracts could degrade the activated Cry toxins, which in turn could reduce their toxicities to *C. medinalis*.

Citation: Yang, Y.; Wu, Z.; He, X.; Xu, H.; Lu, Z. Processing Properties and Potency of *Bacillus thuringiensis* Cry Toxins in the Rice Leaffolder *Cnaphalocrocis medinalis* (Guenée). *Toxins* **2023**, *15*, 275. https://doi.org/10.3390/toxins15040275

Received: 27 February 2023
Revised: 30 March 2023
Accepted: 4 April 2023
Published: 6 April 2023

1. Introduction

Bacillus thuringiensis (Bacillales: Bacillaceae; [Bt]) is a Gram-positive bacterium that generates parasporal crystals (formed mainly by Cry and Cyt proteins), which are activated after solubilization and enzymatic digestion and have insecticidal activity against a variety of insects [1–3]. It has been widely used in pest management for several decades due to its insecticidal activity [4–7]. Upon ingestion by insect larvae, Cry proteins are solubilized and proteolyzed into activated toxins in the alkaline environment of the midgut [8,9]. The activated toxin binds to the receptor on the brush border membrane vesicles (BBMVs); then, an oligomer of the toxin forms, binds with the other receptors on the BBMVs, and is inserted into the membrane, resulting in the formation of pores [10,11], or signal transduction involving Ac/PKA is induced, leading to subsequent cell death [12–15].

The midgut, an essential insect organ with various proteases, is important in food digestion, utilization, and detoxification [16,17]. Enzymes in midgut juices were reported in the mechanism of Bt action [18]. Cry protoxins are generally processed in the midgut fluids of lepidopteran larvae from 130–140 kDa to 60–70 kDa [19]. However, the activated toxins are digested into smaller molecules and may even be fully destroyed after prolonged contact between toxins and midgut extracts. The digestive activity of insect pests is critical to toxin action and influences toxin toxicity and specificity. When midgut extracts interacted with Cry proteins, two consequences were observed: activation and degradation, which might result in two different toxicity effects [20]. Variations in midgut juices between susceptible and resistant *Plutella xylostella* (Linnaeus) (Lepidoptera: Plutellidae) strains may play an important role in *P. xylostella* resistance [21]. The activated toxin's stability was shown to be proportional to its toxicity against the target insect [22]. *Anomala cuprea* (Hope) (Coleoptera: Scarabaeidae) neat gut juice demonstrated the capacity to breakdown Cry toxin into smaller, atoxic particles in one minute [23]. Brunet et al. [24] proposed that *Manduca sexta* (Linnaeus) (Lepidoptera: Sphingidae) midgut juice contains protease inhibitors, which may have an essential function in Bt toxin action. Accelerated Cry toxin degradation leads to loss of Cry1C vulnerability in fifth-instar larvae of *Spodoptera littoralis* (Boisduval) (Lepidoptera: Noctuidae) [25], while serine protease inhibitors can increase the insecticidal efficacy of some Cry proteins by up to 20-fold [26].

Rice, *Oryza sativa* L. (Poales: Poaceae), is one of the world's most important essential foodstuffs [27,28]. It is consumed by roughly half of the world's population, the majority of whom live in Asia, the primary rice-producing region [28,29]. With the growing human population and consumption, it is critical to enhance the current rice yields [30]. However, rice can be damaged by insect pests, leading to losses in rice yields [31]. *Cnaphalocrocis medinalis* (Guenée) (Lepidoptera: Pyralidae), a significant rice insect pest, is widely distributed in China, Japan, Korea, India, Vietnam, Thailand, the Philippines, Malaysia, and other Asian countries [32,33]. It causes chlorophyll loss by folding and feeding leaves [32,33], and its heavy outbreaks could cause significant losses in rice production [34]. In the worst cases, this pest can cause 63–80% rice yield losses [35]. In 2015, *C. medinalis* damaged 15.5 million ha of rice plants, resulting in yield losses of 0.47 million tons in China [34]. For a long time, the control of *C. medinalis* relied on chemical insecticides. However, the overuse and misuse of chemicals can cause many negative issues, such as environmental pollution and insect resistance. Recently, with the proposal of "reductions in chemicals" in China, an increasing number of nonchemical measures were recommended for pest control in paddy fields. Bt sprays have been used as biological pesticides to control *C. medinalis* for decades [36]. Bt rice cultivars on trial could suppress the population of Lepidoptera insect pests such as *C. medinalis* and mitigate their damage in paddy fields [29]. Several Cry toxins and Bt rice lines have both been shown to be effective against *C. medinalis* [37–39]. However, insect tolerance and resistance to Bt toxins may impede their implementation, and many insects were found to be resistant to Bt toxins in laboratory or field populations [40–43]. A report from Wu et al. [44] suggested that *C. medinalis* has the potential to develop resistance to low amounts of Cry toxin. It is crucial to understand the interaction between Cry toxins and *C. medinalis*. The midgut is an essential organ in which the Cry toxin functions, and midgut extracts play a role in its activity [7,45]. Yang et al. [46,47] reported that pH and inhibitors could influence the protease profiles and the degradation of activated Cry toxins in the midgut juices of *C. medinalis*. Moreover, variations in the toxicities of Cry toxins were observed in *C. medinalis* [38]. Degradation of Cry toxins by insect midgut extracts might be involved in the action of toxins in *C. medinalis*. In this work, we explored the processing patterns of different Cry-activated toxins in *C. medinalis* midgut extracts and evaluated the impact of Cry toxins digestion on their efficacy against *C. medinalis* to better understand the function of midgut extracts in the action of different Cry toxins. Our findings may help to explain variations in the potency of Cry toxins against *C. medinalis*, as well as their interaction with Cry toxins, and will enhance the safe use of Cry toxins.

2. Results

2.1. Processing of Cry Toxins with Different Concentrations of *C. medinalis* Midgut Extracts

Three Cry-activated toxins, Cry1Aa, Cry1Ac, and Cry1C, were processed in *C. medinalis* midgut extracts at four different ratios (10:1, 1:1, 1:10, and 1:100) (Cry toxin:extracts, *w*/*w*) at 30 °C for 8 h, respectively. The results demonstrated that all three Cry toxins in this study could be degraded by *C. medinalis* midgut extracts; however, the degradation levels varied among the Cry toxins. Cry1Aa degradation levels rose with increasing concentrations of *C. medinalis* midgut extracts, and Cry1Aa toxins were entirely degraded into small fragments at a ratio of 1:100 (Cry toxin:extracts, *w*/*w*) (Figure 1). The Cry1Ac and Cry1C toxins showed a similar pattern of degradation as Cry1Aa, whereas Cry1Ac was not totally degraded at a ratio of 1:100 (Cry toxin:extracts, *w*/*w*) (Figure 1).

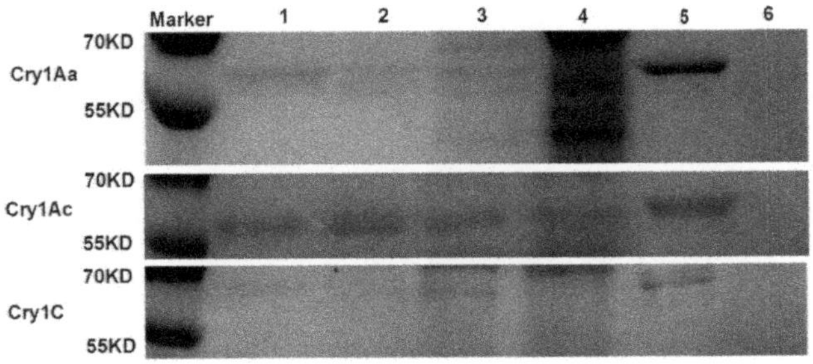

Figure 1. In vitro processing of Cry toxins with different concentrations of *C. medinalis* midgut extracts. For in vitro processing, Cry1Aa, Cry1Ac, and Cry1C (1 μg) were incubated with *C. medinalis* midgut extracts at 30 °C for 8 h at ratios (Cry toxin:extracts, *w*/*w*) of 10:1, 1:1, 1:10, and 1:100, respectively. Cry toxins and midgut extracts were incubated at 30 °C for 8 h as controls, respectively. Processed Cry toxins, after incubation with midgut extracts, were separated by SDS–PAGE gel. Lanes 1–4: 10:1, 1:1, 1:10, and 1:100; Lane 5: Cry toxin; Lane 6: midgut extracts.

2.2. *Processing of Cry Toxins with C. medinalis Midgut Extracts over Time*

Three Cry-activated toxins, Cry1Aa, Cry1Ac, and Cry1C, were processed in *C. medinalis* midgut extracts (1:10, Cry toxin:extracts, *w*/*w*) at 30 °C over various times (2, 4, 8, 12, and 24 h). The results revealed that when the incubation period was prolonged, the degradation levels of Cry toxins rose. The Cry1Ac, Cry1Aa, and Cry1C toxins began to break down into small fragments after four hours of incubation (Figure 2). After 24 h of incubation, Cry1Ac and Cry1C toxins were entirely degraded into small fragments; however, Cry1Aa was not completely degraded, but the small fragments were further degraded (Figure 2).

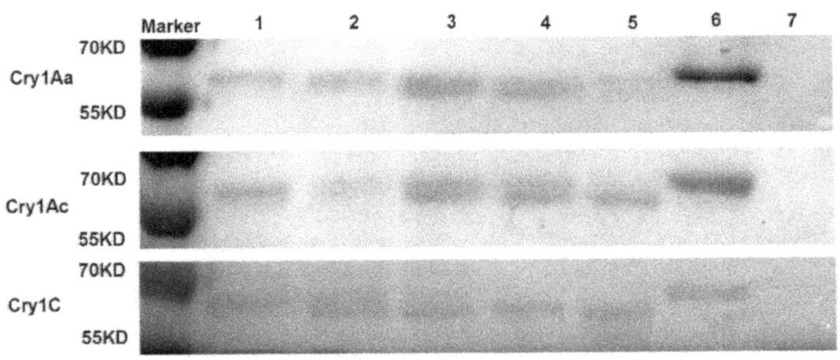

Figure 2. In vitro processing of Cry toxins with *C. medinalis* midgut extracts over various times. Cry1Aa, Cry1Ac, and Cry1C (10 μg) toxins were incubated with *C. medinalis* midgut extracts (1:10, Cry toxin:extracts, *w*/*w*) at 30 °C for 2, 4, 8, 12, and 24 h, respectively. Cry toxin and midgut extracts were incubated at 30 °C for 8 h as controls, respectively. Processed Cry toxins, after incubation with midgut extracts, were separated by SDS–PAGE gel. Lanes 1–5: 2, 4, 8, 12, 24 h; Lane 6: Cry toxin; Lane 7: midgut extracts.

2.3. *Potency of Cry Toxin Processed by C. medinalis Midgut Extracts*

The toxicities of Cry1Aa, Cry1Ac, and Cry1C toxins, with and without digestion by *C. medinalis* midgut extracts, were evaluated in *C. medinalis* larvae through detached leaf-dipping methods. The Cry-activated toxins processed by midgut extracts were prepared by incubating activated Cry toxin and *C. medinalis* midgut extracts at 30 °C for 8 h at a ratio of 1:10 (*w*/*w*). The LC_{50} values of activated Cry1Aa, Cry1Ac, and Cry1C toxins against

C. *medinalis* were 1.981, 0.673, and 1.207 μg/mL, respectively (Table 1). The LC_{50} values of activated Cry1Aa, Cry1Ac, and Cry1C toxins processed by midgut extracts against C. *medinalis* were 3.498, 1.068, and 2.186 μg/mL, respectively (Table 1). After being processed by midgut extracts, the toxicities of activated Cry1Aa, Cry1Ac, and Cry1C toxins decreased. The toxicity regression lines of activated Cry toxins were not equal but rather parallel with those of activated Cry toxins digested by midgut extracts (Figure 3). The ratios of the LC_{50} values of activated Cry toxins processed by midgut extracts to activated Cry toxins varied from 1.586 to 1.811.

Table 1. Median lethal concentrations of Cry toxins and Cry toxins processed by midgut extracts against C. *medinalis*.

Toxins	*n*	Slope (SE)	LC_{50} (95%FL) (μg/mL)
Cry1Aa	300	1.684 (0.181)	1.981 (1.556–2.472)
Cry1Aa processed by midgut extracts	300	1.562 (0.174)	3.498 (2.438–5.136)
Cry1Ac	300	1.511 (0.180)	0.673 (0.397–0.989)
Cry1Ac processed by midgut extracts	300	1.427 (0.168)	1.068 (0.664–1.609)
Cry1C	300	1.695 (0.180)	1.207 (0.961–1.506)
Cry1C processed by midgut extracts	300	1.498 (0.174)	2.186 (1.463–3.534)

n = number of larvae in the probit analysis. LC_{50} (median lethal concentration): concentration of toxins (μg/mL) required to kill 50% of larvae over 48 h. 95% FL = 95% fiducial limits.

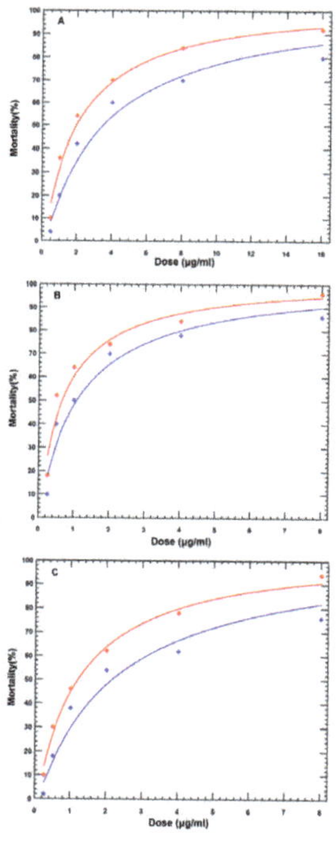

Figure 3. Equality and parallelism of the toxicities of Cry toxins compared with those of Cry toxins processed by midgut extracts against C. *medinalis*. Red and blue lines represent the treatments of Cry toxins and Cry toxins processed by midgut extracts, respectively. (**A**) Cry1Aa toxins with or without midgut extracts digestion (equality: chi-square = 12.39, $p < 0.05$; parallelism: chi-square = 0.24, $p > 0.05$); (**B**) Cry1Ac toxins with or without midgut extracts digestion (equality: chi-square: 7.24, $p < 0.05$; parallelism: chi-square = 0.12, $p > 0.05$); (**C**) Cry1C toxins with or without midgut extracts digestion (equality: chi-square = 12.92, $p < 0.05$; parallelism: chi-square = 0.62, $p > 0.05$).

3. Discussion

Over 200 distinct Cry toxins, derived from Bt, were identified, with the majority being harmful to insect pests and nematodes [48,49]. Numerous studies have revealed differences in the toxicities of Cry toxins [2,36,50–55]. The difference in toxicity is affected by many factors, and predigestion treatment by solubilization or enzymatic processing has a great effect [2]. Degradation of Cry toxins by insect midgut extracts might be involved in the action of toxins, and accelerated Cry toxin degradation could reduce or eliminate insecticidal activity [25]. In this study, we explored the degradation properties of Cry toxins in *C. medinalis* midgut extracts and evaluated the potency of the Cry toxins processed by the midgut extracts. All three activated Cry toxins could be degraded by midgut extracts in *C. medinalis*, and the digestion of Cry toxins by midgut extracts could reduce their toxicities against *C. medinalis*. Our findings will help researchers better understand the variations in the toxicity of Cry toxins against *C. medinalis* and aid in the investigation of interactions between *C. medinalis* and the Cry protein.

In insects, the midgut is an essential organ for metabolism and food usage, and enzymes in the midgut play a critical role in these functions [16,17]. Midgut extracts are rich in enzymes that can activate the Cry protein as well as degrade it into smaller peptides [20]. Our results suggested that all three Cry toxins in this study could be degraded by *C. medinalis* midgut extracts, and they began to break down into small fragments after four hours of incubation. However, the degradation levels varied among the Cry toxins. At a ratio of 1:100 (Cry toxin:extracts, *w*/*w*), Cry1Aa and Cry1C toxins were entirely degraded into smaller fragments, whereas Cry1Ac was not totally degraded at a ratio of 1:100 (Cry toxin:extracts, *w*/*w*). After 24 h of incubation, Cry1Ac and Cry1C toxins were entirely degraded into small fragments; however, Cry1Aa was not completely degraded, but the small fragments were further degraded. Tomimoto et al. reported that pronase in *Bombyx mori* (Lepidoptera: Bombycidae) could degrade Cry protein into several tiny fragments [56]. Cry1Ca toxin is completely destroyed when incubated with midgut extracts from high larval instars of *S. littoralis* (Boisduval) (Lepidoptera: Noctuidae) [25]. Cry3Aa toxin is digested into smaller pieces than the 55-kDa activated fragments in the red palm weevil, *Rhynchophorus ferrugineus* (Coleoptera: Curculionidae), under distinct conditions [57]. However, no degradation of any of the toxins was observed in the proteolytic processing of Bt toxins Cry3Bb1 and Cry34Ab1/Cry35Ab1 by western corn rootworm midgut extracts [58]. Our previous studies indicated that pH and inhibitors could influence the degradation of Cry toxins in *C. medinalis* [46,47]. Apart from these factors, protease activity, developmental stages of insects, and the structure of Cry toxins may be connected to Cry toxin degradation in midgut extracts.

Toxicity variations were found in many Cry toxins against *C. medinalis* [38]. We examined the efficacy of Cry toxins against *C. medinalis* with and without digestion by midgut extracts in this study. The results showed that a Cry toxin digested by *C. medinalis* midgut extracts had reduced toxicity. A protein complex in the midgut of the spruce budworm, *Choristoneura occidentalis* Freeman (Lepidoptera: Tortricidae), might inactivate the Cry toxin by precipitation [59]. Previous studies indicated that overdigestion of Cry toxins by lepidopteron midgut juice was normally associated with a loss of toxicity [20,60,61]. Pang et al. [62] discovered that an increase in midgut juice content was associated with a reduction in the insecticidal efficacy of Cry toxin due to the generation of nonactive pieces. Smaller fragments of Cry1Ab toxin degraded by midgut extracts of *M. sexta* and *Spodoptera frugiperda* (Lepidoptera: Noctuidae) were correlated with a decrease in pore formation and insecticidal activities, and cleavage in domain II of Cry1Ab toxin may be involved in toxin inactivation [20]. The toxicity of *B. thuringiensis* var. *thuringiensis* to *Pieris brassicae* (Lepidoptera: Pieridae) and *Mamestra brassicae* (Lepidoptera: Noctuidae) was shown to be closely linked to protein content and activity in the midgut [63]. Moreover, in resistant strains of insects, decreased toxicity was associated with alterations in midgut juice [20,61,64]. The initial stage influencing the variations in the toxicity of Cry toxins was the distinct digestive levels of midgut juice for various Cry toxins. Brunet et al. [24]

proposed that *M. sexta* midgut fluid components might influence pore formation by Cry9Ca toxin. Yamazaki et al. [21] discovered that midgut extracts of *P. xylostella* (Lepidoptera: Plutellidae), which are highly resistant to Cry1Ac, possess three times larger amounts of glucosinolate sulfatase, which binds to Cry1Ac, compared to susceptible strains. Tetreau et al. [65] found that midgut extract alterations were involved in the process of Bt resistance in the yellow fever mosquito *Aedes aegypti* (Linnaeus) (Diptera: Culicidae) using proteomic and transcriptomic approaches. Changes in the enzymes in the midgut juice also influence Cry1Ac toxicity against *Heliothis virenscens* (Fabricius) (Lepidoptera: Noctuidae) through a proteinase inhibition assay [66]. Engineering multiple trypsin/chymotrypsin sites in the Cry3A toxin could enhance its activity against *Monochamus alternatus* (Coleoptera: Cerambycidae) larvae [67].

Cry toxin activity was linked not only to midgut enzymes but also to a series of receptors on the BBMV in the midgut [68]. Karim and Dean [39] found that Cry1Ac, Cry1Ab, and Cry1Aa had distinct high binding affinities to *C. medinalis* and were linked to an essential step in the Cry toxin's action. Yang et al. [69] discovered that several genes may be implicated in the *C. medinalis* reaction with the Cry toxin. Recently, at least seven ABC proteins were reported to be associated with the *C. medinalis'* response to the Cry1C toxin [70]. Our results indicated the importance of midgut extracts in the degradation of Cry toxins. Interestingly, *C. medinalis* possesses the potential to develop resistance to a low amount of Cry toxin by increasing the activities of the main enzymes in the midgut [44]. Prevention of the further degradation of Cry-activated toxins might maintain their toxicity against *C. medinalis*. In the case of *Ephestia kuehniella* (Lepidoptera: Phycitidae), Cry1Ac toxicity was enhanced toward this lepidopteran pest through the toxin's protection against excessive proteolysis [71]. As an important material for bioagents, Cry toxins play a crucial role in sustainable agriculture. Delaying insect resistance or maintaining the toxicity of Cry toxin is important for the application of Bt toxins. The findings in our study provide novel insights into the potential threat of *C. medinalis* resistance to Cry toxins and promote the development of sustainable agriculture. Our results in this research only provide one perspective on the interplay of Cry toxins and *C. medinalis* via Cry toxin digestion. More investigations on the interaction of Cry toxins and *C. medinalis* will elucidate the mechanism of Cry toxins in *C. medinalis*, boosting the application of Cry toxins in *C. medinalis* management.

4. Conclusions

Herein, we investigated the degradation properties of Cry toxins in *C. medinalis* midgut extracts and tested the efficacy of the Cry toxins processed by the midgut extracts. The results suggested that midgut extracts from *C. medinalis* could degrade the activated Cry toxins, and the degradation levels of Cry toxins by midgut extracts differed depending on the time or concentration effects. In addition, further degradation of Cry toxins by midgut extracts could reduce their toxicities to *C. medinalis*. The findings here will facilitate the understanding of Bt action on *C. medinalis* and promote Bt application in the *C. medianlis* control.

5. Materials and Methods

5.1. Insects

C. medinalis adults were gathered using a sweep net from paddy fields (30.7° N, 120.9° E) in Jiaxing, Zhejiang, China. The moths were given a 10% honey solution in a plastic cup covered with nylon mesh. Eggs laid on the mesh were removed and placed in a box with a detached leaf from a 45-day-old Taichung Native 1 (TN1) rice plant. The insect cultures were maintained at 27 $\pm$ 1 °C with a relative humidity of 70–80% and a photoperiod of 14:10 (L:D) h.

5.2. Toxins

Activated Cry1Ac, Cry1C, and Cry1Aa toxins (MP, Cavey, CWRU, US) derived from *Bacillus thurigensis* were purchased from Youlong BioTech Co., Ltd. (Shanghai, China). They were produced from Cry protoxins through proteolysis using trypsin and refined using ion exchange HPLC, and the activated Cry toxins were really 97% pure.

5.3. Preparation of Midgut Extracts

C. medinalis fifth-instar larvae were cooled on ice for 30 min before their midgut tissues were dissected. The midgut fluids were separated from the solids by centrifugation at 10,000× *g* for 20 min and then filtered through 0.22-m filters. The total protein content of midgut extracts was measured through Bradford's method [72] with a microplate reader (Tecan Trading AG, Mannedorf, Switzerland). The midgut extracts were aliquoted and kept at −70 °C until needed.

5.4. Processing of Cry Toxins with Different Concentrations of C. medinalis Midgut Extracts

One microgram of activated Cry toxin was combined with midgut extracts at various ratios (Cry toxin:midgut extracts, *w*/*w*: 10:1, 1:1, 1:10, and 1:100) and incubated at 30 °C for 8 h. Toxin digestion was halted using a 1 mM solution of phenylmethylsulfonyl fluoride (PMSF, Sigma-Aldrich®, Sigma Aldrich (Shanghai) Trading Co., Ltd., Shanghai, China).

Protein was separated using an 8–10% SDS-PAGE and stained with Coomassie brilliant blue.

5.5. Processing of Cry Toxins with C. medinalis Midgut Extracts over Time

In vitro testing of Cry toxin degradation levels by midgut extracts over time was performed. Ten micrograms of activated Cry toxins were incubated at 30 °C with midgut extracts at a concentration of 1:10 (Cry toxin:midgut extracts, *w*/*w*) for 2, 4, 8, 12, and 24 h, respectively. Toxin digestion was halted by 1 mM PMSF. Protein was separated using an 8–10% SDS-PAGE and stained with Coomassie brilliant blue.

5.6. Bioassays

Activated Cry toxins and Cry toxins processed by midgut extracts were employed to determine insecticidal activity against *C. medinalis* larvae. The Cry-activated toxins processed by midgut extracts were prepared by incubating activated Cry toxin and *C. medinalis* midgut extracts at 30 °C for 8 h at a ratio of 1:10 (*w*/*w*). The bioassays were delivered using the detached leaf-dip technique with modifications [73]. A final 0.1% Triton X-100 solution was prepared in solutions for diluting and spreading over the rice leaf. The leaves of the main rice stalks were chopped into 3–4 cm portions and soaked in every solution for 1 min before being placed in Petri dishes coated with damp absorbent cotton (5 cm in diameter). 0.01 M PBS (containing 0.1% Triton X-100) was used to treat control leaves. Ten second-instar larvae were placed into each Petri dish using a camel hair brush, and the dishes were then sealed with Parafilm® (Sigma Aldrich (Shanghai) Trading Co., Ltd., Shanghai, China) to keep the larvae from escaping and covered with wet mesh to keep the moisture in. Each treatment was carried out five times. After 48 h, the survival rate was calculated.

5.7. Data analysis

Bioassay data were analyzed through probit analysis using POLO Plus software (version 2.0) (LeOra Software, Berkeley, CA, USA), with a natural response (control mortality) included as a model parameter, and the equality and parallelism analysis of probit-regression lines were also treated with POLO Plus software [74,75].

Author Contributions: Conceptualization, Y.Y. and Z.L.; methodology, Y.Y.; validation, Y.Y.; formal analysis, Y.Y. and H.X.; investigation, Y.Y., Z.W. and X.H.; resources, Z.W.; data curation, Y.Y., Z.W. and X.H.; writing—original draft preparation, Y.Y.; writing—review and editing, Y.Y.; visualization,

Y.Y.; supervision, Z.L.; funding acquisition, Z.L. All authors have read and agreed to the published version of the manuscript.

Funding: This research was funded by the earmarked fund for the China Agriculture Research System (CARS-01).

Institutional Review Board Statement: Not applicable.

Informed Consent Statement: Not applicable.

Data Availability Statement: Not applicable.

Conflicts of Interest: The authors declare no conflict of interest.

References

1. Schnepf, E.; Crickmore, N.; Van Rie, J.; Lereclus, D.; Baum, J.; Feitelson, J.; Zeigler, D.R.; Dean, D.H. *Bacillus thuringiensis* and its pesticidal crystal proteins. *Microbiol. Mol. Biol. Rev.* **1998**, *62*, 775–806. [CrossRef]
2. Van Frankenhuyzen, K. Insecticidal activity of *Bacillus thuringiensis* crystal proteins. *J. Insect Physiol.* **2009**, *101*, 1–16. [CrossRef] [PubMed]
3. Adang, M.J.; Crickmore, N.; Jurat-Fuentes, J.L. Diversity of *Bacillus thuringiensis* crystal toxins and mechanism of action. *Adv. Insect Physiol.* **2014**, *47*, 39–87.
4. Ali, S.; Zafar, Y.; Ali, G.M.; Nazir, F. *Bacillus thuringiensis* and its application in agriculture. *Afr. J. Biotechnol.* **2010**, *9*, 2022–2031.
5. Sanahuja, G.; Banakar, R.; Twyman, R.; Capell, T.; Christou, P. *Bacillus thuringiensis*: A century of research, development and commercial applications. *Plant Biotech. J.* **2011**, *9*, 283–300. [CrossRef] [PubMed]
6. Sanchis, V. From microbial sprays to insect-resistant transgenic plants: History of the biopesticide *Bacillus thuringiensis*. A review. *Agron. Sustain. Dev.* **2011**, *31*, 217–231. [CrossRef]
7. Bravo, A.; Gómez, I.; Porta, H.; García-Gómez, B.I.; Rodriguez-Almazan, C.; Pardo, L.; Soberón, M. Evolution of *Bacillus thuringiensis* Cry toxins insecticidal activity. *Microb. Biotechnol.* **2013**, *6*, 17–26. [CrossRef]
8. Heckel, D.G. How do toxins from *Bacillus thuringiensis* kill insects? An evolutionary perspective. *Arch. Insect. Biochem. Physiol.* **2020**, *104*, e21673. [CrossRef]
9. Bel, Y.; Ferré, J.; Hernández-Martínez, P. *Bacillus thuringiensis* toxins: Functional characterization and mechanism of action. *Toxins* **2020**, *12*, 785. [CrossRef]
10. Ferré, J.; Van Rie, J. Biochemistry and genetics of insect resistance to *Bacillus thuringiensis*. *Annu. Rev. Entomol.* **2002**, *47*, 501–533. [CrossRef]
11. Bravo, A.; Gomez, I.; Conde, J.; Muñoz-Garay, C.; Sánchez, J.; Miranda, R.; Zhuang, M.; Gill, S.S.; Soberón, M. Oligoerization triggers binding of a *Bacillus thuringiensis* Cry1Ab pore-forming toxin to aminopeptidase N receptor leading to insertion into membrane microdomains. *Biochim. Biophys. Acta* **2004**, *1667*, 38–46. [CrossRef] [PubMed]
12. Zhang, X.; Candas, M.; Griko, N.B.; Rose-Young, L.; Bulla, L.A., Jr. Cytotoxicity of *Bacillus thuringiensis* Cry1Ab toxin depends on specific binding of the toxin to the cadherin receptor BT-R1 expressed in insect cells. *Cell Death Differ.* **2005**, *12*, 1407–1416. [CrossRef]
13. Zhang, X.; Candas, M.; Griko, N.B.; Taussig, R.; Bulla, L.A., Jr. A mechanism of cell death involving an adenylyl cyclase/PKA signaling pathway is induced by the Cry1Ab toxin of *Bacillus thuringiensis*. *Proc. Natl. Acad. Sci. USA* **2006**, *103*, 9897–9902. [CrossRef]
14. Zhang, X.; Griko, N.B.; Corona, S.K.; Bulla, L.A., Jr. Enhanced exocytosis of the receptor BT-R(1) induced by the Cry1Ab toxin of *Bacillus thuringiensis* directly correlates to the execution of cell death. *Comp. Biochem. Physiol. B Biochem. Mol. Biol.* **2008**, *149*, 581–588. [CrossRef]
15. Guo, Z.; Kang, S.; Chen, D.; Wu, Q.; Wang, S.; Xie, W.; Zhu, X.; Baxter, S.W.; Zhou, X.; Jurat-Fuentes, J.L.; et al. MAPK signaling pathway alters expression of midgut *ALP* and *ABCC* genes and causes resistance to *Bacillus thuringiensis* Cry1Ac toxin in diamondback moth. *PLoS Genet.* **2015**, *11*, e1005124. [CrossRef] [PubMed]
16. Terra, W.R.; Ferriera, C. Biochemistry of Digestion. In *Insect Molecular Biology and Biochemistry*; Gilbert, L.I., Ed.; Elsevier: New York, NY, USA, 2012; pp. 365–418.
17. Linser, P.J.; Dinglasan, R.R. Chapter One–Insect Gut Structure, Function, Development, and Target of Biological Toxins. In *Advances in Insect Physiology*; Dhadialla, T.S., Gill, S.S., Eds.; Academic Press: Oxford, UK, 2014; pp. 1–13.
18. Oppert, K.; Kramer, J.; Beeman, R.; Johnson, D.; McGaughey, W.H. Proteinase-mediated insect resistance to *Bacillus thuringiensis* toxins. *J. Biol. Chem.* **1997**, *272*, 23473–23476. [CrossRef] [PubMed]
19. Pardo-López, L.; Soberón, M.; Bravo, A. *Bacillus thuringiensis* insecticidal three-domain Cry toxins:mode of action, insect resistance and consequences for crop protection. *FEMS Microbiol. Rev.* **2013**, *37*, 3–22. [CrossRef]
20. Miranda, R.; Zamudio, F.Z.; Bravo, A. Processing of Cry1Ab delta-endotoxin from *Bacillus thuringiensis* by *Manduca sexta* and *Spodoptera frugiperda* midgut proteases: Role in protoxin activation and toxin inactivation. *Insect Biochem. Mol. Biol.* **2001**, *31*, 1155–1163. [CrossRef]

21. Yamazaki, T.; Ishikawa, T.; Pandian, G.N.; Okazaki, K.; Haginoya, K.; Tachikawa, Y.; Mitsui, T.; Miyamoto, K.; Angusthanasombat, C.; Hori, H. Midgut juice of *Plutella xylostella* highly resistant to *Bacillus thuringiensis* Cry1Ac contains a three times larger amount of glucosinolate sulfatase which binds to Cry1Ac compared to that of susceptible strain. *Pestic. Biochem. Phys.* **2011**, *101*, 125–131. [CrossRef]
22. Pang, A.S.D.; Gringorten, J.L. Degradation of *Bacillus thuringiensis* δ-endotoxin in host insect gut juice. *FEMS Microbiol. Lett.* **1998**, *167*, 281–285. [CrossRef]
23. Sugimura, M.; Iwahana, H.; Sato, R. Unusual proteolytic processing of a δ-endotoxin from *Bacillus thuringiensis* strain Buibui by larval midgut-juice of *Anomala cuprea* Hope (Coleoptera: Scarabaeidae). *Appl. Entomol. Zool.* **1997**, *32*, 533–540. [CrossRef]
24. Brunet, J.F.; Vachon, V.; Marsolais, M.; Van Rie, J.; Schwartz, J.L.; Laprade, R. Midgut juice components affect pore formation by the *Bacillus thuringiensis* insecticidal toxin Cry9Ca. *J. Invertebr. Pathol.* **2010**, *104*, 203–208. [CrossRef] [PubMed]
25. Keller, M.; Sneh, B.; Strizhov, N.; Prudovsky, E.; Regev, A.; Koncz, C.; Schell, J.; Zilberstein, A. Digestion of δ-entotoxin by gut proteases may explain reduced sensitivity of advanced instar larvae of *Spodoptera littoralis* to Cry1C. *Insect Biochem. Mol. Biol.* **1996**, *26*, 365–373. [CrossRef] [PubMed]
26. MacIntosh, S.C.; Kishore, G.M.; Perlak, F.J.; Marrone, P.G.; Stone, T.B.; Sims, S.R.; Fuchs, R. Potentiation of *Bacillus thuringiensis* insecticidal activity by serine protease inhibitor. *J. Agr. Food Chem.* **1990**, *8*, 1145–1152. [CrossRef]
27. Yuan, L.P. Development of hybrid rice to ensure food security. *Rice Sci.* **2014**, *21*, 1–2. [CrossRef]
28. Zhang, Q.F. Strategies for developing green super rice. *Proc. Natl. Acad. Sci. USA* **2007**, *104*, 16402–16409. [CrossRef]
29. Chen, M.; Shelton, A.; Ye, G.Y. Insect-resistant genetically modified rice in China: From research to commercialization. *Annu. Rev. Entomol.* **2011**, *56*, 81–101. [CrossRef]
30. Fan, M.; Shen, J.; Yuan, L.; Jiang, R.; Chen, X.; Davies, W.J.; Zhang, F. Improving crop productivity and resource use efficiency to ensure food security and environmental quality in China. *J. Exp. Agric.* **2012**, *63*, 13–24. [CrossRef]
31. Savary, S.; Horgan, F.; Willocquet, L.; Heong, K.L. A review of principles for sustainable pest management in rice. *Crop. Prot.* **2012**, *32*, 54–63. [CrossRef]
32. Shepard, B.M.; Barrion, A.T.; Litsinger, J.A. *Rice Feeding Insects of Tropical Asia*; IRRI: Manila, Philippines, 1995.
33. Cheng, J.A. *Rice Pests*; China Agricultural Press: Beijing, China, 1996.
34. Yang, Y.J.; Xu, H.X.; Zheng, X.S.; Tian, J.C.; Lu, Y.H.; Lu, Z.X. Progresses in management technology of rice leaffolders in China. *J. Plant Prot.* **2015**, *42*, 691–701.
35. Rajendran, R.; Rajendran, S.; Sandra, B.P.C. Varietal resistance of rice to leaffolder. *Int. Rice Res. Newsl.* **1986**, *11*, 17–18.
36. Wang, D.G.; Zhou, S.F.; Yang, A.M.; Yang, S.X. Application of bio-pesticide Bt agent in the green technology of rice pest management. *Plant Dr.* **2005**, *18*, 14–15.
37. Zheng, X.S.; Yang, Y.J.; Xu, H.X.; Chen, H.; Wang, B.J.; Lin, Y.J.; Lu, Z.X. Resistance performances of transgenic Bt rice lines T2A-1 and T1c-19 against *Cnaphalocrocis medinalis* (Lepidoptera: Pyralidae). *J. Econ. Entomol.* **2011**, *104*, 1730–1735. [CrossRef] [PubMed]
38. Yang, Y.J.; Xu, H.X.; Zheng, X.S.; Lu, Z.X. Susceptibility and selectivity of *Cnaphalocrocis medinalis* (Lepidoptera: Pyralidae) to different Cry toxins. *J. Econ. Entomol.* **2012**, *105*, 2122–2128. [CrossRef] [PubMed]
39. Karim, S.; Dean, D.H. Toxicity and receptor binding properties of *Bacillus thuringiensis* δ–endotoxins to the midgut brush border membrane vesicles of the rice leaf folders, *Cnaphalocrocis medinalis* and *Marasmia patnalis*. *Curr. Microbiol.* **2000**, *41*, 276–283. [CrossRef]
40. Jurat-Fuentes, J.L.; Heckel, D.G.; Ferré, J. Mechanisms of resistance to insecticidal proteins from *Bacillus thuringiensis*. *Annu. Rev. Entomol.* **2021**, *66*, 121–140. [CrossRef]
41. Liu, L.; Li, Z.; Luo, X.; Zhang, X.; Chou, S.H.; Wang, J.; He, J. Which is stronger? A continuing battle between cry toxins and insects. *Front Microbiol.* **2021**, *12*, 665101. [CrossRef]
42. Fabrick, J.A.; Li, X.; Carrière, Y.; Tabashnik, B.E. Molecular genetic basis of lab- and field-selected Bt resistance in pink bollworm. *Insects* **2023**, *14*, 201. [CrossRef]
43. Gassmann, A.J.; Shrestha, R.B.; Kropf, A.L.; St Clair, C.R.; Brenizer, B.D. Field-evolved resistance by western corn rootworm to Cry34/35Ab1 and other *Bacillus thuringiensis* traits in transgenic maize. *Pest Manag. Sci.* **2020**, *76*, 268–276. [CrossRef]
44. Wu, Z.H.; Yang, Y.J.; Xu, H.X.; Zheng, X.S.; Tian, J.C.; Lu, Y.H.; Lu, Z.X. Changes in growth and development and main enzyme activities in midgut of *Cnaphalocrocis medinalis* intermittently treated with low amount of Bt rice leaves over generations. *Chin. J. Rice Sci.* **2015**, *29*, 417–423. (In Chinese)
45. Xu, L.; Pan, Z.Z.; Zhang, J.; Liu, B.; Zhu, Y.J.; Chen, Q.X. Proteolytic activation of *Bacillus thuringiensis* Cry2Ab through a belt-and-braces approach. *J. Agri. Food Chem.* **2016**, *64*, 7195–7200. [CrossRef]
46. Yang, Y.J.; Xu, H.X.; Wu, Z.H.; Lu, Z.X. pH influences the profiles of midgut extracts in *Cnaphalocrocis medinalis* (Guenée) and its degradation of activated Cry toxins. *J. Integr. Agri.* **2020**, *19*, 775–784. [CrossRef]
47. Yang, Y.J.; Xu, H.X.; Wu, Z.H.; Lu, Z.X. Effect of protease inhibitors on the profiles of midgut juices in *Cnaphalocrocis medinalis* (Lepidoptera: Pyralidae) and its degradation to activated Cry toxins. *J. Integr. Agri.* **2021**, *20*, 2195–2203. [CrossRef]
48. Roh, Y.J.; Choi, J.Y.; Li, M.S.; Jin, B.R.; Je, Y.H. *Bacillus thuringiensis* as a specific, safe, and effective tool for insect pest control. *J. Microbiol. Biotechnol.* **2007**, *17*, 547–559. [PubMed]
49. Crickmore, N.; Baum, J.; Bravo, A.; Lereclus, D.; Narva, K.; Sampson, K.; Schnepf, E.; Sun, M.; Zeigler, D.R. Bacillus Thuringiensis Toxin Nomenclature. 2013. Available online: https://www.btnomenclature.info/ (accessed on 1 March 2023).

50. Xu, X.; Yu, L.; Wu, Y. Disruption of a cadherin gene associated with resistance to Cry1Ac δ-endotoxin of *Bacillus thuringiensis* in *Helicoverpa armigera*. *Appl. Environ. Microbiol.* **2005**, *71*, 948–954. [CrossRef] [PubMed]
51. Avilla, C.; Vargas-Osuna, E.; Gonzalez-Cabrera, J.; Ferre, J.; Gonzalez-Zamora, J.E. Toxicity of several δ-endotoxins of *Bacillus thuringiensis* against *Helicoverpa armigera* (Lepidoptera: Noctuidae) from Spain. *J. Invertebr. Pathol.* **2005**, *90*, 51–54. [CrossRef]
52. Bird, L.J.; Akhurst, R.J. Variation in susceptibility of *Helicoverpa armigera* (Hubner) and *Helicoverpa punctigera* (Wallengren) (Lepidoptera: Noctuidae) in Australia to two *Bacillus thuringiensis* toxins. *J. Invertebr. Pathol.* **2007**, *94*, 84–94. [CrossRef]
53. Hernandez-Martinez, P.; Ferre, J.; Escriche, B. Susceptibility of *Spodoptera exigua* to 9 toxins from *Bacillus thuringiensis*. *J. Invertebr. Pathol.* **2007**, *97*, 245–250. [CrossRef]
54. Gao, Y.L.; Hu, Y.; Fu, Q.; Zhang, J.; Oppert, B.; Lai, F.X.; Peng, Y.F.; Zhang, Z.T. Screen of *Bacillus thuringiensis* toxins for transgenic rice to control *Seamia inferens* and *Chilo suppressalis*. *J. Invertebr. Pathol.* **2011**, *105*, 11–15. [CrossRef]
55. Li, H.; Bouwer, G. The larvicidal activity of *Bacillus thuringiensis* Cry proteins against *Thaumatotibia leucotreta* (Lepidoptera: Tortricidae). *Crop. Prot.* **2012**, *32*, 47–53. [CrossRef]
56. Tomimoto, K.; Hayakawa, T.; Hori, H. Pronase digestion of brush border membrane-bound Cry1A shows that almost the whole activated Cry1Aa molecule penetrates into the membrane. *Comp. Biochem. Physiol. B Biochem. Mol. Biol.* **2006**, *144*, 413–422. [CrossRef]
57. Guo, Y.; Sun, Y.; Liao, Q.; Carballar-Lejarazú, R.; Sheng, L.; Wang, S.; Zhou, J.; Zhang, F.; Wu, S. Proteolytic activation of *Bacillus thuringiensis* Cry3Aa toxin in the red palm weevil (Coleoptera: Curculionidae). *J. Econ. Entomol.* **2021**, *114*, 2406–2411. [CrossRef] [PubMed]
58. Kaiser-Alexnat, R.; Büchs, W.; Huber, J. Studies on the Proteolytic Processing and Binding of Bt Toxins Cry3Bb1 and Cry34Ab1/Cry35Ab1 in the Midgut of Western Corn Rootworm (*Diabrotica virgifera virgifera* LeConte). In *Insect Pathogens and Insect Parasitic Nematodes, Proceedings of the 12th European Meeting of the IOBC/wprs Working Group, Pamplona, Spain, 22-25 June 2009*; Ehlers, R.U., Crickmore, N., Enkerli, J., Glazer, I., Lopez-Ferber, M., Tkaczuk, C., Eds.; IOBC/wprs Bulletin: Zurich, Switzerland, 2009; Volume 45, pp. 235–238.
59. Milne, R.E.; Pang, A.S.; Kaplan, H. A protein complex from *Choristoneura fumiferana* gut-juice involved in the precipitation of delta-endotoxin from *Bacillus thuringiensis* subsp. *sotto*. *Insect Biochem. Mol. Biol.* **1995**, *25*, 1101–1114. [CrossRef] [PubMed]
60. Tojo, A.; Aizawa, K. Dissolution and degradation of *Bacillus thuringiensis* delta-endotoxin by gut juice protease of the silkworm *Bombyx mori*. *Appl. Environ. Microbiol.* **1983**, *45*, 576–580. [CrossRef] [PubMed]
61. Lightwood, D.J.; Ellar, D.J.; Jarrett, P. Role of proteolysis in determining potency of *Bacillus thuringiensis* Cry1Ac δ-endotoxin. *Appl. Environ. Microbiol.* **2000**, *66*, 5174–5181. [CrossRef] [PubMed]
62. Pang, A.S.; Gringorten, J.L.; Bai, C. Activation and fragmentation of *Bacillus thuringiensis* delta-endotoxin by high concentrations of proteolytic enzymes. *Can. J. Microbiol.* **1999**, *45*, 816–825. [CrossRef]
63. Bai, C.; Yi, S.X.; Degheele, D. Determination of protease activity in regurgitated gut juice from larvae of *Pieris brassicae*, *Mamestra brassicae* and *Spodoptera littoralis*. *Med. Fac Landbouww Rijksuniv. Gent.* **1990**, *55*, 439–798.
64. Rajagopal, R.; Arora, N.; Sivakumar, S.; Rao, N.G.; Nimbalkar, S.A.; Bhatnagar, R.K. Resistance of *Helicoverpa armigera* to Cry1Ac toxin from *Bacillus thuringiensis* is due to improper processing of the protoxin. *Biochem. J.* **2009**, *419*, 309–316. [CrossRef]
65. Tetreau, G.; Bayyareddy, K.; Jones, C.M.; Stalinski, R.; Riaz, M.A.; Paris, M.; David, J.P.; Adang, M.J.; Després, L. Larval midgut modifications associated with *Bti* resistance in the yellow fever mosquito using proteomic and transcriptomic approaches. *BMC Genom.* **2012**, *13*, 248–262. [CrossRef]
66. Zhu, Y.C.; West, S.; Liu, F.X.; He, Y. Interaction of proteinase inhibitors with Cry1Ac toxicity and the presence of 15 chymotrypsin cDNAs in the midgut of the tobacco budworm, *Heliothis virescens* (F.) (Lepidoptera: Noctuidae). *Pest Manag. Sci.* **2012**, *68*, 692–701. [CrossRef] [PubMed]
67. Guo, Y.; Wang, Y.; O'Donoghue, A.J.; Jiang, Z.; Carballar-Lejarazú, R.; Liang, G.; Hu, X.; Wang, R.; Xu, L.; Guan, X.; et al. Engineering of multiple trypsin/chymotrypsin sites in Cry3A to enhance its activity against *Monochamus alternatus* Hope larvae. *Pest Manag. Sci.* **2020**, *76*, 3117–3126. [CrossRef]
68. Pigott, C.R.; Ellar, D.J. Role of receptors in *Bacillus thuringiensis* crystal toxin activity. *Microbiol. Mol. Biol. Rev.* **2007**, *71*, 255–281. [CrossRef] [PubMed]
69. Yang, Y.J.; Xu, H.X.; Lu, Y.H.; Wang, C.Y.; Lu, Z.X. Midgut transcriptomal response of the rice leaffolder, *Cnaphalocrocis medinalis* (Guenée) to Cry1C toxin. *PLoS ONE* **2018**, *13*, e0191686. [CrossRef] [PubMed]
70. Yang, Y.; Lu, K.; Qian, J.; Guo, J.; Xu, H.; Lu, Z. Identification and characterization of ABC proteins in an important rice insect pest, *Cnaphalocrocis medinalis* unveil their response to Cry1C toxin. *Int. J. Biol. Macromol.* **2023**, *237*, 123949. [CrossRef] [PubMed]
71. Elleuch, J.; Jaoua, S.; Tounsi, S.; Zghal, R.Z. Cry1Ac toxicity enhancement towards lepidopteran pest *Ephestia kuehniella* through its protection against excessive proteolysis. *Toxicon* **2016**, *120*, 42–48. [CrossRef] [PubMed]
72. Bradford, M.M. A rapid and sensitive method for the quantitation of microgram quantities of protein utilizing the principle of dye-protein binding. *Anal. Biochem.* **1976**, *72*, 248–254. [CrossRef]
73. Ye, G.Y.; Hu, C.; Shu, Q.Y.; Cui, H.R.; Gao, M.W. The application of detached- leaf bioassay for evaluating the resistance of Bt transgenic rice to stem borers. *Acta Phytophy. Sin.* **2000**, *27*, 1–6.

74. LeOra Software. *PoloPlus: A User's Guide to Probit and Logit Analysis*; LeOra Software: Berkeley, CA, USA, 2003.
75. Finney, D.J. *Probit Analysis*, 3rd ed.; Cambridge University Press: Cambridge, UK, 1971.

toxins

MDPI

Article

Bacillus thuringiensis Cyt Proteins as Enablers of Activity of Cry and Tpp Toxins against *Aedes albopictus*

Liliana Lai [1,2], Maite Villanueva [2], Ane Muruzabal-Galarza [3], Ana Beatriz Fernández [2], Argine Unzue [2], Alejandro Toledo-Arana [3], Primitivo Caballero [1] and Carlos J. Caballero [2,*]

[1] Institute of Multidisciplinary Research in Applied Biology-IMAB, Universidad Pública de Navarra, 31192 Mutilva, Spain
[2] Departamento de Investigación y Desarrollo, Bioinsectis SL, Plaza Cein 5, Nave A14, 31110 Noáin, Spain
[3] Instituto de Agrobiotecnología (IDAB), CSIC-Gobierno de Navarra, 31192 Mutilva, Spain
* Correspondence: carlos.caballero@bioinsectis.com

Abstract: *Aedes albopictus* is a species of mosquito, originally from Southeast Asia, that belongs to the Culicidae family and the Dipteran insect order. The distribution of this vector has rapidly changed over the past decade, making most of the temperate territories in the world vulnerable to important human vector-borne diseases such as dengue, yellow fever, zika or chikungunya. *Bacillus thuringiensis* var. *israeliensis* (Bti)-based insecticides represent a realistic alternative to the most common synthetic insecticides for the control of mosquito larvae. However, several studies have revealed emerging resistances to the major Bti Crystal proteins such as Cry4Aa, Cry4Ba and Cry11Aa, making the finding of new toxins necessary to diminish the exposure to the same toxicity factors overtime. Here, we characterized the individual activity of Cyt1Aa, Cry4Aa, Cry4Ba and Cry11Aa against *A. albopictus* and found a new protein, Cyt1A-like, that increases the activity of Cry11Aa more than 20-fold. Additionally, we demonstrated that Cyt1A-like facilitates the activity three new Bti toxins: Cry53-like, Cry56A-like and Tpp36-like. All in all, these results provide alternatives to the currently available Bti products for the control of mosquito populations and position Cyt proteins as enablers of activity for otherwise non-active crystal proteins.

Keywords: *Bacillus thuringiensis*; Cyt toxins; *Aedes albopictus*; synergy; mosquitocidal; Cry toxins; Tpp toxins

Key Contribution: A new function for Cyt proteins as enablers of activity of non-active proteins.

Citation: Lai, L.; Villanueva, M.; Muruzabal-Galarza, A.; Fernández, A.B.; Unzue, A.; Toledo-Arana, A.; Caballero, P.; Caballero, C.J. *Bacillus thuringiensis* Cyt Proteins as Enablers of Activity of Cry and Tpp Toxins against *Aedes albopictus*. *Toxins* **2023**, *15*, 211. https://doi.org/10.3390/toxins15030211

Received: 10 February 2023
Revised: 27 February 2023
Accepted: 8 March 2023
Published: 10 March 2023

1. Introduction

Aedes (Stegomya) albopictus (Diptera, Culicidae) (Skuse 1894), commonly known as the Asian tiger mosquito, and originally from Southeast Asia, experimented a quick expansion across the Pacific and Indian Ocean [1], reaching the Americas, Africa and Europe in the past few decades [2]. In Europe, *A. albopictus* was first reported in Albania in 1979 and, after its introduction in Italy, it rapidly colonized the rest of the Mediterranean countries [3–5]. It is thought that the expansion of this invasive species was mainly favored by the shipment of used tires infested with eggs, which were able to survive until they reached their destination. Additionally, the increase in the temperatures due to climate change may have facilitated its establishment in temperate European countries [6]. Although, *A. albopictus* is generally considered thermophilic; however, because of its ecological plasticity, tolerance to cold temperatures and the ability of its eggs to enter diapause have allowed it to be present in every inhabited continent [7]. For this reason, recent studies predict its establishment in every suitable environment beyond 2050, affecting up to 197 territories of the world [8,9].

Because of its opportunistic blood feeding behavior, *A. albopictus* can be exposed to different pathogens. Although *A. albopictus* can feed on a large number of animals, they prefer humans as a blood source, which results in the infection of more than one billion people with vector-borne diseases every year [10]. This makes it one of the most important vectors of arbovirus diseases, such as chikungunya (CHIKV) [11,12], dengue (DENV) [13] and Zika (ZIKV) [14,15], which are a major threat to public health, as they can cause symptoms like fever, hemorrhages and neurological diseases, among others [16]. In 2007, the first outbreak of chikungunya in Italy causing high fever, joint pain and an itchy skin rash caused public alarm and concern about the reemergence of previously eradicated mosquito-borne diseases in Europe [6].

Today, the control of dipteran vectors of disease is addressed through source reduction, chemical pesticides, biological control, genetic control or combinations of these [17,18]. *A. albopictus* larvae can grow in natural and artificial water containers, either outdoors or in peridomestic environments that can be eliminated when possible or treated with larvicidal insecticides [18]. Larvicide and source reduction are the most efficient methods because they give long-term results. The control of larvae can be achieved through chemical insect growth regulators, organophosphates or *Bacillus thuringiensis* (Bt)-based insecticides. Chemicals such as pyrethroids and organophosphates can be highly toxic if used in great amounts, and they are most likely to produce insecticide resistance in mosquitoes [19–21]. These have been consistently used as an efficient method to control the spread of mosquito-borne diseases; however, the development of resistances has led to outbreaks and an increased vector competence of mosquitoes [22]. The most common mechanisms of resistance in mosquitoes are genetic mutations on the target sites of the active ingredients or changes in metabolism [21].

Bt-based solutions represents a good alternative to chemicals since they have a relatively low environmental impact and a high target specificity that make them eligible for the treatment of drinkable water due to its non-toxicity to humans and animals [23,24]. Bt is a gram-positive and spore-forming bacterium capable of producing a parasporal crystal composed of insecticidal proteins [25]. Strains from the serovar *israeliensis*, Bti, are mainly used for the control of dipteran pests. Their parasporal crystals are mainly composed of four major δ-endotoxins (Cry4Aa, Cry4Ba, Cry11Aa and Cyt1Aa), which are highly toxic to insects of the Dipteran order, including mosquitoes. However, recent studies showed that species of the *Aedes* genus had started to develop resistances to the individual Bti proteins Cry4Aa, Cry4Ba and Cry11Aa [26–30]. Among the Bti-based products, Vectobac-*12AS*, a liquid formulation of strain AM 65–52®, is probably the most popular [31–33]. Although this particular strain had already been evaluated against *A. albopictus* larvae, the effects of each of its individual proteins remained unknown [34]. In this study, we characterized the toxicity of the individual Cry4Aa, Cry4Ba, Cry11Aa and Cyt1Aa proteins against *A. albopictus* larvae. Additionally, we focused on identifying new Cry and Cyt proteins for the control of *A. albopictus* larvae that represent an alternative to the classic Bti toxins and, hence, serve as a preventive measure for upcoming resistances. Finally we found a new Cyt protein, Cyt1A-like, which behaved as a synergistic factor by enhancing the activity of Cry11Aa and as an enabler of activity of three new non-active mosquitocidal proteins.

2. Results

2.1. Activity of Cry and Cyt Proteins from Strain AM 65-52 against A. albopictus Larvae

In previous studies, the genome sequencing of strain AM 65-52 revealed its pesticidal gene content, which included the *cry4Aa*, *cry4Ba*, *cry10Aa*, *cry11Aa*, *cry60Aa/cry60Ab*, *cyt1Aa*, *cyt2Ba* and *cyt1Ca* genes [35–39]. Since there was no publicly available data on the activity of each of the top four components of its crystals, Cry4Aa, Cry4Ba, Cry11Aa and Cyt1Aa against *A. albopictus* larvae, we decided to evaluate them individually. For this purpose, the genes required for expressing each of the proteins were cloned in vectors optimized for the expression of δ-endotoxins in Bt [40,41], and the resulting plasmids transformed into the acrystalliferous Bt strain BMB171, as previously described [40]. The *cry4Aa* (3543 bp), *cry4Ba* (3426 bp), *cry11Aa* (1941 bp) and *cyt1Aa* (750 bp) genes were independently cloned, and the resulting recombinant strains expressing *cry4Aa*, *cry4Ba*, *cry11Aa* and *cyt1Aa* were grown in CCY medium for 48–72 h. The mixtures of spores and crystals were run in an SDS-PAGE. Figure 1 shows that the observed bands are coincidental with the predicted molecular weight of each of the four proteins: 134 kDa for Cry4Aa, 128 kDa for Cry4Ba, 27 kDa for Cyt1Aa and 73 kDa for Cry11Aa.

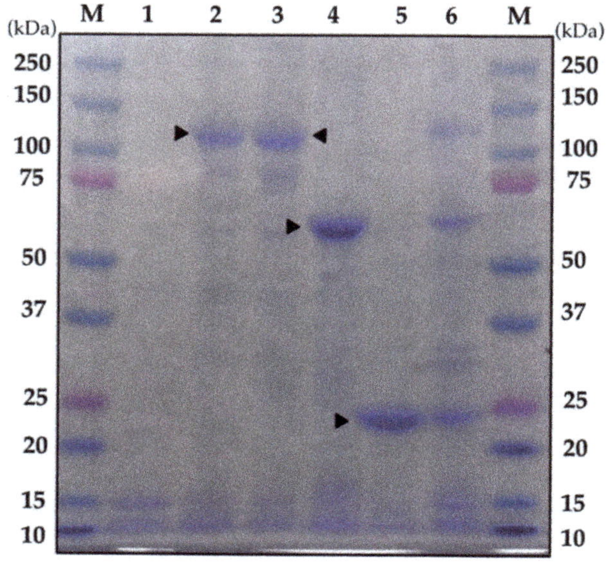

Figure 1. SDS-PAGE showing the protein profile of the BMB171 recombinant strains expressing the AM 65-52 major crystal proteins. Lane M, molecular weight; lane 1, BMB171 carrying an empty plasmid; lane 2, BMB171-Cry4Aa (134 kDa); lane 3, BMB171-Cry4Ba (128 kDa); lane 4, BMB171-Cry11Aa (73 kDa); lane 5, BMB171-Cyt1Aa (27 kDa); lane 6, AM 65-52. Triangles point at major protein bands.

The mosquitocidal activity of the single δ-endotoxins was evaluated on second instar larvae of *A. albopictus* at two different concentrations of spores and crystals (1000 ng/mL and 1.00 ng/mL). Cry4Ba showed the highest activity at 1000 ng/mL, with a mortality of 100%. Conversely, Cry11Aa seemed to be the least active of the tested proteins, with a mortality of 73.33% $\pm$ 0.11 at 1000 ng/mL (Table S1). Once the activity scale for each toxin was defined, the LC_{50} for each of them was calculated, which resulted in 178 ng/mL, 46 ng/mL, 228 ng/mL and 171 ng/mL for Cry4Aa, Cry4Ba, Cry11Aa and Cyt1Aa, respectively (Table 1). Strains AM 65-52 and BMB171 were used as positive and negative controls, respectively. As expected, AM 65-52 showed high mosquitocidal activity, with a LC_{50} of 0.019 ng/mL.

Table 1. Mean lethal concentration (LC_{50}) value of the AM 65-52 Cry4Aa, Cry4Ba, Cry11Aa and Cyt1Aa proteins for second instar larvae of *A. albopictus*.

Treatment	Observed LC_{50} (ng/mL)	Lower Limits	Upper Limits	χ^2	df	Slope	SE Slope	Intercept
Cry4Aa	178	142	226	6.66	4	1.80	0.202	−4.04
Cry4Ba	46	30.4	65.7	2.45	4	1.17	0.182	−1.95
Cry11Aa	228	144	324	3.14	4	1.11	0.166	−2.61
Cyt1Aa	171	133	219	4.55	4	1.85	0.223	−4.12
AM 65-52	0.019	0.013	0.024	4.20	4	1.66	0.191	2.90

Treatment: Spore-and-crystal mixtures; LC_{50}: median lethal concentration; χ^2: chi-square; df: degree of freedom; SE: standard error. Control insects experienced no mortality in all cases.

2.2. Potential Mosquitocidal Cry, Cyt and Tpp Genes from the BST Collection

To expand the number of potential mosquitocidal proteins with activity against *A. albopictus*, we conducted an in silico screening of 36 Bt wild type strains from our private collection. These were previously isolated from soil samples from several regions and habitats from Spain. Considering that the literature on active Bt proteins against *A. albopictus* larvae is scarce, we used previously described toxins against *Aedes aegypti* (Diptera, Culicidae) and other *Aedes* species as a query to select potential new toxic proteins and synergistic factors [31,33,42–44]. Based on our sequencing data, we chose strains BST059.3 and BST-230 as a source of potentiators of toxicity and new mosquitocidal toxins, respectively. Although BST059.3 did not show any activity against *A. albopictus* larvae ($LC_{50} > 1 \times 10^5$ ng/mL of spore-and-crystal mixture), the strain carried two *cyt1-like* genes: *cyt1A-like* and *cyt1D-like*. Cyt proteins have been extensively described as synergistic factors that are able to potentiate the activity of Bti toxins in different genera of mosquitoes, such as *Aedes*, *Anopheles* and *Culex* [45–47]. In particular, Cyt toxins were shown to increase the activity of Cry4Aa, Cry4Ba, Cry10Aa and Cry11Aa when combined together [40,42,45,48]. In agreement with this, we hypothesized that *cyt1A-like* and *cyt1D-like* might be able to produce proteins with a similar function to that of the previously described Cyt toxins. In the case of BST230, several new genes of interest were found, namely, *cry4-like*, *cry53-like*, *cry56-like* and *tpp36-like*. Table 2 contains a list of the new δ-endotoxins and their identity percentages with each of their closest matches, including Cyt1Aa5, Cyt1Da1, Cry4Aa4, Cry53Ab1, Cry56Aa2 and Tpp36Aa1.

Table 2. List of new *cry* and *cyt* genes selected for this study.

Target Database	Pairwise Identity %	MW (kDa)	Accession Number of Reference	Accession Number	Strain
cyt1Aa5	65	31	CAD30079	OQ397557	BST059.3
cyt1Da1	48	59	ADV33305	OQ397558	BST059.3
cry4Aa4	38	132	AFB18317	OQ397551	BST230
cry53Ab1	40	76	ACP43734	OQ397553	BST230
cry56Aa2	54	73	ADK38584	OQ397555	BST230
tpp36Aa1	30	56	AAK64558	OQ397552	BST230

MW: Molecular weight.

To study the mosquitocidal properties of the newly found crystal proteins, genes *cry4-like*, *cry53-like*, *cry56A*-like and *Tpp36*-like were amplified from strain BST-230, cloned into pTBT02, and the resulting vectors electroporated into strain BMB171. In the case of Cry53-like, we were not able to observe any crystals at first (Figure 2A). In order for it to crystallize, we had to include an additional open reading frame (ORF) of 1623 bp downstream of the *cry53-like* coding sequence (CDS) (3004 bp), mimicking its original architecture in the genome. The sequence of this ORF matched the typical Domain V from the crystallization region (C terminal) of Cry1Ac, according to the Conserved Domains Database (CDD) [49]. These C-terminal domain ORFs are often found in mosquitocidal protoxins [50] and, in this case, were required for crystal formation (Figure 2B). Likewise, Cry56A-like also required a fragment of 1692 bp, namely *orf1*, upstream of its CDS (1974 bp) for effective crystal formation (Figure 2B). Analogously, the *cyt1A*-like and *cyt1D-like* genes were amplified from strain BST059.3 and independently cloned into pTBT02. However, neither of them were able to form crystals (Figure 2A). To solve this, both genes were cloned including the *p20* orf, and the resulting plasmids were used to transform strain BMB171. The reason for including *p20* was to promote crystal formation in the Cyt proteins during sporulation (Figure 2B) [51,52].

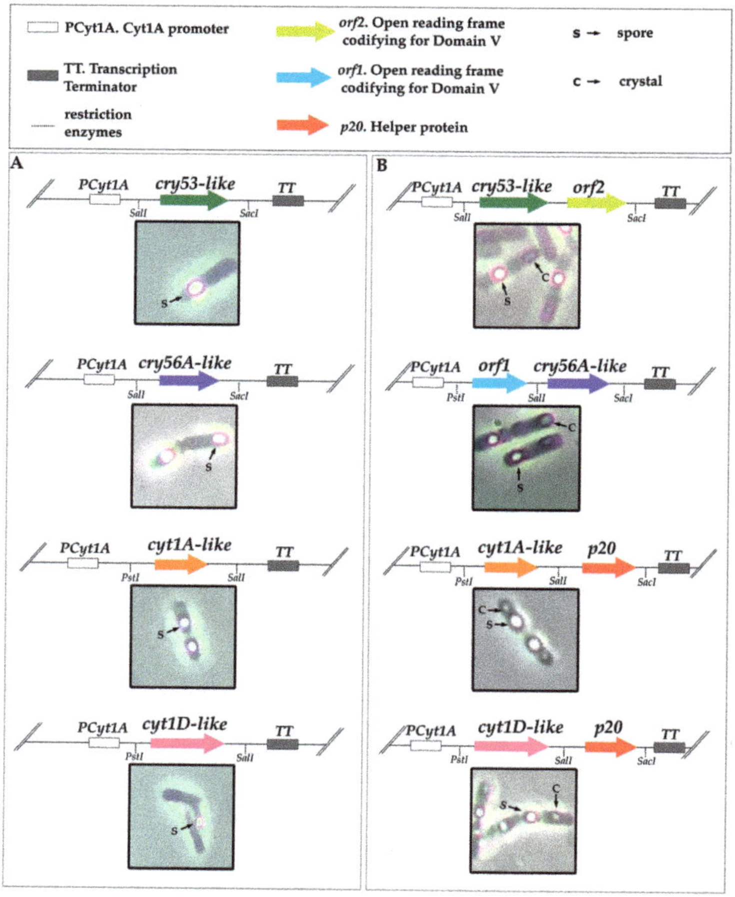

Figure 2. Graphical representation of the impact of helper proteins on the crystallization of Cry53-like, Cry56A-like, Cyt1A-like and Cyt1D-like and of their phenotypes under the microscope after their expression in strain BMB171. (**A**) Images of BMB171 expressing *cry53-like*, *cry56A-like*, *cyt1A-like* and *cyt1D-like* (produce only spores). (**B**) *cry53-like* + *orf2*, *orf1:cry56A-like*, *cyt1A-like:p20* and *cyt1D-like:p20* are able to form both spores and crystals.

2.3. Characterization of the Bt Recombinant Strains Expressing the New Potential Mosquitocidal Proteins

The BMB171 recombinant strains carrying *cry4-like*, *cry53-like*, *cry56A-like*, *cyt1A-like*, *cyt1D-like* and *tpp36-like* were able to produce spores and crystals when grown in CCY medium for 48–72 h. An SDS-PAGE analysis showed the expected molecular weights for most of the new recombinant proteins. The predicted sizes of Cyt1A-like (~31 kDa), Cyt1D-like (~59kDa) and Cry4-like (~130 kDa) correlated with their observed bands (Figure 3). Cry53-like:orf2 (lane 6) showed two bands: one of ~70 kDa (Cry53-like), which was a little lower than the expected size (76 kDa), and a band of ~62kDa (orf2). For Cry56A-like:orf1 (lane 7), we were able to detect both expected band sizes ~73 kDa (Cry56A-like) and 64 kDa (orf1). Tpp36-like produced a clear band of ~40 kDa (lane 8), although the expected size was about ~55kDa. All of the plasmids were sequenced and did not have any changes and/or mismatches in their sequences.

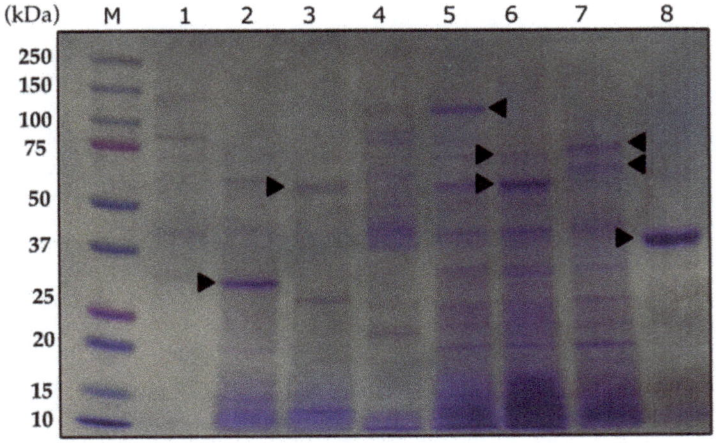

Figure 3. SDS-PAGE showing the protein profile of the BMB171 recombinant strains carrying the new recombinant proteins. Lane M, molecular weight; lane 1, strain TF059.3; lane 2, BMB171-Cyt1A-like (31 kDa); lane 3, BMB171-Cyt1D-like (59 kDa); lane 4, strain BST-230; lane 5, BMB171-Cry4-like (130 kDa); lane 6, BMB171-Cry53-like:orf2 (76 kDa:62 kDa); lane 7, BMB171-orf1:Cry56A-like (64 kDa:73 kDa); lane 8, BMB171-Tpp36-like (55 kDa). Triangles point at major protein bands that correspond with the expressed proteins.

The mosquitocidal activity of the new δ-endotoxins was evaluated on second instar larvae of *A. albopictus*. For these experiments, 1×10^5 ng/mL and 1×10^3 ng/mL of spore-and-crystal mixtures were used as high and low concentrations, respectively. Surprisingly, when analyzing the results, none of the Cry-like and Tpp-like individual proteins showed activity, indicating that they might need to act in partnership with others to produce toxicity.

2.4. Synergies between Cyt and AM 65-52 Cry Proteins

To address the potential of Cyt1Aa, Cyt1A-like and Cyt1D-like to produce synergistic interactions with Cry4Aa, Cry4Ba and Cry11Aa against *A. albopictus* larvae, we decided to evaluate the effect of 1:1 mixtures on the aforementioned proteins. The preliminary results of the potential synergistic interactions are shown in Table S2. A potential synergy was considered when the activity of the 1:1 mix was higher than the sum of the individual activities of the two proteins. Finally, LC_{50} values were calculated. As previously shown in other mosquito species, Cyt1Aa enhanced the activity of Cry4Aa, Cry4Ba and Cry11Aa. However, in the case of the two new Cyt proteins, only Cyt1A-like increased the activity of Cry11Aa, with a synergism factor of 23.14. Cyt1D-like did not potentiate the activity of any of the toxins. Table 3 shows the LC_{50} for each of the binary combinations that produced synergistic interactions.

Table 3. Mean lethal concentration (LC_{50}) value of the 1:1 mixture calculated for L2 larvae of *A. albopictus*.

Treatment	Observed LC_{50} (ng/mL)	Expected LC_{50} (ng/mL)	Lower Limits	Upper Limits	χ^2	df	Slope	SE Slope	Intercept	Synergism Factor
Cyt1Aa + Cry4Aa	6.04	170	4.68	8.13	1.91	4	1.66	0.212	−1.30	28.14
Cyt1Aa + Cry4Ba	3.08	100	2.05	5.29	7.42	4	2.14	0.243	−1.05	32.46
Cyt1Aa + Cry11Aa	17.1	195.6	13.8	21.0	1.92	4	2.37	0.257	−2.93	11.44
Cyt1A-like + Cry11Aa	19.7	456	16.1	24	0.704	4	2.04	0.215	−2.69	23.14

Treatment: Spore-and-crystal mixtures; Expected LC_{50}: Expected median lethal concentration calculated with the method Tabashnik (1992); χ^2: chi-square; df: degree of freedom; SE: standard error; Synergism Factor: the ratio of the expected LC_{50} and the observed LC_{50}. Control insects experienced no mortality in all cases.

2.5. Cyt Proteins as Enablers of Activity of Cry and Tpp Toxins

Because none of the new potential mosquitocidal proteins found in strain BST-230 showed activity on their own, we wondered if Cyt1Aa, Cyt1A-like and Cyt1D-like might be able to make them active against *A. albopictus* larvae. Considering that Cry56A-like is non-active, we would have expected the LC_{50} of the Cyt1Aa + Cry56A-like mix to equal the LC_{50} of Cyt1Aa (171 ng/mL). However, the activity of the mix was higher by 7.3-fold, with an observed LC_{50} of 23.3 ng/mL. For the Cyt1A-like combinations, the expected LC_{50} was greater than 10^5 ng/mL since no activity could be observed for them when tested individually. Nevertheless, when mixing Cyt1A-like with Cry53-like, Cry56A-like and Tpp36-like, the LC_{50} values were 1331, 186 and 1053 ng/mL, respectively (Table 4). These results indicated that although Cyt1 proteins are considered synergistic factors of Cry proteins, in some cases, they may behave as enablers of activity of otherwise non-toxic proteins in a specific manner.

Table 4. Mean lethal concentration (LC_{50}) value of the 1:1 mixture (enabler+Cry/Tpp) and the toxins inoculated individually, calculated for L2 larvae of *A. albopictus*.

Treatment	Observed LC50 (ng/mL)	Lower Limits	Upper Limits	χ^2	df	Slope	SE Slsope	Intercept
Cyt1Aa	171	133	219	4.55	4	1.85	0.223	−4.12
Cry56A-like	>10^5							
Cyt1Aa+Cry56A-like	23.3	19.2	27.9	3.71	4	2.20	0.220	−3.01
Cyt1A-like	>10^5							
Cry53-like	>10^5							
Cyt1A-like+Cry53-like	1331	1091	1649	2.87	4	2.03	0.202	−6.35
Cyt1A-like	>10^5							
Cry56A-like	>10^5							
Cyt1A-like+Cry56A-like	186	138	239	6.26	4	1.59	0.192	−3.60
Cyt1A-like	>10^5							
Tpp36-like	>10^5							
Cyt1A-like+Tpp36-like	1053	860	1298	5.85	4	2.04	0.208	−6.18

Treatment: Spore-and-crystal mixtures; χ^2: chi-square; df: degree of freedom; SE: standard error. Control insects experienced no mortality in all cases.

3. Discussion

Mosquito larvae are highly susceptible to Bti crystals, for which the typical composition is usually a combination of the Cry4Aa, Cry4Ba, Cry11Aa and Cyt1Aa proteins in variable proportions. Interestingly, when analyzing the relative amount of each of the proteins in the parasporal crystal, only a limited number of the anticipated pesticidal proteins are represented and with variable abundance: Cyt1Aa (38–61%), Cry60Ba (5–12%), Cry11Aa (10–27%), Cry4Ba (10–28%), Cry60Aa (2–4%) and Cry4Aa (2–4%) [53]. Although the activity of the major components of the AM 65-52 crystal were previously characterized against *A. aegypti* and species from the genera Culex and Anopheles, little information is available on their effect on *A. albopictus* larvae [54–58]. Additionally, the synergistic interactions between the components of the crystals received attention from different research groups due to the activity of the single Bti proteins being below the toxicity of the complex crystal. Despite this, *A. albopictus* has rarely been used as a model organism for these kind of studies. The reason for this may be the fact that it was traditionally associated with wild animals and territories, and hence, it was a less dangerous species for humans [59]. *A. aegypti*, on the contrary, has often been regarded as the primary vector of arborviruses in human health, becoming the most studied mosquito within its genus. However, in recent times, the *A. albopictus* gained notoriety due to its establishment in Western countries, specifically those in the Mediterranean coast. This led the European authorities to increase the resources allocated to monitoring its spread and to enforce measures to avoid the reemergence of previously eradicated diseases, such as Dengue, especially in countries like Italy and France, where outbreaks of this mosquito have been reported [10,60,61]. For this reason, our research group deemed it appropriate to expand the knowledge on the Bti toxin susceptibility of other Aedes species to *A. albopictus*. We began by characterizing the insecticidal activity of individual Bti toxins on tiger mosquito L2 larvae. The LC_{50} values for Cry4Aa, Cry4Ba, Cry11Aa and Cyt1Aa of 178, 46, 228 and 171 ng/mL, respectively, were similar to the previously described ones for *A. aegypti*. These results suggested that both species may have a similar susceptibility to Bti toxins and crystals [42,44]. When analyzing the synergistic interactions of Cyt1Aa, we found that it was able to potentiate the activity of Cry4Aa, Cry4Ba and Cry11Aa by 28.14-, 32.46- and 11.44-fold, respectively. This synergistic effect on the activity of the individual toxins was considerably higher than previously reported in *A. aegypti*, for which the corresponding synergistic factors were 15.5, 10.91 and 3.15 for the Cyt1A+Cry4A, Cyt1A+Cry4B and Cyt1A+Cry11 combinations, respectively [42,45].

Crystal toxins differ in structure and sequence identity, those differences being that Cry proteins are characterized by a three-domain structure [62], Cyt proteins by a single domain constituted by a β-sheet in the middle and surrounded by two α-helical layers [63] and Tpp proteins by a single domain named Toxin_10 (Bin-like) [54]. The mode of action of Cry and Cyt δ-endotoxins is somehow similar at the beginning of the infection, when the crystals are ingested by a susceptible host. Once they reach the midgut, they are solubilized due to the alkaline conditions and processed by proteases into their active form [64]. However, whereas Cyt toxins directly interact with membrane lipids and insert themselves into the membrane of the epithelial host cells, Cry toxins interact with specific receptors of the surface of said cells and oligomerize before the insertion and pore formation occurs [64]. In the case of Tpp proteins, the mode of action is not completely clear, but it was observed that their toxin form binds to the mosquito midgut, specifically the posterior midgut and the gastric caecum [65].

Synergistic interactions between Cry and Cyt toxins have been extensively described in mosquitoes, but the precise mechanism of action remains poorly understood. The most studied combination is the one between Cyt1Aa and Cry11Aa. One of the major assumptions is that Cyt1Aa may function as a receptor for Cry11Aa, facilitating its insertion in the peritrophic membrane of the insect midgut [66]. In agreement with this, Cyt1Aa was shown to delay the development of resistances in mosquitoes by, hypothetically, habilitating new binding sites to other toxins that do not necessarily need to be from the same crystals, such as the *Lysinibacillus sphaericus* Mtx1 and Mtx2 toxins [67]. Currently, there is no evidence of mosquito resistances to Bti crystals as a whole, probably due to the interactions that occur between the proteins within the crystal [68,69].

Although resistance to Bti crystals seems unlikely, results showing the existence of resistant biotypes to single Bti proteins open the possibility of specific populations of mosquitoes becoming resistant to Bti-based solutions over time if used irresponsibly. Therefore, we decided to look for new Bt toxins and synergistic factors that represent an alternative to the most common Bti proteins for the control of *A. albopictus* larvae. Acknowledging that resistances are less likely to develop when synergies take place, one of our focuses was to find new synergistic proteins and characterize them. For this purpose, we selected strain BST059.3, which, despite not showing any activity against *A. albopictus* larvae ($LC_{50} > 1 \times 10^5$ ng/mL of spore-and-crystal mixture), carried two *cyt1-like* genes: *cyt1A-like* and *cyt1D-like*. In addition, BST230, came across as an interesting strain due to its novel insecticidal content: *cry4-like*, *cry53-like*, *cry56-like* and *tpp36-like*. Cyt1A-like and Cyt1D-like were selected as possible synergistic factors, and the rest of the proteins were selected as hypothetical mosquitocidal proteins. Cry4 toxins have been widely described as one of the major mosquitocidal toxins [54,70,71], whereas, in a previous study, Cry56 showed activity against *A. aegypti* larvae [72]. Despite Cry53 and Tpp36 not been previously reported as potential mosquitocidal toxins, we decided to include them in the study because strain BST-230 showed toxicity against *A. albopictus* larvae (LC_{50} = 39.5 ng/mL) (Table S4). Interestingly, the Cyt1A-like toxin showed high specificity as a synergistic factor. As opposed to Cyt1Aa, which had activity against *A. albopictus* larvae and interacted synergistically with Cry4Aa, Cry4Ba and Cry11Aa, Cyt1A-like showed no activity (not active at a concentration of 1×10^5 ng/mL) and was only able to potentiate Cry11Aa when tested in combination with the aforementioned proteins. Interestingly, such synergistic effect increased the activity of Cry11Aa 23.14 fold, which was at least two times greater than the one provided by Cyt1Aa (synergistic factor of 11.44).

Possibly the most remarkable feature of Cyt1A-like was the ability to activate the otherwise non-active and newly described Cry53-like, Cry56A-like and Tpp36-like, which showed no activity on their own at a concentration of 1×10^5 ng/mL. This phenomenon was shared with Cyt1Aa for Cry56-like, for which the mixture produced a similar LC_{50}. The combination of Cyt1A-like+Cry56A-like was the most effective, with an LC_{50} of 186 ng/mL of spore-and-crystal mixture, a concentration that is close to the ones found in some of the major Bti toxins alone such as Cry4Aa and Cyt1Aa. Cyt1A-like+Cry53-like and Cyt1A-like+Tpp36-like had considerably higher LC_{50} values of 1331 ng/mL and 1053 ng/mL, respectively. Although the combination Cyt1A-like+Tpp36-like was among the least active ones, it confirmed that Cyt1A-like was also able to activate proteins that have a different structure compared to the classic three-domain Cry proteins (Tpp proteins have a typical structure that includes only one domain, named Toxin_10 [54]. In this study, Cyt1A and Cyt1Aa-like were able to activate toxins that were non-active on their own, came from different strains, and had different structures, opening the possibility for them to enable the activity of proteins that have been, until now, considered non-toxic. The biological explanation for this could be that communities of Bt strains in the wild act in a cooperative manner by activating and potentiating the crystal toxins of their fellow Bt neighbors in the pursuit of higher efficacies and more varied mechanisms of actions for when infecting their hosts. Although we were unable to decipher the mechanism of action of said interactions, it may be similar to the one described for Cyt1Aa and Cry11Aa [64]. Here we propose that the characterized Cyt1A and Cyt1A-like toxins could function as enablers of the activity of otherwise non-active Cry and Tpp proteins by habilitating binding sites for them on the lipidic membrane of the insect midgut epithelial cells.

Considering that Cyt1A toxins may delay the appearance of resistance and that the proteins that we described are new, we believe they could represent a great alternative for the control of *A. albopictus* larvae. Additionally, the capability of Cyt proteins to activate non-toxic proteins in a specific manner may be a common characteristic among Cyt1A proteins. To the best of our knowledge, this represents a new function for Cyt proteins, rendering them not only as capable of potentiating the activity of Cry proteins but also as activators of otherwise non-active proteins.

4. Materials and Methods

4.1. Total DNA Extraction and Genomic Sequencing of the Bacterial Strains

Bacillus thuringiensis var. *israeliensis* was isolated from the commercial Bt-based product Vectobac-*12AS*®. TF059.3 and BST-230 were obtained from Spanish soils and belong to the BST collection. Total genomic DNA (chromosome and plasmid) was extracted from the strains using the Wizard® Genomic DNA purification kit (Promega, Madison, WI, USA). A sequencing library was prepared for Illumina sequencing by using a NextSeq500 sequencer (Genomics Research Hub Laboratory, School of Biosciences, Cardiff University, Cardiff, UK).

4.2. Identification of the Potential Mosquitocidal Genes in the BST Collection

CLC Genomic Workbench 10.1.1 (QIAGEN, Aarhus, Denmark) was used to process and assemble the genomic raw data. Reads were trimmed and filtered to remove those of low quality, and reads shorter than 50 bp were removed. Processed reads were de novo assembled using a stringent criterion of overlap of at least 95 bp of the read and 95% identity, and reads were then mapped back to the contigs for assembly correction. Genes were predicted using GeneMark v2.5 (Georgia Institute of Technology, Atlanta, GA,

USA) [73]. To assist the identification process of potential mosquitocidal toxin proteins, the Basic Local Alignment Search Tool [74,75] was deployed against a database built in our laboratory, including the amino acid sequences of known Bt toxins with pesticidal activity from the bacterial pesticidal protein database (https://www.bpprc.org, accessed on 18 July 2022) [76,77]. The pairwise sequence alignment comparison was calculated by using needle v6.6.0 [78]. The prediction of structurally conserved domains was carried out using CD-search [49].

4.3. Bacterial Strains and Plasmids Used in the Cloning Process

The recombinant plasmids pHT606:*cry4Aa*, pHT618:*cry4Ba* and pWF45:*cyt1Aa:p20* were provided by Dr. Colin Berry (Cardiff University, Cardiff, UK) and electroporated into the acrystalliferous BMB171 strain. BMB171 was used as a host vector to express all of the proteins used in this study. *Escherichia coli* XL1 blue was used for transformation. *Cry4Aa4-like, cry53Ab1-like, cry56Aa2-like, cyt1Aa5-like, cyt1Da1-like* and *tpp36Aa1-like* were expressed in vector pTBT02. *Cry11Aa* (1941 bp) was cloned alongside *p19* (540 bp) and *p20* (549 bp) [40]. Both P19 and P20 helped crystalize Cry11Aa as well as increase its mosquitocidal activity [79].

4.4. Amplification, Cloning and Sequencing of Cyt1a-like, Cyt1d-like, Cry4-like, Cry53-like, Cry56a-like and Tpp36-like

SnapGene® software (GSL Biotech, Chicago, IL, USA) was used to design plasmids and simulate the cloning process.

To clone each of the selected toxin genes, these were first amplified using a specific set of oligonucleotides listed in Table 5.

- cyt1-like

Primers harboring the restriction enzymes *PstI and SalI* recognition sites at their extremes were used to amplify the full coding sequence of the *cyt1-like* genes. *SalI* and *SacI* were used for the amplification of *p20*.

- cry4-like

The coding sequence of cry4-like was amplified by using primers harboring SalI and SacI recognition sites.

- cry53-like

Primers harboring the restriction enzymes SalI and SacI recognition sites at their extremes were used to amplify the full coding sequence of the two contiguous genes cry53-like and orf2.

- cry56A-like

Primers harboring the restriction enzymes SalI and SacI recognition sites were used to amplify the gene cry56A-like, whereas PstI and SalI were used for the orf1 gene.

- tpp36-like

Primers used for the amplification of tpp36-like harbored SalI and SacI recognition sites.

The PCR reactions were performed using a Q5® High-Fidelity 2X Master Mix (New England Biolabs, Ipswich, SD, USA). PCR products were gel-purified by using NucleoSpin® Gel and a PCR Clean Up kit (Macherey-Nagel Inc., Bethlehem, PA, USA). After the first ligation into pJET-blunt plasmid using a CloneJET PCR Cloning Kit (Thermo Scientific, Waltham, MA, USA), the ligation products were electroporated into *E. coli* XL1 blue cells. Colonies were checked via PCR in order to isolate the ones carrying the plasmid. Plasmids from positive clones were purified using the NucleoSpin® Plasmid Kit (Macherey-Nagel Inc., Bethlehem, PA, USA), following the manufacturer's instructions. Finally, pJET plasmids sequences were confirmed via sequencing (StabVida, Caparica, Portugal). Once the sequences were verified, the plasmids were digested with the specific set restriction enzymes for each fragment of interest and run in agarose gels, and the corresponding bands were excised and purified. These were then ligated to pre-digested expression vectors using the Rapid DNA ligation kit (ThermoScientific, Vilnius, Lithuania). *Cyt1A-like:p20*, *cyt1D-like:p20*, *cry4-like*, *cry53-like:orf2*, *orf1:cry56A-like* and *Tpp36-like* were cloned in the pTBT02 vector. The final plasmids were then electroporated into *E. coli* XL1 blue cells. Positive clones were verified via colony-PCR, and plasmids were purified and verified via digestion. *pTBT02-cyt1A-like:p20*, *pTBT02:cyt1D-like:p20*, *pTBT02:cry4-like*, *pTBT02:cry53-like:orf2*, *pTBT02:orf1:cry56A-like* and *pTBT02:Tpp36-like* were finally introduced into the BMB171 Bt strain. The pTBT02 expression vector was created using the pSTAB backbone and by adding a more versatile multicloning site (MCS) as well as a terminator of transcription (TT), which was not present in the previously utilized plasmid. The new MCS and the TT were amplified from the pCN47 plasmid [80] using primers MCS-TT_MfeI and MCS-TT_AatII. The resulting amplicon was cloned in pJET and excised utilizing the MfeI and Aatll enzymes. Next, the digested fragment was inserted in the pSTAB plasmid. Additionally, the *cyt1A* promoter was relocated in pSTAB by amplifying it with new primers carrying the sequence for the Sphl and SalI restriction sites. The resulting fragment was subcloned in pJET, digested using the corresponding restriction enzymes and inserted into the pSTAB.

Table 5. Primers used for PCR and sequencing.

Primer Name	Sequence (5′-3′)
Cyt1Aa_like_FW_PstI	GTGTCGACCAAAGGCAGTGGTGTTTTAAG
Cyt1Aa_like_RV_SalI	CTCTGCAGGGGCTACCCAATTATAATCG
p20-Fw-SalI	CCTGCAGGGATAAAATTGGAGGATAATTGATG
p20-Rv-SacI	GGCATGCGTTTCCAGTGCATTCAATTTAC
Cry4Aa4_FW_SalI_BST230	GTCGACGAAATTCAATTGGAAATGGAGGAAC
Cry4Aa_RV_SacI_BST230	GAGCTCCTTTTTTCCAAATTTGTAATAGAAT
Orf_Cry56_FW_PstI	CTGCAGCAGCAAAAAATACGCAGAAAAGGTA
Orf_Cry56_RV_SalI	GTCGACGAATCGTTAACGGTTATATCTTTG
Cry56Aa2_FW_SalI_BST230	GTCGACGGACTACATAAGGAGTGAAA
Cry56Aa2_RV_SacI_BST230	GAGCTCCTATAGAACTGGCCGCTTGA
Cry53+Orf2_FW_SalI	GTCGACGGACTACATAAGGAGTGAAAAAT
Cry53+Orf2_RV_SacI	GAGCTCCTAATTCTCATTTGGAATCGT
Tpp36Aa1_FW_SalI_BST230	GTCGACGAAAAAAATCACATAAGGAGTG
Tpp36Aa1_RV_SacI_BST230	GAGCTCCCCTTACTTCGTTCTACTTAC
Cyt1Aa4like_FW_PstI	CTGCAGCAAAGGCAGTGGTGTTTTAAG
Cyt1Aa4Like_RV_SalI	GTCGACGGGCTACCCAATTATAATCG
Cyt1Da1_like_FW_PstI	CTGCAGCGAGAGAGGTATAAATATGAACC
Cyt1Da1_like_RV_SalI	GTCGACGTAAGAACCCTACGACTAGG
MCS-TT_MfeI	CAATTGGCATGCCTGCAGGTCGACTCTAGAA GATCTCCCGGGTACCGAGC
MCS-TT_AatII	GACGTCAAAGGCGCCTGTCACTTTGCTTG
Sequencing	
Cry4Aa_seq_BST230	CTAGTGAATAATGTAGGTTCTTTA
Cry4Aa_seq1_BST230	CAAGTATGCAATACTGCTTAC
Cry4Aa_seq2_BST230	GATATGGTTTCTATTTCACTTG
Cry4Aa_seq3_BST230	GTCAATCAAGAAATTTACTTCAAA
Cry56orf_seq_FW	GAAGTGTCACGATCGCCAT
Cry56orf_seq_RV	TTCACATGTTCCAATGCTTCA
Cry56orf_seq_FW_1	ATTCCGGCTGCACATGTAAC
Cry56orf_seq_RV_1	GAGCTGTTTGGTGAAGTATCCA
Cry56orf_seq_FW_2	CCATAACATTATATACTAACGTGG
Cry56orf_seq_RV_2	TACTGCTCAGATGCCACGTT
Cry53like_FW_seq1	GTAGAGAAATGACCATAACAG
Cry53like_RV_seq2	GCAGGAAATAGAGCAACTATATCT
Cry53like_FW_seq3	GCTTTGTCACTAAATAATTTGCG

Table 5. *Cont.*

Primer Name	Sequence (5′-3′)
Cry53like_RV_seq4	GTAAGCAAAATTCTCATTTCGCAA
Cry53like_RV_seq5	CATACCTAAGTTTGTATTTGTATCT
Cry53like_FW_seq6	GATTTTCATATTGACACAGGAGA
Tpp36like_FW_seq1	CATTAATTCCGTGTATACTTGTAAA
Tpp36like_RV_seq2	CTGCTAATGAATATTGATAATCA
Seq pCyt1A F (59)	CATATATTTGCACCGTCTAATGG
MCS-TT_AatII	GACGTCAAAGGCGCCTGTCACTTTGCTTG

Underlined nucleotides represent restriction enzyme sites.

4.5. Nucleotide Sequence Accession Numbers

The nucleotide sequence data reported in this paper were deposited in the GeneBank database under the following accession numbers: OQ397557 for *cyt1A-like*, OQ397558 for *cyt1D-like*, OQ397551 for *cry4-like*, OQ397553 for *Cry53-like*, OQ397554 for *orf2*, OQ397555 for *cry56A-like*, OQ397556 for *orf1* and OQ397552 for *tpp36-like*.

4.6. Spore-and-Crystal Mixture Production, Protein Quantification and SDS-PAGE Analysis

BMB171 recombinant strains carrying pHT606:cry4Aa, pHT618:cry4Ba, pWF45: cyt1Aa:p20, pTBT02-cyt1Aa-like:p20, pTBT02:cyt1Da1-like:p20, pTBT02:cry4Aa4-like, pTBT02:cry53Ab1-like:orf2, pTBT02:orf1:cry56Aa-like and pTBT02:Tpp36Aa1-like were grown in 50 mL of CCY medium (supplemented with 20 μg/mL erythromycin) after inoculating single colonies from LB plates [81]. The strains were grown constantly at 28 °C with shaking at 200 rpm. Crystal formation was observed daily at the microscope. Once the cells lysed, after 48–72 h, spore-and-crystal mixtures were washed first with 1M NaCl and 10 mM EDTA, resuspended in 1 mL of dH_2O water and kept at 4 °C until use. The mixtures were solubilized in carbonate buffer (50 mM Na_2CO_3 and 100 mM NaCl, pH 11.3) and quantified for their total amounts of protein by using the Bradford method [82] and by using bovine serum albumin as a standard. For protein profile analysis, the washed spore-and-crystal mixtures were mixed with 2× sample buffer (Bio-Rad, Hercules, CA, USA), boiled at 100 °C for 5 min and then subjected to electrophoresis with a previously described method [83] using Criterion TGX™ 4–20% Precast Gel (Bio-Rad, Laboratories Inc., Hercules, CA, USA). Gels were stained with Coomassie brilliant blue R-250 (Bio-Rad, Laboratories Inc., Hercules, CA, USA) and then destained in a solution of 30% ethanol and 10% acetic acid.

4.7. Bioassays on L2 Larvae of A. Albopictus

The toxicities of the single proteins and mixtures were determined on second instar larvae of *A. albopictus*. Eggs of *A. albopictus* were provided by BioGenius GmbH (Friedrich-Ebert-Straße 75, 51429 Bergisch Gladbach, Germany). The bioassays were performed by placing between 10–15 larvae (L2) in each well of a 6-well plate Corning® Costar® (Corning™, Corning, NY, USA). The bioassays were performed following a previously described method [84]. Each well contained a known concentration of spores and crystals in a total volume of 5 mL, with 0.5 mg of brewer's yeast as food source for the larvae. In order to calculate the median lethal concentration LC_{50}, 6 concentrations (from high to low) of Bt suspension were chosen for each recombinant strain: 1000, 500, 250, 125, 62.5 and 31.5 ng/mL for Cry4Aa and Cyt1Aa, 500, 250, 125, 62.5, 31.5 and 15.75 ng/mL for Cry4Ba and 2500, 1250, 625, 312.5, 156.25 and 78.12 ng/mL for Cry11Aa. The highest concentrations (C1) were defined as a dose that produces between 90–100% of mortality, whereas the lower doses were simply 1:2 serial dilutions of the C1 dose.

4.8. Statistical Analysis

Concentration–mortality data were subjected to logit regression to estimate the LC_{50} for individual toxins and mixtures of toxins [85].The observed and expected LC_{50} values for the individual toxins and the toxin mixture in *A. albopictus* were used to evaluate the interaction of Cyt1Aa with Cry4Aa, Cry4Ba, Cry11Aa and Cry56Aa2 and the interaction between Cyt1Aa and Cyt1A-like with Cry11Aa, Cry53Ab-like, Cry56Aa-like and Tpp36Ab-like. To calculate the expected LC_{50} values for the toxin mixture under the null hypothesis of no interaction, the "simple similar action" model was used [86]. This model assumes that the concentration–response regression lines for different components of a mixture are parallel and suitable for testing synergism in chemically compounds that are alike, such as Bt toxins. All synergies were evaluated by first calculating the expected LC_{50}, as follows, as there were no synergisms between them:

$$LC_{50(m)}\left[\frac{ra}{LC_{50(A)}}+\frac{rb}{LC_{50(B)}}\right]^{-1}$$

where $LC_{50(m)}$ is the expected LC_{50} of the mixture of toxin A and toxin B, $LC_{50(A)}$ is the observed LC_{50} for toxin A alone, $LC_{50(B)}$ is the observed LC_{50} for toxin B alone and rA and rB represent the relative proportions of toxin A and toxin B in the mixture, respectively. All statistical procedures were performed using R software (v.4.1.1) (R Foundation for Statistical Computing, Vienna, Austria).

Supplementary Materials: The following supporting information can be downloaded at: https://www.mdpi.com/article/10.3390/toxins15030211/s1, Table S1: Mean lethal concentration (LC50) value of BST-230; Table S2: Raw data of the synergies between Cry and Cyt toxins; Table S3: Raw data of the activated Cry and Tpp toxins; Table S4: Mean lethal concentration (LC50) value of BST-230.

Author Contributions: Conceptualization, C.J.C., P.C. and M.V.; methodology, L.L., M.V., A.M.-G., A.T.-A. and A.U.; software, L.L. and A.B.F.; formal analysis, L.L., C.J.C. and P.C.; writing—original draft preparation, L.L., C.J.C., M.V., A.B.F. and P.C.; writing—review and editing, L.L., C.J.C., M.V., A.B.F., A.U., A.M.-G., A.T.-A. and P.C.; funding acquisition, C.J.C. All authors have read and agreed to the published version of the manuscript.

Funding: This research was funded by Gobierno de Navarra with grant numbers 0011-1365-2020-000297 to C.J.C. and 0011-1365-2020-000033 to A.T.-A. L.L. received a doctoral grant from the European Union's H2020 research and innovation programme under Marie Sklodowska-Curiegrant agreement N°801586. A.M.-G. was supported by a FPU PhD grant from the Spanish Ministry of Science and Innovation (FPU20/05496). A.B. Fernández's contract was cofounded by the Program Torres Quevedo (Ministerio de Ciencia, Innovación y Universidades, Gobierno de España) (PTQ-2018-010091) and Bioinsectis SL.

Institutional Review Board Statement: Not applicable.

Informed Consent Statement: Not applicable.

Data Availability Statement: Not applicable.

Acknowledgments: The authors thank Colin Berry (Cardiff University, Cardiff, UK) for providing the Bt recombinant strains 4Q2-81 pHT606:*cry4Aa* and 4Q2-81 pHT611:*cry4Ba* used in this study.

Conflicts of Interest: The authors declare no conflict of interest.

References

1. Kambhampati, S.; Black Iv, W.C.; Rai, K.S. Geographic Origin of the US and Brazilian Aedes Albopictus Inferred from Allozyme Analysis. *Heredity* **1991**, *67*, 85–94. [CrossRef]
2. Bonizzoni, M.; Gasperi, G.; Chen, X.; James, A.A. The Invasive Mosquito Species Aedes Albopictus: Current Knowledge and Future Perspectives. *Trends Parasitol.* **2013**, *29*, 460–468. [CrossRef]
3. Adhami, J.; Reiter, P. Introduction and Establishment of Aedes (Stegomyia) Albopictus Skuse (Diptera: Culicidae) in Albania. *J. Am. Mosq. Control. Assoc.* **1998**, *14*, 340–343. [PubMed]
4. Tippelt, L.; Werner, D.; Kampen, H. Tolerance of Three Aedes Albopictus Strains (Diptera: Culicidae) from Different Geographical Origins towards Winter Temperatures under Field Conditions in Northern Germany. *PLoS ONE* **2019**, *14*, e0219553. [CrossRef] [PubMed]

5. Sabatini, A.; Raineri, V.; Trovato, G.; Coluzzi, M. Aedes Albopictus in Italy and Possible Diffusion of the Species into the Mediterranean Area. *Parassitologia* **1990**, *32*, 301–304.
6. Rezza, G. Chikungunya Is Back in Italy: 2007–2017. *J. Travel Med.* **2018**, *25*, tay004. [CrossRef]
7. Lwande, O.W.; Obanda, V.; Lindström, A.; Ahlm, C.; Evander, M.; Näslund, J.; Bucht, G. Globe-Trotting Aedes Aegypti and Aedes Albopictus: Risk Factors for Arbovirus Pandemics. *Vector-Borne Zoonotic Dis.* **2020**, *20*, 71–81. [CrossRef]
8. Battaglia, V.; Agostini, V.; Moroni, E.; Colombo, G.; Lombardo, G.; Rambaldi Migliore, N.; Gabrieli, P.; Garofalo, M.; Gagliardi, S.; Gomulski, L.M.; et al. The Worldwide Spread of Aedes Albopictus: New Insights from Mitogenomes. *Front. Genet.* **2022**, *13*. [CrossRef] [PubMed]
9. Kraemer, M.U.G.; Reiner, R.C.; Brady, O.J.; Messina, J.P.; Gilbert, M.; Pigott, D.M.; Yi, D.; Johnson, K.; Earl, L.; Marczak, L.B.; et al. Past and Future Spread of the Arbovirus Vectors Aedes Aegypti and Aedes Albopictus. *Nat. Microbiol.* **2019**, *4*, 854–863. [CrossRef] [PubMed]
10. Leta, S.; Beyene, T.J.; de Clercq, E.M.; Amenu, K.; Kraemer, M.U.G.; Revie, C.W. Global Risk Mapping for Major Diseases Transmitted by Aedes Aegypti and Aedes Albopictus. *Int. J. Infect. Diseases* **2018**, *67*, 25–35. [CrossRef]
11. Sasayama, M.; Benjathummarak, S.; Kawashita, N.; Rukmanee, P.; Sangmukdanun, S.; Masrinoul, P.; Pitaksajjakul, P.; Puiprom, O.; Wuthisen, P.; Kurosu, T.; et al. Chikungunya Virus Was Isolated in Thailand, 2010. *Virus Genes* **2014**, *49*, 485–489. [CrossRef]
12. da Cunha, R.V.; Trinta, K.S. Chikungunya Virus: Clinical Aspects and Treatment. *Mem. Inst. Oswaldo Cruz* **2017**, *112*, 523–531. [CrossRef] [PubMed]
13. Rezza, G. Aedes Albopictus and the Reemergence of Dengue. *BMC Public Health* **2012**, *12*, 72. [CrossRef]
14. Gardner, L.M.; Chen, N.; Sarkar, S. Global Risk of Zika Virus Depends Critically on Vector Status of Aedes Albopictus. *Lancet Infect. Dis* **2016**, *16*, 522–523. [CrossRef] [PubMed]
15. Ciota, A.T.; Bialosuknia, S.M.; Zink, S.D.; Brecher, M.; Ehrbar, D.J.; Morrissette, M.N.; Kramer, L.D. Effects of Zika Virus Strain and Aedes Mosquito Species on Vector Competence. *Emerg. Infect. Dis* **2017**, *23*, 1110–1117. [CrossRef]
16. Gómez, M.; Martinez, D.; Muñoz, M.; Ramírez, J.D. Aedes Aegypti and Ae. Albopictus Microbiome/Virome: New Strategies for Controlling Arboviral Transmission? *Parasit Vectors* **2022**, *15*, 287. [CrossRef]
17. Rivero, A.; Vézilier, J.; Weill, M.; Read, A.F.; Gandon, S. Insecticide Control of Vector-Borne Diseases: When Is Insecticide Resistance a Problem? *PLoS Pathog* **2010**, *6*, 5–6. [CrossRef]
18. Medlock, J.M.; Hansford, K.M.; Schaffner, F.; Versteirt, V.; Hendrickx, G.; Zeller, H.; van Bortel, W. A Review of the Invasive Mosquitoes in Europe: Ecology, Public Health Risks, and Control Options. *Vector-Borne Zoonotic Dis.* **2012**, *12*, 435–447. [CrossRef] [PubMed]
19. Rodríguez, M.M.; Bisset, J.A.; Fernández, D. Levels of Insecticide Resistance and Resistance Mechanisms in Aedes Aegypti from Some Latin American Countries. *J. Am. Mosq. Control Assoc.* **2007**, *23*, 420–429. [CrossRef]
20. Hemingway, J.; Hawkes, N.J.; McCarroll, L.; Ranson, H. The Molecular Basis of Insecticide Resistance in Mosquitoes. *Insect Biochem. Mol. Biol. Insect Biochem. Mol. Biol.* **2004**, *34*, 653–665. [CrossRef] [PubMed]
21. Marcombe, S.; Farajollahi, A.; Healy, S.P.; Clark, G.G.; Fonseca, D.M. Insecticide Resistance Status of United States Populations of Aedes Albopictus and Mechanisms Involved. *PLoS ONE* **2014**, *9*, e101992. [CrossRef]
22. Liu, N. Insecticide Resistance in Mosquitoes: Impact, Mechanisms, and Research Directions. *Annu. Rev. Entomol.* **2015**, *60*, 537–559. [CrossRef] [PubMed]
23. Department of Control of Neglected Tropical Diseases WHO Pesticide Evaluation Scheme (WHOPES). *Pesticides and Their Application for the Control of Vectors and Pests of Public Health Importance*; WHO: Geneva, Switzerland, 2006.
24. Kache, P.A.; Eastwood, G.; Collins-Palmer, K.; Katz, M.; Falco, R.C.; Bajwa, W.I.; Armstrong, P.M.; Andreadis, T.G.; Diuk-Wasser, M.A. Environmental Determinants of Aedes Albopictus Abundance at a Northern Limit of Its Range in the United States. *Am. J. Trop. Med. Hyg.* **2020**, *102*, 436–447. [CrossRef] [PubMed]
25. Raymond, B.; Federici, B.A. In Defense of Bacillus Thuringiensis, the Safest and Most Successful Microbial Insecticide Available to Humanity—A Response to EFSA. *FEMS Microbiol. Ecol.* **2017**, *93*, fix084. [CrossRef] [PubMed]
26. Wirth, M.C.; Park, H.W.; Walton, W.E.; Federici, B.A. Cyt1A of *Bacillus thuringiensis* Delays Evolution of Resistance to Cry11A in the Mosquito Culex Quinquefasciatus. *Appl. Environ. Microbiol.* **2005**, *71*, 185–189. [CrossRef] [PubMed]
27. Stalinski, R.; Tetreau, G.; Gaude, T.; Després, L. Pre-Selecting Resistance against Individual Bti Cry Toxins Facilitates the Development of Resistance to the Bti Toxins Cocktail. *J. Invertebr. Pathol.* **2014**, *119*, 50–53. [CrossRef]
28. Stalinski, R.; Laporte, F.; Tetreau, G.; Després, L. Receptors Are Affected by Selection with Each *Bacillus thuringiensis* Israelensis Cry Toxin but Not with the Full Bti Mixture in Aedes Aegypti. *Infect. Genet. Evol.* **2016**, *44*, 218–227. [CrossRef]
29. Lee, S.B.; Aimanova, K.G.; Gill, S.S. Alkaline Phosphatases and Aminopeptidases Are Altered in a Cry11Aa Resistant Strain of Aedes Aegypti. *Insect. Biochem. Mol. Biol.* **2014**, *54*, 112–121. [CrossRef]
30. Cadavid-Restrepo, G.; Sahaza, J.; Orduz, S. Treatment of an Aedes Aegypti Colony with the Cry11Aa Toxin for 54 Generations Results in the Development of Resistance. *Mem. Inst. Oswaldo Cruz* **2012**, *107*, 74–79. [CrossRef]
31. Boyce, R.; Lenhart, A.; Kroeger, A.; Velayudhan, R.; Roberts, B.; Horstick, O. *Bacillus thuringiensis* Israelensis (Bti) for the Control of Dengue Vectors: Systematic Literature Review. *Trop. Med. Int. Health* **2013**, *18*, 564–577. [CrossRef]
32. Russell, T.L.; Brown, M.D.; Purdie, D.M.; Ryan, P.A.; Kay, B.H. Efficacy of VectoBac (*Bacillus thuringiensis* Variety Israelensis) Formulations for Mosquito Control in Australia. *J. Econ. Entomol.* **2003**, *96*, 1786–1791. [CrossRef]

33. Federici, B.A.; Park, H.W.; Bideshi, D.K.; Wirth, M.C.; Johnson, J.J. Recombinant Bacteria for Mosquito Control. *J. Exp. Biol.* **2003**, *206*, 3877–3885. [CrossRef] [PubMed]
34. Kamgang, B.; Marcombe, S.; Chandre, F.; Nchoutpouen, E.; Nwane, P.; Etang, J.; Corbel, V.; Paupy, C. Insecticide Susceptibility of Aedes Aegypti and Aedes Albopictus in Central Africa. *Parasit Vectors* **2011**, *4*, 79. [CrossRef]
35. Guerchicoff, A.; Ugalde, R.A.; Rubinstein, C.P. Identification and Characterization of a Previously Undescribed Cyt Gene in *Bacillus thuringiensis* subsp. Israelensis. *Appl. Environ. Microbiol.* **1997**, *63*, 2716–2721. [CrossRef] [PubMed]
36. Waalwijk, C.; Dullemans, A.M.; van Workum, E.S.; Visser, B. Molecular Cloning and the Nucleotide Sequence of the Mr 28000 Crystal Protein Gene of Bacilus Thuringiensis Subsp. Israelensis. *Nucleic Acids Res.* **1985**, *13*, 8207–8217. [CrossRef] [PubMed]
37. Anderson, I.; Sorokin, A.; Kapatral, V.; Reznik, G.; Bhattacharya, A.; Mikhailova, N.; Burd, H.; Joukov, V.; Kaznadzey, D.; Walunas, T.; et al. Comparative Genome Analysis of Bacillus Cereus Group Genomes with Bacillus Subtilis. *FEMS Microbio.l Lett.* **2005**, *250*, 175–184. [CrossRef] [PubMed]
38. Donovan, W.P.; Dankocsik, C.; Pearce Gilbert, M. Molecular Characterization of a Gene Encoding a 72-Kilodalton Mosquito-Toxic Crystal Protein from *Bacillus thuringiensis* subsp. Israelensis. *J. Bacteriol.* **1988**, *170*, 4732–4738. [CrossRef] [PubMed]
39. Berry, C.; O'Neil, S.; Ben-Dov, E.; Jones, A.F.; Murphy, L.; Quail, M.A.; Holden, M.T.G.; Harris, D.; Zaritsky, A.; Parkhill, J. Complete Sequence and Organization of PBtoxis, the Toxin-Coding Plasmid of *Bacillus thuringiensis* Subsp. Israelensis. *Appl. Environ. Microbiol.* **2002**, *68*, 5082–5095. [CrossRef] [PubMed]
40. Valtierra-De-Luis, D.; Villanueva, M.; Lai, L.; Williams, T.; Caballero, P. Potential of Cry10Aa and Cyt2Ba, Two Minority δ-Endotoxins Produced by *Bacillus thuringiensis* Ser. Israelensis, for the Control of Aedes Aegypti Larvae. *Toxins* **2020**, *6*, 355. [CrossRef]
41. Park, H.-W.; Bideshi, D.K.; Federici, B.A. Overview of the Basic Biology of *Bacillus thuringiensis* with Emphasis on Genetic Engineering of Bacterial Larvicides for Mosquito Control. *Open Toxinol. J.* **2010**, *3*, 83–100.
42. Valtierra-de-Luis, D.; Villanueva, M.; Berry, C.; Caballero, P. Potential for *Bacillus thuringiensis* and Other Bacterial Toxins as Biological Control Agents to Combat Dipteran Pests of Medical and Agronomic Importance. *Toxins* **2020**, *12*, 773. [CrossRef]
43. Silva-Filha, M.; Romão, T.; Rezende, T.; Carvalho, K.; de Menezes, H.G.; Nascimento, N.A.D.; Soberón, M.; Bravo, A. Bacterial Toxins Active against Mosquitoes: Mode of Action and Resistance. *Toxins* **2021**, *13*, 523. [CrossRef]
44. Ben-Dov, E. *Bacillus thuringiensis* subsp. Israelensis and Its Dipteran-Specific Toxins. *Toxins* **2014**, *6*, 1222–1243. [CrossRef] [PubMed]
45. Crickmore, N.; Bone, E.J.; Williams, J.A.; Ellar, D.J. Contribution of the Individual Components of the Î´-Endotoxin Crystal to the Mosquitocidal Activity of *Bacillus thuringiensis* Subsp. Israelensis. *FEMS Microbiol. Lett.* **1995**, *131*, 249–254. [CrossRef]
46. Fernández-Luna, M.T.; Tabashnik, B.E.; Lanz-Mendoza, H.; Bravo, A.; Soberón, M.; Miranda-Ríos, J. Single Concentration Tests Show Synergism among *Bacillus thuringiensis* Subsp. Israelensis Toxins against the Malaria Vector Mosquito Anopheles Albimanus. *J. Invertebr. Pathol.* **2010**, *104*, 231–233. [CrossRef]
47. Georghiou, G.P.; Wirth, M.C. Influence of Exposure to Single versus Multiple Toxins of *Bacillus thuringiensis* Subsp. Israelensis on Development of Resistance in the Mosquito Culex Quinquefasciatus (Diptera: Culicidae). *Appl. Environ. Microbiol.* **1997**, *63*, 1095–1101. [CrossRef]
48. Wirth, M.C.; Georghiou, G.P.; Federici, B.A. CytA Enables CryIV Endotoxins of *Bacillus thuringiensis* to Overcome High Levels of CryIV Resistance in the Mosquito, Culex Quinquefasciatus. *Proc. Natl. Acad. Sci. USA* **1997**, *94*, 10536–10540. [CrossRef] [PubMed]
49. Marchler-Bauer, A.; Bo, Y.; Han, L.; He, J.; Lanczycki, C.J.; Lu, S.; Chitsaz, F.; Derbyshire, M.K.; Geer, R.C.; Gonzales, N.R.; et al. CDD/SPARCLE: Functional Classification of Proteins via Subfamily Domain Architectures. *Nucleic Acids Res.* **2017**, *45*, D200–D203. [CrossRef] [PubMed]
50. Adalat, R.; Saleem, F.; Crickmore, N.; Naz, S.; Shakoori, A.R. In Vivo Crystallization of Three-Domain Cry Toxins. *Toxins* **2017**, *9*, 80. [CrossRef]
51. Sazhenskiy, V.; Zaritsky, A.; Itsko, M. Expression in Escherichia Coli of the Native Cyt1Aa from *Bacillus thuringiensis* subsp. Israelensis. *Appl. Environ. Microbiol.* **2010**, *76*, 3409–3411. [CrossRef]
52. Wu1, D.; Federici', B.A. A 20-Kilodalton Protein Preserves Cell Viability and Promotes CytA Crystal Formation during Sporulation in *Bacillus thuringiensis*. *J. Bacteriol.* **1993**, *175*, 5276–5280.
53. Caballero, J.; Jiménez-Moreno, N.; Orera, I.; Williams, T.; Fernández, A.B.; Villanueva, M.; Ferré, J.; Caballero, P.; Ancín-Azpilicueta, C.; Johnson, K.N. Unraveling the Composition of Insecticidal Crystal Proteins in *Bacillus thuringiensis*: A Proteomics Approach. *Appl. Environ. Microbiol.* **2020**, *86*, e00476-20. [CrossRef]
54. Crickmore, N.; Berry, C.; Panneerselvam, S.; Mishra, R.; Connor, T.R. Bacterial Pesticidal Protein Resource Center, Viewed (02/12/2022). Available online: https://www.Bpprc.Org (accessed on 12 November 2022).
55. de Barros Moreira Beltrão, H.; Silva-Filha, M.H.N.L. Interaction of *Bacillus thuringiensis* Svar. Israelensis Cry Toxins with Binding Sites from Aedes Aegypti (Diptera: Culicidae) Larvae Midgut. *FEMS Microbiol. Lett.* **2007**, *266*, 163–169. [CrossRef]
56. Delécluse, A.; Bourgouin, C.; Klier, A.; Rapoport, G. Specificity of Action on Mosquito Larvae of *Bacillus thuringiensis* Israelensis Toxins Encoded by Two Different Genes. *Mol. Genet. Genom.* **1988**, *214*, 42–47. [CrossRef]
57. Chilcott, C.N.; Ellar, D.J. Comparative Toxicity of *Bacillus thuringiensis* Var. Israelensis Crystal Proteins in Aiuo and in Tritvo. *Microbiology* **1988**, *134*, 2551–2558. [CrossRef] [PubMed]

58. Otieno-Ayayo, Z.N.; Zaritsky, A.; Wirth, M.C.; Manasherob, R.; Khasdan, V.; Cahan, R.; Ben-Dov, E. Variations in the Mosquito Larvicidal Activities of Toxins from *Bacillus thuringiensis* ssp. Israelensis. *Environ. Microbiol.* **2008**, *10*, 2191–2199. [CrossRef] [PubMed]
59. Hawley, W. The Biology of Aedes Albopictus. *J. Am. Mosq. Control Assoc. Suppl.* **1988**, *1*, 1–39. [PubMed]
60. Albieri, A.; Carrieri, M.; Angelini, P.; Baldacchini, F.; Venturelli, C.; Zeo, S.M.; Bellini, R. Quantitative Monitoring of Aedes Albopictus in Emilia-Romagna, Northern Italy: Cluster Investigation and Geostatistical Analysis. *Bull. Insectology* **2010**, *63*, 209–216.
61. Caputo, B.; Ienco, A.; Cianci, D.; Pombi, M.; Petrarca, V.; Baseggio, A.; Devine, G.J.; della Torre, A. The "Auto-Dissemination" Approach: A Novel Concept to Fight Aedes Albopictus in Urban Areas. *PLoS Negl. Trop. Dis.* **2012**, *6*, e1793. [CrossRef]
62. Li, J.; Carroll, J.; Ellar, D.J. Crystal Structure of Insecticidal 6-Endotoxin 0 from *Bacillus thuringiensis* at 2.5 A Resolution. *Nature* **1991**, *353*, 815–821. [CrossRef] [PubMed]
63. Xu, C.; Wang, B.C.; Yu, Z.; Sun, M. Structural Insights into *Bacillus thuringiensis* Cry, Cyt and Parasporin Toxins. *Toxins* **2014**, *6*, 2732–2770. [CrossRef]
64. Bravo, A.; Gill, S.S.; Soberón, M. Mode of Action of *Bacillus thuringiensis* Cry and Cyt Toxins and Their Potential for Insect Control. *Toxicon* **2007**, *49*, 423–435. [CrossRef] [PubMed]
65. Best, H.L.; Williamson, L.J.; Lipka-Lloyd, M.; Waller-Evans, H.; Lloyd-Evans, E.; Rizkallah, P.J.; Berry, C. The Crystal Structure of *Bacillus thuringiensis* Tpp80Aa1 and Its Interaction with Galactose-Containing Glycolipids. *Toxins* **2022**, *14*, 863. [CrossRef] [PubMed]
66. Pérez, C.; Fernandez, L.E.; Sun, J.; Folch, J.L.; Gill, S.S.; Soberó, M.; Bravo, A. *Bacillus thuringiensis* subsp. Israelensis Cyt1Aa Synergizes Cry11Aa Toxin by Functioning as a Membrane-Bound Receptor. *Proc. Natl. Acad. Sci. USA* **2005**, *102*, 18303–18308. [CrossRef]
67. Wirth, M.C.; Delécluse, A.; Walton, W.E. Cyt1Ab1 and Cyt2Ba1 from *Bacillus thuringiensis* Subsp. Medellin and *B. thuringiensis* Subsp. Israelensis Synergize *Bacillus sphaericus* against Aedes Aegypti and Resistant Culex Quinquefasciatus (Diptera: Culicidae). *Appl. Environ. Microbiol.* **2001**, *67*, 3280–3284. [CrossRef]
68. Tetreau, G.; Stalinski, R.; David, J.P.; Després, L. Monitoring Resistance to *Bacillus thuringiensis* subsp. Israelensis in the Field by Performing Bioassays with Each Cry Toxin Separately. *Mem. Inst. Oswaldo Cruz.* **2013**, *108*, 894–900. [CrossRef]
69. Paris, M.; Tetreau, G.; Laurent, F.; Lelu, M.; Despres, L.; David, J.P. Persistence of *Bacillus thuringiensis* Israelensis (Bti) in the Environment Induces Resistance to Multiple Bti Toxins in Mosquitoes. *Pest Manag. Sci.* **2011**, *67*, 122–128. [CrossRef] [PubMed]
70. Delecluse, A.; Poncet, S.; Klier, A.; Rapoport, G. Expression of CryIVA and CryIVB Genes, Independently or in Combination, in a Crystal-Negative Strain of *Bacillus thuringiensis* Subsp. Israelensis. *Appl. Environ. Microbiol.* **1993**, *59*, 3922–3927. [CrossRef] [PubMed]
71. Abdullah, M.A.F.; Alzate, O.; Mohammad, M.; McNall, R.J.; Adang, M.J.; Dean, D.H. Introduction of Culex Toxicity into *Bacillus thuringiensis* Cry4Ba by Protein Engineering. *Appl. Environ. Microbiol.* **2003**, *69*, 5343–5353. [CrossRef]
72. Zhu, J.; Zheng, A.P.; Tan, F.R.; Wang, S.Q.; Deng, Q.M.; Li, S.C.; Wang, L.X.; Li, P. Characterisation and Expression of a Novel Holotype Crystal Protein Gene, Cry56Aa1, from *Bacillus thuringiensis* Strain Ywc2-8. *Biotechnol. Lett.* **2010**, *32*, 283–288. [CrossRef]
73. Borodovsky Mark; McIninch James GENMARK: Parallel Gene Recognition for Both DNA Strands. *Comput. Chem.* **1993**, *17*, 123–133. [CrossRef]
74. Altschul, S.F.; Gish, W.; Miller, W.; Myers, E.W.; Lipman, D.J. Basic Local Alignment Search Tool. *J. Mol. Biol.* **1990**, *215*, 403–410. [CrossRef] [PubMed]
75. Available online: Https://Blast.Ncbi.Nlm.Nih.Gov/Blast.Cgi (accessed on 12 November 2022).
76. Crickmore, N.; Berry, C.; Panneerselvam, S.; Mishra, R.; Connor, T.R.; Bonning, B.C. A Structure-Based Nomenclature for *Bacillus thuringiensis* and Other Bacteria-Derived Pesticidal Proteins. *J. Invertebr. Pathol.* **2020**, *186*, 107438. [CrossRef] [PubMed]
77. Crickmore, N.; Zeigler, D.R.; Feitelson, J.; Schnepf, E.; van Rie, J.; Lereclus, D.; Baum, J.; Dean, D.H. Revision of the Nomenclature for the *Bacillus thuringiensis* Pesticidal Crystal Proteins. *Microbiol. Mol. Biol. Rev.* **1998**, *62*, 807–813. [CrossRef]
78. Madeira, F.; Pearce, M.; Tivey, A.R.N.; Basutkar, P.; Lee, J.; Edbali, O.; Madhusoodanan, N.; Kolesnikov, A.; Lopez, R. Search and Sequence Analysis Tools Services from EMBL-EBI in 2022. *Nucleic Acids Res.* **2022**, *50*, W276–W279. [CrossRef]
79. Xu, Y.; Nagai, M.; Bagdasarian, M.; Smith, T.W.; Walker, E.D. Expression of the P20 Gene from *Bacillus thuringiensis* H-14 Increases Cry11A Toxin Production and Enhances Mosquito-Larvicidal Activity in Recombinant Gram-Negative Bacteria. *Appl. Environ. Microbiol.* **2001**, *67*, 3010–3015. [CrossRef]
80. Charpentier, E.; Anton, A.I.; Barry, P.; Alfonso, B.; Fang, Y.; Novick, R.P. Novel Cassette-Based Shuttle Vector System for Gram-Positive Bacteria. *Appl. Environ. Microbiol.* **2004**, *70*, 6076–6085. [CrossRef]
81. Stewart, G.S.; Johnstone, K.; Hagelberg, E.; Ellar, D.J. Commitment of Bacterial Spores to Germinate A Measure of the Trigger Reaction. *Biochem. J.* **1981**, *198*, 101–106. [CrossRef]
82. Bradford, M.M. A Rapid and Sensitive Method for the Quantitation of Microgram Quantities of Protein Utilizing the Principle of Protein-Dye Binding. *Anal. Biochem.* **1976**, *72*, 248–254. [CrossRef]
83. Laemmli, U.K. Cleavage of Structural Proteins during the Assembly of the Head of Bacteriophage T4. *Nature* **1970**, *227*, 680–685. [CrossRef] [PubMed]

84. Mclaughlin, R.E.; Dulmage, H.T.; Alls, R.; Couch, T.L.; Dame, D.A.; Hall, I.M.; Rose, R.I.; Versoi, P.L. Standard Bioassay for the Potency Assessment of *Bacillus thuringiensis* Serotype H-14 against Mosquito Larvae [Aedes Aegypti, Biological Control]. *Bull. Entomol. Soc. Am.* **1984**, *30*, 26–29.
85. Finney, D.J. Probit Analysis. *J. Pharm. Sci.* **1971**, *60*, 1432.
86. Tabashnikt, B.E. Evaluation of Synergism among *Bacillus thuringiensis* Toxins. *Appl. Environ. Microbiol.* **1992**, *58*, 3343–3346. [CrossRef] [PubMed]

toxins

MDPI

Communication

Mpp23Aa/Xpp37Aa Insecticidal Proteins from *Bacillus thuringiensis* (Bacillales: Bacillaceae) Are Highly Toxic to *Anthonomus grandis* (Coleoptera: Curculionidae) Larvae

Jéssica A. de Oliveira [1], Bárbara F. Negri [2], Patricia Hernández-Martínez [3], Marcos F. Basso [4] and Baltasar Escriche [3,*]

1 Laboratório de Prospecção de Cepas e Genes, Instituto Mato-Grossense do Algodão (IMAmt), Rondonópolis 78740-970, Mato Grosso, Brazil
2 Laboratório de Biologia Molecular e Transformação de Plantas, Instituto Mato-Grossense do Algodão (IMAmt), Rondonópolis 78740-970, Mato Grosso, Brazil
3 Departamento de Genética, Instituto de Biotecnología y Biomedicina (BIOTECMED), Universitat de València, 46100 Burjassot, Valencia, Spain
4 Dipartimento di Biologia e Incubatore Universitario Fiorentino, Dipartimento di Biologia, Università degli Studi di Firenze, 50019 Sesto Fiorentino, Firenze, Italy
* Correspondence: baltasar.escriche@uv.es; Tel.: +34-963543401

Abstract: The beetle *Anthonomus grandis* Boheman, 1843, is the main cotton pest, causing enormous losses in cotton. The breeding of genetically modified plants with *A. grandis* resistance is seen as an important control strategy. However, the identification of molecules with high toxicity to this insect remains a challenge. The susceptibility of *A. grandis* larvae to proteins (Cry1Ba, Cry7Ab, and Mpp23Aa/Xpp37Aa) from *Bacillus thuringiensis* Berliner, 1915, with toxicity reported against Coleopteran, has been evaluated. The ingestion of different protein concentrations (which were incorporated into an artificial diet) by the larvae was tested in the laboratory, and mortality was evaluated after one week. All Cry proteins tested exhibited higher toxicity than that the untreated artificial diet. These Cry proteins showed similar results to the control Cry1Ac, with low toxicity to *A. grandis*, since it killed less than 50% of larvae, even at the highest concentration applied (100 $\mu g \cdot g^{-1}$). Mpp/Xpp proteins provided the highest toxicity with a 0.18 $\mu g \cdot g^{-1}$ value for the 50% lethal concentration. Importantly, this parameter is the lowest ever reported for this insect species tested with *B. thuringiensis* proteins. This result highlights the potential of Mpp23Aa/Xpp37Aa for the development of a biotechnological tool aiming at the field control of *A. grandis*.

Keywords: cotton boll weevil; Cry proteins; Mpp/Xpp proteins; Cry23Aa; Cry37Aa; Coleoptera; Curculionidae

Key Contribution: Mpp23Aa/Xpp37Aa Bt proteins are promising molecular agents to control the *A. grandis* cotton pest.

Citation: de Oliveira, J.A.; Negri, B.F.; Hernández-Martínez, P.; Basso, M.F.; Escriche, B. Mpp23Aa/Xpp37Aa Insecticidal Proteins from *Bacillus thuringiensis* (Bacillales: Bacillaceae) Are Highly Toxic to *Anthonomus grandis* (Coleoptera: Curculionidae) Larvae. *Toxins* **2023**, *15*, 55. https://doi.org/10.3390/toxins15010055

Received: 14 December 2022
Revised: 2 January 2023
Accepted: 3 January 2023
Published: 8 January 2023

1. Introduction

Cotton (*Gossypium hirsutum* L.) is one of the major crops of the global agricultural economy. During the 2020/2021 growing season, more than 24 million tons of cotton fiber were produced worldwide, with an output of more than 25 million tons in the 2021/2022 growing season. Brazil is the world's fourth-largest cotton producer after China, India, and the USA, with an expected increase of around 19% for the 2021/2022 growing season [1,2]. One of the greatest challenges for cotton cultivation in Brazil is the high demand for agricultural inputs, particularly insecticides. Data indicate that the investment in insecticides, mainly to control the *Anthonomus grandis* Boheman, 1843 (cotton boll weevil, CBW) accounts for approximately 21% of the costs of cotton farming in Mato Grosso, Brazil [3].

The cotton boll weevil is a Coleopteran insect, considered the most harmful cotton pest, and its stereotypic behavior makes control difficult [4]. In addition to feeding on the flower buds, after mating, females lay eggs inside the fruiting structures of the cotton plant, where the development of all immature stages of the insect occurs, causing abscission or reduction in fiber quality [5]. In addition, newly emerged or old adults are commonly protected by the flower bud bracts or refugees in plants outside of the cotton crops, decreasing exposure to mortality factors, i.e., insecticides applied by spraying [6].

Once the insect is detected in pheromone traps, insecticide applications are performed, using mainly malathion. These applications can be performed weekly until the insect is no longer detected. Consequently, the field may receive as many as 25 applications in a single growing season, boosting expenses regarding agricultural inputs, and negatively affecting the environment and entomofauna [7–9]. However, the low effectiveness of the available synthetic agents has contributed to inefficient control measures, an increase in the insect population, and its genetic diversity. In addition, the non-existence of conventional or transgenic commercial cultivars with some resistance to CBW has stimulated the search for new biotechnological tools for the effective control of this insect pest.

The development of genetically modified crops expressing Cry and Vip toxins from *Bacillus thuringiensis* (Bt) has been widely studied due to the toxic effect of these proteins, mainly against Lepidoptera [10]. Interestingly, transgenic plants expressing insecticidal proteins from Bt are effective in controlling stem borers, ear feeders, and rootworms [10]. Therefore, cotton plants expressing Bt proteins (toxic to CBW) may also control this devastating insect pest. Currently, transgenic cotton plants expressing *cry* and *vip* genes to control Lepidoptera are already being commercially launched in Brazil, but to date, no variety or cultivar capable of controlling the beetle *A. grandis* has been released [11–13].

The insect order Coleoptera comprises numerous species considered pests for the world's major crops, few of which have proved susceptible to Bt toxins [14]. Despite causing significant economic losses around the world, the number of studies carried out to identify Bt proteins active against beetles is lower than those performed in Lepidopteran insect species. In the case of CBW, only a few studies have been performed to assess the insecticidal activity of either Bt strains or individual Bt insecticidal proteins. Therefore, more research is needed to determine which Bt proteins are able to effectively control CBW.

The application of Bt toxins to control *A. grandis* in cotton crops, especially as a transgenic plant, is of considerable agronomic interest, since this biotechnological tool is promising for reducing production and yield losses and containing CBW infestation, which represents a major problem for the cotton farmers in the Americas. Herein, in the present work, four Bt proteins known to be toxic to Coleopterans have been tested for the first time on *A. grandis* larvae. Two of the proteins tested belong to the 3-domain toxin class (Cry1Ba and Cry7Ab), and the other two mixed proteins (Mpp23Aa and Xpp37Aa, previously known as Cry23Aa and Cry37Aa, respectively) to the Mpp/Xpp class. Specifically, these last two proteins are expressed in a single operon and are mentioned as Mpp23Aa/Xpp37Aa. Therefore, results from this study could assess the potential of these proteins to effectively control CBW in the field.

2. Results

The survival of newly hatched larvae of *A. grandis* was affected by the ingestion of an artificial diet containing the different Bt proteins (Figure 1). The decreased survival of Bt-treated individuals increased with the increase in Bt protein concentration. For example, the analysis of the number of dead larvae after treatment at the highest dose (100 $\mu g \cdot g^{-1}$) was significantly different from that in the untreated group (one-way ANOVA, n = 269, F = 78, $p < 0.0001$). At this dose, the Cry1Ac, Cry1Ba, and Cry7Ab proteins had similar effects (Tuckey's post-test, $p > 0.99$ for any combination), but the results were statistically different from those regarding the effect of Mpp23Aa/Xpp37Aa (Tuckey's post-test, $p > 0.0001$). Cry proteins showed corrected mortality below 50%, which shows the low susceptibility of these insects to these proteins. In contrast, when the toxicity of the Mpp23Aa/Xpp37Aa proteins was tested at the same concentrations, 100% mortality was observed.

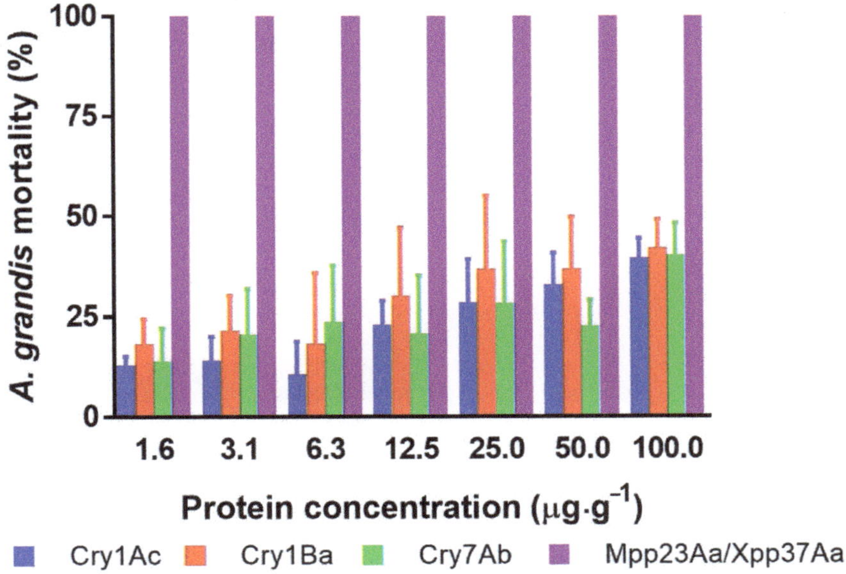

Figure 1. Corrected mortality [15] of *A. grandis* larvae after seven days of protein intake in an artificial diet at seven concentrations of different *B. thuringiensis* Cry proteins. Error bars represent the mean standard error. The error bars for Mpp23Aa/Xpp37Aa are not shown because of the death of all treated larvae in all replicates.

Moreover, the death of all larvae tested was found in all assayed concentrations, even at the lower level (1.56 $\mu g \cdot g^{-1}$). Based on these results, Mpp23Aa/Xpp37Aa proteins were selected for additional bioassay testing with lower toxin concentrations (ranging from 0.13 to 4 $\mu g \cdot g^{-1}$) to obtain dose–response data suitable for analysis. An example of the effect of these proteins can be observed in Figure 2.

Figure 2. Effect of Mpp23Aa/Xpp37Aa on *A. grandis* larvae with 0.5 $\mu g \cdot g^{-1}$ treatment. (**A**) Neonate larva at the start of the treatment (1 mm size), (**B**) larva on control untreated media after 7 days of treatment (5 mm size), (**C**) larva on Mpp/Xpp treated media after 7 days of treatment (1 mm size).

The probit analysis was applied to the data obtained, estimating the bioassay parameters (Table 1). In addition, the analysis provided some extrapolated data of lethal concentration (LC) parameters useful for toxicity comparisons. As expected, Cry1Ac protein (described as a Coleopteran non-toxic) was marginally toxic (indicated by a high LC_{50} value). Based on the LC_{50} values obtained, Cry1Ba and Cry7Aa were slightly more toxic than Cry1Ac; however, the test of the hypothesis that slopes and intercepts are the same, discarded significant differences (χ^2 = 7.04, degrees of freedom, d.f. = 4).

Table 1. Parameters estimated by probit analysis from the bioassays with *A. grandis* larvae tested with several Bt proteins. Toxin concentrations are given in μg of protein per·g^{-1} of artificial diet.

Bt Protein	n	Slope (±SE)	LC_{10} (FL_{95})	LC_{50} (FL_{95})	χ^2 (d.f.)
Cry1Ac	841	0.58 (0.16)	2.01 (0.04–7.70)	317 (188–3378) *	2.08 (5)
Cry1Ba	840	0.43 (0.11)	0.33 (0.01–1.84)	293 (92–4486) *	2.05 (5)
Cry7Aa	841	0.35 (0.12)	0.38 (0.01–3.05)	1598 (233–ND) *	4.60 (5)
Mpp23Aa/Xpp37Aa	735	3.80 (0.47)	0.08 (0.04–0.12) *	0.18 (0.13–0.22)	4.43 (4)

n = number of evaluated larvae; slope = dose–response quickness; SE = standard error of the mean; LC_{10} and LC_{50} = concentration that killed 10% and 50%, respectively; FL_{95} = 95% fiducial limits; d.f. = degrees of freedom; ND = not determined; * value interpolated because they were out of the range of assayed concentrations.

On the other hand, the Mpp23Aa/Xpp37Aa proteins showed significantly different toxicity dynamics (slope parameter) to *A. grandis* than to the Cry proteins tested, according to the previous test applied (χ^2 = 532, d.f. = 2). The Mpp23Aa/Xpp37Aa toxicity was higher than that of the Cry proteins tested, since they provided a higher response to the concentration (a slope value higher), and the LC_{10} and LC_{50} values were lower. Only the LC parameters ratio (LC Cry1Ac protein/LC Mpp/Xpp proteins) could be provided, since the slopes were not parallel between Mpp/Xpp and the Cry proteins. The ratio of LC_{10} and LC_{50} values were 25 and 1761, respectively.

The concentration of Mpp/Xpp proteins required to kill 90% of the treated larvae (LC_{90}), which is significant for pest control, was estimated as 0.4 μg·g^{-1} (fiducial limits, FL_{95} = 0.32–0.56), increasing the evidence of its high toxicity.

3. Discussion

Although *A. grandis* is currently one of the most devastating cotton insect pests in Brazil and other cotton crop-producing countries, there is little information about which Bt insecticidal proteins are effective against this insect pest [12–14]. The toxins usually described as active against insects of the order Coleoptera are those of the Cry3, Cry7, Cry8, Cry10, and Gpp34/Tpp35 classes [9,14], although other classes can be toxic as well, including the Cry1 class. Particularly, the Cry1Ba6 [16], Cry8Ka5 [17], Cry1Ia [18,19], Cry1Ia12 [20,21], and Cry10Aa [22] proteins have already been proven as toxic to CBW larvae in bioassays using purified recombinant protein or virus-infected insect extracts, while the Cry1Ia12 [20], Cry1Ia [19], and Cry10Aa [11,13] proteins were also confirmed to show some toxic activity when used in transgenic cotton lines.

It is important to highlight that studies evaluating the toxicity of single insecticidal proteins from Bt are required to show which proteins are effective against *A. grandis*. Furthermore, additional studies indicating whether it is suitable to combine the selected proteins in the same transgenic crop will be desirable to avoid the occurrence of cross-resistance among these proteins [23].

Results from the selective bioassay showed that although the survival of the larvae was affected after the ingestion of an artificial diet containing Cry1Ac, Cry1Ba, and Cry7Ab proteins, the highest dose used (100 μg·g^{-1}) caused less than 50% mortality. The results obtained indicate that these proteins have low toxicity towards *A. grandis* larvae. The Cry1Ac is a protein known to be toxic to lepidopteran agricultural pest species such as *Helicoverpa zea*, *Heliothis virescens* [24], *Helicoverpa armigera* [25], *Anticarsia gemmatalis*, and *Chrysodeixis includens* [26], among others. For Coleopterans, although toxicity has been reported for some species [27], studies demonstrating high susceptibility to Cry1Ac are rare. Therefore, given the expression levels of Cry1Ac in commercial varieties, with a maximum of 7 and 20 μg·g^{-1} of fresh and dry leaf tissue, respectively [28,29], our results indicate that these current commercial varieties would not be able to control *A. grandis*.

Cry1Ba was included in this study because it has been described as a protein that has dual activity (toxicity to both Lepidopteran and Coleopteran insects) [14,30]. A previous study that tested the toxicity of Cry1Ba on *A. grandis* established an LC_{50} of

380.8 μg·g^{-1} [16], which is in line with our results. Interestingly, the authors reported that the toxicity of the original Bt strains that contained the *cry1B* gene was higher, suggesting that other virulence factors were causing the toxicity of these strains on *A. grandis*.

The Cry7 group is known to be poisonous to different Coleopteran insect species [10,14]. Here, we showed that the mortality of *A. grandis* caused by Cry7Ab was not high (40% as a mean in response to 100 μg·g^{-1}), suggesting that this protein is not active against this insect pest. Similarly, *Anomala corpulenta* and *Pyrrhalta aenescens* larvae were not affected after ingestion of protein Cry7Ab3, and low susceptibility to Cry7Ab was observed in *Acanthoscelides obtectus* [30,31]. However, Cry7Ab was highly toxic to *Xylotrechus arvicola* and *Henosepilachna vigintioctomaculata* larvae, respectively [31,32].

On the other hand, Mpp23Aa/Xpp37Aa were extremely active against *A. grandis* larvae, with an estimated LC_{50} = 0.18 μg·g^{-1} (Table 1). These two proteins were described as toxic to *A. grandis* in a previous work [33], although this is the first study that establishes LC_{50} values. The toxic effect of Mpp23Aa/Xpp37Aa was also observed on larvae of *Tribolium castaneum* [34], *Cylas* spp. [35], and *Xylotrechus arvicola* [32], suggesting that these proteins are able to control different Coleopteran pests.

The Mpp23Aa/Xpp37Aa proteins were first described as a binary toxin based on the fact that both genes were located in the same operon. However, recently, it was shown that these proteins were toxic for *C. puncticollis* larvae when tested individually [36]. In our study, we have tested the toxicity against *A. grandis* larvae by using a *B. thuringiensis* strain that expresses both proteins; thus, we cannot determine whether both proteins are required to exert their toxic effect. Further research will be required to test whether both proteins are required for the toxic effect against *A. grandis* larvae.

Concentrations of some Cry proteins considered viable for introgression in cotton range up to about 20 μg·g^{-1}, since the levels of protein expression in plants differ between flower buds and leaf tissues, ranging from approximately 3.0 to 19.0 μg·g^{-1} of fresh tissue [10,13]. Thus, according to our results, the potentially possible control of *A. grandis* by Mpp23Aa/Xpp37Aa toxins can be satisfactory. More experiments will be conducted to deepen the understanding of the toxicity and mode of action of this protein complex on *A. grandis*; however, the results presented here may be useful to outline several lines of research.

4. Materials and Methods

4.1. Bt Proteins Preparation

The source and preparation of the Bt proteins (Cry1Ba, Cry7Ab, and Mpp23Aa/Xpp37Aa) tested in this study were the *B. thuringiensis* strains described by Rodríguez-González et al. [30]. An exception was Cry1Ac, but it was prepared in the same way. In detail, Cry1Ac and Cry7Ab proteins were obtained from the *B. thuringiensis* HD-73 and HD867 strains, respectively, provided by the Bacillus Genetic Stock Center, USA. Cry1Ba was obtained from the recombinant *B. thuringiensis* strain EG11916 (Ecogen, Inc., Langhorne, PA, USA) and Mpp23Aa/Xpp37Aa were obtained from the EG10327 strain (Ref. No. NRRL B-21365) obtained from the Agricultural Research Culture Collection, Northern Regional Research Laboratory (NRRL), USA, respectively.

The *Bacillus thuringiensis* strains were grown in CCY medium supplemented with the appropriate antibiotic for 48 h at 29 °C, under constant agitation. Spores and crystals were separated by centrifugation at 16,000× *g* for 10 min at 4 °C and then washed three times with 1 M NaCl, 10 mM ethylenediaminetetraacetic acid (EDTA), and twice with 10 mM KCl. The final pellets were frozen using liquid nitrogen and lyophilized. The presence of the proteins in the lyophilized powder was determined using sodium dodecyl sulfate 12% polyacrylamide gel electrophoresis (SDS-PAGE) (Figure S1).

4.2. Rearing of Anthonomus grandis

Approximately 500 adult boll weevils collected from cotton fields in the region of Rondonópolis City, Mato Grosso, Brazil, were placed in cages at a 1:1 male–female ratio and maintained at 27 °C, under 40% relative humidity, and with a photoperiod of 12:12 h light:dark. In small, cube-shaped cages, the insects were fed an artificial diet [37].

The insect cages were serviced every two days, when the artificial diet was replaced by a fresh one, and insect eggs were collected. The eggs were superficially disinfected with a 20% copper sulfate and 0.2% benzalkonium chloride solution and separated from impurities such as diet leftovers and feces. Then, they were incubated in Petri dishes lined with filter paper and moistened with sterile water until hatching, and the newly hatched larvae (<48 h) were used for the bioassays.

4.3. Bioassays of *A. grandis* with Bt Proteins

The bioassays were based on the method of incorporating the test substances into an artificial diet [16]. The proteins, supplied as a lyophilized powder, were resuspended in sterile deionized water, and the concentration in this solution was determined by the Bradford method, using bovine serum albumin (BSA) as a standard. The aliquots required to obtain the doses corresponding to concentrations of 100; 50; 25; 12.5; 6.25, 3.13, and 1.56 $\mu g \cdot g^{-1}$ were incorporated into the artificial CBW diet in a total volume of 30 mL. After blending, the mixture was poured into Petri dishes (Ø 90 × 15 mm) for solidification, and then 35 newly hatched *A. grandis* larvae were placed at equal distances on the diet surface. The control for natural mortality consisted of an artificial diet, without the protein addition. The plates were incubated at 27 °C, with 40% relative humidity, and a photoperiod of 12:12 h light:dark for seven days, at which time the number of live and dead larvae on each plate was counted. All assays were performed in triplicate.

Mpp23Aa/Xpp37Aa proteins that induced very high mortality (100%) at the lowest dose (1.56 $\mu g \cdot g^{-1}$) were selected for an additional bioassay at concentrations of 4, 2, 1, 0.5, 0.25, and 0.13 $\mu g \cdot g^{-1}$, according to the methodology described above.

4.4. Data Analysis

Mortality results for the figure were corrected based on mortality observed in the control using Abbott's formula [15], obtaining the mean value and the standard error of the mean. All data analyses were performed with uncorrected mortality data. Differences in dead larvae among different protein treatments at 100 $\mu g \cdot g^{-1}$ were tested by one-way ANOVA, followed by a Tukey's post-test, considering $p < 0.05$ as a significant difference. GraphPad Prism 7.0 (GraphPad Software, La Jolla, CA, USA) was used.

Bioassay data were analyzed through probit analysis [38] using the POLO-PC software program (LeOra Software, Berkeley, CA, USA, 1987) in order to obtain the toxicological parameters and to compare the obtained results.

Supplementary Materials: The following supporting information can be downloaded at: https://www.mdpi.com/article/10.3390/toxins15010055/s1, Figure S1: SDS–polyacrylamide gel electrophoresis of insecticidal proteins used in the toxicity tests: Cry1Ac, Cry1Ba, Cry7Ab, and Mpp23Aa/Xpp37Aa. BlueStar pre-stained protein marker (bands in kDa: 180, 130, 100, 75, 63, 48, 35, 28, 17, 10). Bands were stained with Coomassie blue. The Bt insecticidal protein bands are indicated by an arrow.

Author Contributions: Conceptualization, B.E. and M.F.B.; statistical software, B.E. and J.A.d.O.; formal analysis, B.E., B.F.N., M.F.B. and J.A.d.O.; investigation, M.F.B., B.F.N., P.H.-M. and J.A.d.O.; writing—original draft preparation, B.E. and J.A.d.O.; writing—review and editing, B.E., P.H.-M., B.F.N. and J.A.d.O.; funding acquisition, B.E. All authors have read and agreed to the published version of the manuscript.

Funding: This research was funded in part by grants from Generalitat Valenciana (grant number PROMETEO/2020/010), grant PID2021-122914OB-100, funded by MCIN/AEI/10.13039/501100011033, and by "ERDF: A way of making Europe," funded by the European Union.

Institutional Review Board Statement: Not applicable.

Informed Consent Statement: Not applicable.

Data Availability Statement: The datasets generated and/or analyzed during the current study are available from the corresponding author upon reasonable request.

Acknowledgments: We would like to thank Guilherme Debiazi Beloni (IMAmt) and Guilherme Gomes Rolim (IMAmt) for the pictures of the adult insect and plant damage in the graphical abstract.

Conflicts of Interest: The authors declare no conflict of interest.

References

1. CONAB. Acompanhamento da Safra Brasileira de Grãos, Safra 2021/22. 2022. Available online: https://www.conab.gov.br/info-agro/safras/graos/boletim-da-safra-de-graos (accessed on 12 December 2022).
2. USDA. Cotton: World Markets and Trade. 2022. Available online: https://www.fas.usda.gov/data/cotton-world-markets-and-trade (accessed on 12 December 2022).
3. IMEA. Custo de Produção do Algodão, Safra 2020/2021. 2022. Available online: https://www.imea.com.br/imea-site/relatorios-mercado (accessed on 12 December 2022).
4. Arruda, L.S.; Torres, J.B.; Rolim, G.G.; Silva-Torres, C.S.A. Dispersal of boll weevil toward and within the cotton plant and implications for insecticide exposure. *Pest Manag. Sci.* **2020**, *77*, 1339–1347. [CrossRef] [PubMed]
5. Grigolli, J.F.J.; Souza, L.A.; Fernandes, M.G.; Busoli, A.C. Spatial distribution of adult *Anthonomus grandis* Boheman (Coleoptera: Curculionidae) and damage to cotton flower buds due to feeding and oviposition. *Neotrop. Entomol.* **2017**, *46*, 442–451. [CrossRef] [PubMed]
6. Neves, R.C.S.; Showler, A.T.; Pinto, E.S.; Bastos, C.S.; Torres, J.B. Reducing boll weevil populations by clipping terminal buds and removing abscised fruiting bodies. *Entomol. Exp. Appl.* **2012**, *146*, 276–285. [CrossRef]
7. Perkin, L.C.; Perez, J.L.; Suh, C.P.-C. The identification of boll weevil, *Anthonomus grandis* grandis (Coleoptera: Curculionidae), genes involved in pheromone production and pheromone biosynthesis. *Insects* **2021**, *12*, 893. [CrossRef] [PubMed]
8. Oliveira, A.A.S.; Araújo, T.A.; Showler, A.T.; Araújo, A.C.A.; Almeida, I.S.; Aguiar, R.S.A.; Miranda, J.E.; Fernandes, F.L.; Bastos, C.S. Spatio-temporal distribution of *Anthonomus grandis* Boh. in tropical cotton fields. *Pest Manag. Sci.* **2022**, *78*, 2492–2501. [CrossRef] [PubMed]
9. Torres, J.B.; Rolim, G.G.; Arruda, L.S.; Dos Santos, M.P.; Leite, S.A.; Neves, R.C.D.S. Insecticides in use and risk of control failure of boll weevil (Coleoptera: Curculionidae) in the Brazilian Cerrado. *Neotrop. Entomol.* **2022**, *51*, 613–627. [CrossRef]
10. Bravo, A.; Gómez, I.; Porta, H.; García-Gómez, B.I.; Rodriguez-Almazan, C.; Pardo, L.; Soberón, M. Evolution of *Bacillus thuringiensis* Cry toxins insecticidal activity. *Microb. Biotechnol.* **2012**, *6*, 17–26. [CrossRef]
11. Ribeiro, T.P.; Arraes, F.B.M.; Lourenço-Tessutti, I.T.; Silva, M.S.; Lisei-de-Sá, M.E.; Lucena, W.A.; Macedo, L.L.P.; Lima, J.N.; Santos Amorim, R.M.; Artico, S.; et al. Transgenic cotton expressing Cry10Aa toxin confers high resistance to the cotton boll weevil. *Plant Biotechnol. J.* **2017**, *15*, 997–1009. [CrossRef]
12. ISAAA. GM Approval Database. 2022. Available online: https://www.isaaa.org/gmapprovaldatabase (accessed on 12 December 2022).
13. Ribeiro, T.P.; Basso, M.F.; Carvalho, M.H.d.; Macedo, L.L.P.d.; Silva, D.M.L.d.; Lourenço-Tessutti, I.T.; Oliveira-Neto, O.B.d.; Campos-Pinto, E.R.d.; Lucena, W.A.; Silva, M.C.M.d.; et al. Stability and tissue-specific Cry10Aa overexpression improves cotton resistance to the cotton boll weevil. *Biotechnol. Res. Innov.* **2019**, *3*, 27–41. [CrossRef]
14. Domínguez-Arrizabalaga, M.; Villanueva, M.; Escriche, B.; Ancín-Azpilicueta, C.; Caballero, P. Insecticidal activity of *Bacillus thuringiensis* proteins against Coleopteran Pests. *Toxins* **2020**, *12*, 430. [CrossRef]
15. Abbott, W.S. A method of computing the effectiveness of on insecticide. *J. Econ. Entomol.* **1925**, *18*, 265–267. [CrossRef]
16. Martins, E.S.; Monnerat, R.G.; Queiroz, P.R.; Dumas, V.F.; Braz, S.V.; Aguiar, R.W.S.; Gomes, A.C.M.M.; Sánchez, J.; Bravo, A.; Ribeiro, B.M. Midgut GPI-anchored proteins with alkaline phosphatase activity from the cotton boll weevil (*A. grandis*) are putative receptors for the Cry1B protein of *Bacillus thuringiensis*. *Insect Biochem. Mol. Biol.* **2010**, *40*, 138–145. [CrossRef] [PubMed]
17. Farias, D.F.; Peijnenburg, A.A.; Grossi-de-Sa, M.F.; Carvalho, A.F. Food safety knowledge on the Bt mutant protein Cry8Ka5 employed in the development of coleopteran-resistant transgenic cotton plants. *Bioengineered* **2015**, *6*, 323–327. [CrossRef] [PubMed]
18. Martins, E.S.; Aguiar, R.W.; Martins, N.F.; Melatti, V.M.; Falcao, R.; Gomes, A.C.; Ribeiro, B.M.; Monnerat, R.G. Recombinant Cry1Ia protein is highly toxic to cotton boll weevil (*Anthonomus grandis* Boheman) and fall armyworm (*Spodoptera frugiperda*). *J. App. Microbiol.* **2008**, *104*, 1363–1371. [CrossRef]
19. Silva, C.R.; Monnerat, R.; Lima, L.M.; Martins, É.S.; Melo Filho, P.A.; Pinheiro, M.P.; Santos, R.C. Stable integration and expression of a *cry1Ia* gene conferring resistance to fall armyworm and boll weevil in cotton plants. *Pest Manag. Sci.* **2016**, *72*, 1549–1557. [CrossRef]
20. de Oliveira, R.S.; Oliveira-Neto, O.B.; Moura, H.F.; de Macedo, L.L.; Arraes, F.B.; Lucena, W.A.; Lourenco-Tessutti, I.T.; de Deus Barbosa, A.A.; da Silva, M.C.; Grossi-de-Sa, M.F. Transgenic cotton plants expressing Cry1Ia12 toxin confer resistance to fall armyworm (*Spodoptera frugiperda*) and cotton boll weevil (*Anthonomus grandis*). *Front. Plant Sci.* **2016**, *7*, 165. [CrossRef]
21. Grossi-de-Sa, M.F.; Quezado de Magalhaes, M.; Silva, M.S.; Silva, S.M.; Dias, S.C.; Nakasu, E.Y.; Brunetta, P.S.; Oliveira, G.R.; Neto, O.B.; Sampaio de Oliveira, R.; et al. Susceptibility of *Anthonomus grandis* (cotton boll weevil) and *Spodoptera frugiperda* (fall armyworm) to a cry1ia-type toxin from a Brazilian *Bacillus thuringiensis* strain. *J. Biochem. Mol. Biol.* **2007**, *40*, 773–782. [CrossRef]

22. Aguiar, R.W.d.S.; Martins, E.S.; Ribeiro, B.M.; Monnerat, R.G. Cry10Aa protein is highly toxic to *Anthonomus grandis* Boheman (Coleoptera: Curculionidae), an important insect pest in Brazilian cotton crop fields. *Bt. Res.* **2012**, *3*, 20–28. [CrossRef]
23. Jakka, S.; Ferré, J.; Jurat-Fuentes, J.L. Cry toxin binding site models and their use in strategies to delay resistance evolution. In *Bt Resistance: Characterization and Strategies for GM Crops Producing Bacillus Thuringiensis Toxins*; Soberón, M., Gao, Y., Bravo, A., Eds.; Oxfordshire, UK, 2015; pp. 138–149.
24. Abdelgaffar, H.M.; Oppert, C.; Sun, X.; Monserrate, J.; Jurat-Fuentes, J.L. Differential heliothine susceptibility to Cry1Ac associated with gut proteolytic activity. *Pestic. Biochem. Physiol.* **2019**, *153*, 1–8. [CrossRef]
25. Wang, P.; Ma, J.; Head, G.P.; Xia, D.; Li, J.; Wang, H.; Yang, M.; Xie, Z.; Zalucki, M.P.; Lu, Z. Susceptibility of *Helicoverpa armigera* to two Bt toxins, Cry1Ac and Cry2Ab, in northwestern China: Toward developing an IRM strategy. *J. Pest. Sci.* **2018**, *92*, 923–931. [CrossRef]
26. Bel, Y.; Sheets, J.J.; Tan, S.Y.; Narva, K.E.; Escriche, B. Toxicity and binding studies of Bacillus thuringiensis Cry1Ac, Cry1F, Cry1C, and Cry2A proteins in the soybean pests *Anticarsia gemmatalis* and *Chrysodeixis (Pseudoplusia) includens*. *Appl. Environ. Microbiol.* **2017**, *83*, e00326-17. [CrossRef] [PubMed]
27. Qi, G.; Lu, J.; Zhang, P.; Li, J.; Zhu, F.; Chen, J.; Liu, Y.; Yu, Z.; Zhao, X. The *cry1Ac gene of Bacillus thuringiensis* ZQ-89 encodes a toxin against long-horned beetle adult. *J. Appl. Microbiol.* **2011**, *110*, 1224–1234. [CrossRef] [PubMed]
28. Singh, A.K.; Paritosh, K.; Kant, U.P.; Burma, K.; Pental, D. High expression of Cry1Ac protein in cotton (*Gossypium hirsutum*) by combining independent transgenic events that target the protein to cytoplasm and plastids. *PLoS ONE* **2016**, *11*, e0158603. [CrossRef]
29. Dhanaraj, A.L.; Willse, A.R.; Kamath, S.P. Stability of expression of Cry1Ac and Cry2Ab2 proteins in Bollgard-II hybrids at different stages of crop growth in different genotypes across cropping seasons and multiple geographies. *Transgenic Res.* **2018**, *28*, 33–50. [CrossRef] [PubMed]
30. Rodríguez-González, A.; Porteous-Álvarez, A.J.; Val, M.; Casquero, P.A.; Escriche, B. Toxicity of five Cry proteins against the insect pest *Acanthoscelides obtectus* (Coleoptera: Chrisomelidae: Bruchinae). *J. Invertebr. Pathol.* **2020**, *169*, 107295. [CrossRef] [PubMed]
31. Song, P.; Wang, Q.; Nangong, Z.; Su, J.; Ge, D. Identification of *Henosepilachna vigintioctomaculata* (Coleoptera: Coccinellidae) midgut putative receptor for *Bacillus thuringiensis* insecticidal Cry7ab3 toxin. *J. Invertebr. Pathol.* **2012**, *109*, 318–322. [CrossRef]
32. Rodríguez-González, A.; Porteous-Álvarez, A.J.; Guerra, M.; González-López, O.; Casquero, P.A.; Escriche, B. Effect of Cry toxins on *Xylotrechus arvicola* (Coleoptera: Cerambycidae) larvae. *Insects* **2021**, *13*, 27. [CrossRef]
33. Guzov, V.M.; Malvar, T.M.; Roberts, J.K.; Sivasupramaniam, S. Insect inhibitory *Bacillus thuringiensis* proteins, fusions, and methods of use therefor. U.S. Patent 7,655,838, 2 February 2010.
34. Contreras, E.; Rausell, C.; Real, M.D. Proteome response of *Tribolium castaneum* larvae to Bacillus thuringiensis toxin producing strains. *PLoS ONE* **2013**, *8*, e0055330. [CrossRef]
35. Ekobu, M.; Solera, M.; Kyamanywa, S.; Mwanga, R.O.; Odongo, B.; Ghislain, M.; Moar, W.J. Toxicity of seven *Bacillus thuringiensis* Cry proteins against *Cylas puncticollis* and *Cylas brunneus* (Coleoptera: Brentidae) using a novel artificial diet. *J. Econ. Entomol.* **2010**, *103*, 1493–1502. [CrossRef]
36. Hernández-Martínez, P.; Khorramnejad, A.; Prentice, K.; Andrés-Garrido, A.; Vera-Velasco, N.M.; Smagghe, G.; Escriche, B. The independent biological activity of *Bacillus thuringiensis* Cry23Aa protein against *Cylas puncticollis*. *Front. Microbiol.* **2020**, *11*, 1734. [CrossRef]
37. Monnerat, R.G.; Dias, S.C.; Oliveira Neto, O.B.; Nobre, S.D.; Silva-Werneck, J.O.; Sá, M.F.G. *Criação Massal do Bicudo do Algodoeiro Anthonomus grandis em Laboratório*; Embrapa Recursos Genéticos e Biotecnologia: Brasília, Brazil, 2000; p. 4.
38. Finney, D.J. *Probit Analysis*, 3rd ed.; Cambridge University Press: Cambridge, UK, 1971.

toxins MDPI

Article

The Crystal Structure of *Bacillus thuringiensis* Tpp80Aa1 and Its Interaction with Galactose-Containing Glycolipids

Hannah L. Best [1], Lainey J. Williamson [1], Magdalena Lipka-Lloyd [1], Helen Waller-Evans [2], Emyr Lloyd-Evans [1], Pierre J. Rizkallah [3] and Colin Berry [1,*]

1 School of Biosciences, Cardiff University, Park Place, Cardiff CF10 3AX, UK
2 School of Pharmacy, Cardiff University, Park Place, Cardiff CF10 3AX, UK
3 School of Medicine, Cardiff University, Heath Campus, Cardiff CF14 3XN, UK
* Correspondence: berry@cardiff.ac.uk

Abstract: AbstractTpp80Aa1 from *Bacillus thuringiensis* is a Toxin_10 family protein (Tpp) with reported action against *Culex* mosquitoes. Here, we demonstrate an expanded target range, showing Tpp80Aa1 is also active against the larvae of *Anopheles gambiae* and *Aedes aegypti* mosquitoes. We report the first crystal structure of Tpp80Aa1 at a resolution of 1.8 Å, which shows Tpp80Aa1 consists of two domains: an N-terminal β-trefoil domain resembling a ricin B lectin and a C-terminal putative pore-forming domain sharing structural similarity with the aerolysin family. Similar to other Tpp family members, we observe Tpp80Aa1 binds to the mosquito midgut, specifically the posterior midgut and the gastric caecum. We also identify that Tpp80Aa1 can interact with galactose-containing glycolipids and galactose, and this interaction is critical for exerting full insecticidal action against mosquito target cell lines.

Keywords: *Bacillus thuringiensis*; Tpp80Aa1 toxin; crystal structure; *Culex*; *Anopheles*; *Aedes*; biocontrol; pesticidal protein

Key Contribution: We present the crystal structure of Tpp80Aa1– to 1.8 Å resolution. We expand the known target range of Tpp80Aa1 to include two more mosquito species: *Anopheles gambiae* and *Aedes aegypti*, increasing the potential utility in the field. We also demonstrate Tpp80Aa1 can bind galactose-containing lipids, and this binding can affect toxicity in cell-based models. This study will underpin future Tpp80Aa1 mode of action investigations and aid in insecticide optimization against mosquito vectors of disease.

Citation: Best, H.L.; Williamson, L.J.; Lipka-Lloyd, M.; Waller-Evans, H.; Lloyd-Evans, E.; Rizkallah, P.J.; Berry, C. The Crystal Structure of *Bacillus thuringiensis* Tpp80Aa1 and Its Interaction with Galactose-Containing Glycolipids. *Toxins* **2022**, *14*, 863. https://doi.org/10.3390/toxins14120863

Received: 2 November 2022
Accepted: 24 November 2022
Published: 8 December 2022

1. Introduction

The *Bacillus thuringiensis* (Bt) is a Gram-positive, sporulating bacterium that can produce a range of insecticidal toxins that are active across an array of invertebrate targets [1]. Bt-produced proteins show highly selective and potent activity and, as such, have revolutionised agriculture by negating the use of hazardous and non-specific chemical pesticides. Additionally, Bt proteins are used for the control of insect vectors of human disease-principally mosquitoes [2]. The importance and commercial success of Bt proteins motivates ongoing searches for novel proteins with new spectra of insecticidal activity.

Tpp80Aa1 (formerly Cry80Aa1) is a recently identified mosquitocidal protein, discovered by whole genome sequencing of Bt strain S3589-1 isolated from a soil sample [3], and it is reported to be active against third instar *Culex pipiens pallens* larvae. Control of disease-spreading Diptera—such as mosquitoes and black-fly—is critical for the control of diseases such as Zika virus, malaria, dengue, yellow fever, and African river blindness [4,5]. To date, the most successful entomopathogenic bacteria for controlling these populations in the field have been Bt serovar. *israelensis* (Bti) and *Lysinibacillus sphaericus*. Bti produces several Cry and Cyt toxins in the form of crystalline inclusions, which allow the bacterium to demonstrate significantly stronger toxicity due to the synergism between individual proteins within the whole Bti crystal [6]. The most utilised *L. sphaericus* toxins are a heterodimer of Tpp1Aa1 and Tpp2Aa1 (formerly BinA and BinB), which are highly potent against *Culex* and *Anopheles* larvae [7], and can also act synergistically with Cyt1Aa to become active against *Ae. aegypti* and Tpp1/2-resistant larvae [8–10]. Tpp1Aa1/Tpp2Aa1-producing *L. sphaericus* strains are heavily utilised in mosquito control in the field, and the emergence of resistance incentivises the discovery of new mosquitocidal proteins—such as Tpp80Aa1—that can be used as stand-alone agents, or in a synergistic approach [11,12].

Bacterial pesticidal proteins belong to one of several distinct structural classes [13]. Tpp proteins have two domains, an N-terminal trefoil domain and a C-terminal pore-formation domain [14]. Conserved domain analysis of Tpp80 indicated the greatest identity to Tpp78Aa1 (38.6%)—a protein with activity against the Hemiptera *Nilaparvata lugens* and *Laodelphax striatellus* and the Tpp78Aa1 structure has recently been published [15]. In this study we report the structure of Tpp80Aa1 (1.8 Å), which consists of an N-terminal ricin B lectin domain and a C-terminal toxin_10 putative pore-forming domain (PFD). We demonstrate that the protein has an affinity for galactose-containing glycans, which may mediate its activity. We also reveal an expanded range of activity against new target mosquito species, *An. gambiae* and *Ae. aegypti*, increasing the protein's value as a novel toxin for mosquito control.

2. Results

2.1. *Tpp80Aa1 Has Activity against C. quinquefasciatus, Ae. aegypti, and An. gambiae Larvae and Mosquito Cell Lines*

In mosquitocidal bioassays, trypsin-activated Tpp80Aa1 (Figure S1A,B) demonstrated activity against *An. gambiae*, *C. quinquefasciatus*, and *Ae. aegypti* larvae, with LC_{50} values of 8.6, 24.6 and 29.6 μg/mL, respectively (Figure 1A). Only *Culex pipiens pallens* (LC_{50} 72 μg/mL) susceptibility has been reported previously [3]. Compared to other mosquitocidal Tpp proteins, Tpp80Aa1 has a lower potency but a broader target range of activity (Table 1). Cellular models were used to investigate Tpp80Aa1 toxicity in vitro. In line with the larval bioassays, 48 h after the addition of Tpp80Aa1, cell viability was substantially reduced in MRA-918 (*C. quinquefasciatus*), C6/36 (*Ae. aegypti*), and Ag55 (*An. gambiae*) cell lines (Figure 1B). Inspection of cellular morphology, via light microscopy, at the 48 h time point shows cells exposed to Tpp80Aa1 have a rounded morphology and are becoming detached from the plate—indicative of cell death. Vacuolisation is also present in the Tpp80Aa1-treated cells (Figure 1C), a cellular phenotype that has been described previously with other Tpp family member two-component toxins, Cry48/Tpp49 [16] and Tpp1/Tpp2 [17–19]. This establishes these cell lines as models for further investigations into the Tpp80Aa1 mechanism of action.

2.2. *Tpp80Aa1 Binding Occurs Predominantly in the Posterior Midgut*

To investigate the gut binding profile of Tpp80Aa1, we fed fluorescently labelled Tpp80Aa1 to *Ae. aegypti* larvae (Figure 1D). Tpp80Aa1 binding is predominantly present in the posterior midgut (PM) and gastric caecum (GC), with substantially weaker binding observable in the anterior midgut (AM). A clear punctate staining pattern is present throughout the midgut—particularly in the PM—suggesting internalisation of Tpp80Aa1 in cytoplasmic vesicles. This binding pattern is very similar to that previously reported with radio-labelled or fluorescently labelled Tpp1/Tpp2 in the *Culex* midgut [7,20,21] and may suggest that the elusive Tpp80Aa1 receptor(s) is specifically localised to brush border membranes of the PM and GC. The localisation pattern of the binding may also be due to the known pH gradient throughout the mosquito midgut (approx. pH 8 in the gastric caecum, >pH 10 in the anterior midgut, falling to pH 7.5 in posterior midgut) [22,23] affecting the processing and binding of Tpp80Aa1.

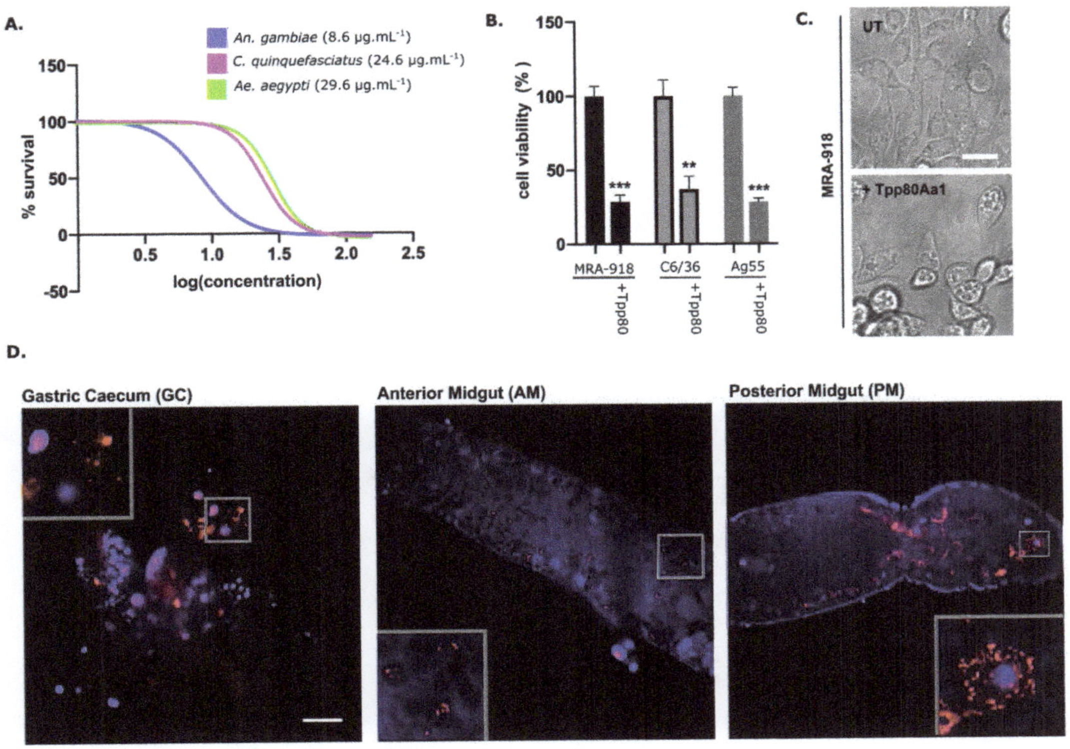

Figure 1. Tpp80Aa1 is active against *Culex quinquefasciatus*, *Aedes aegypti*, and *Anopheles gambiae*. (**A**) Dose response curve of Tpp80Aa1 added to larvae in water plotting log concentration (μg/ mL) against % of surviving larvae. The 50% lethal concentrations (LC_{50}) were determined as 8.6, 24.6, and 29.6 μg/mL for *An. gambiae*, *C. quinquefasciatus*, and *Ae. aegypti*, respectively. (**B**) Cell lines isolated from *C. quinquefasciatus* (MRA-918), *Ae. aegypti* (C6/36), and *An. gambiae* (Ag55) show significantly reduced viability (as determined via resazurin assay) 48 h post addition of Tpp80Aa1 at 50 μg/mL. Data are presented as % of control ± SD and were analysed using unpaired *t*-tests (*** $p \leq 0.0001$, ** $p = 0.0014$) (**C**) Light microscopy images of untreated (UT) or Tpp80Aa1 treated *C. quinquefasciatus* cells (MRA-918), 48 h post addition. Representative scale bar in the UT image = 10 μm. (**D**) Tpp80Aa1 fluorescently labelled with Alexa Fluor® 555 (red) was fed to *Ae. aegypti* larvae via addition to water. Post-ingestion, larvae were transferred to fresh water for 30 min before guts were dissected, labelled with Hoecsht 33342 (blue), and imaged with a single plan illumination microscope. Grey insets showing punctate binding pattern of Tpp80Aa1. Representative scale bar in GC image = 100 μm.

Table 1. Summary of LC_{50} values (μg/mL) of mosquitocidal Tpp proteins.

	Anopheles gambiae	*Culex quinquefasciatus*	*Aedes aegypti*
Tpp80Aa1	8.6	24.6	29.6
Cry48/Tpp49	NT *	0.02/0.006 [24]	NT *
Tpp1/Tpp2	0.013–0.03 [20]	0.013–0.03 [24]	No toxicity to very low toxicity–depending on variant [25,26]

* NT = reported nontoxic.

2.3. *Tpp80Aa1 Structure Description*

Our final model had an R_{work}/R_{free} of 0.17/0.21 at 1.8 Å resolution and showed Tpp80Aa1 packs into the crystal lattice as homodimers, which could be indexed in the monoclinic space group C 1 2 1 (Table S1). The electron density map showed continuous density for residues 5–355 of the 373 aa wild-type Tpp80 protein sequence (Figure S2). Within each Tpp80Aa1 monomer, two distinct conserved domains appear: an N-terminal ricin B-lectin domain (IPR035992) spanning residues 5–155, and a C-terminal Toxin_10 domain (IPR008872) spanning residues 156–355 (Figure 2A). In the Tpp80Aa1 homodimer structure we also see the presence of two calcium ions (1 per monomer) and five buffer molecules present in the crystallization solution; four bis-tris propane and one citrate (Figure 2).

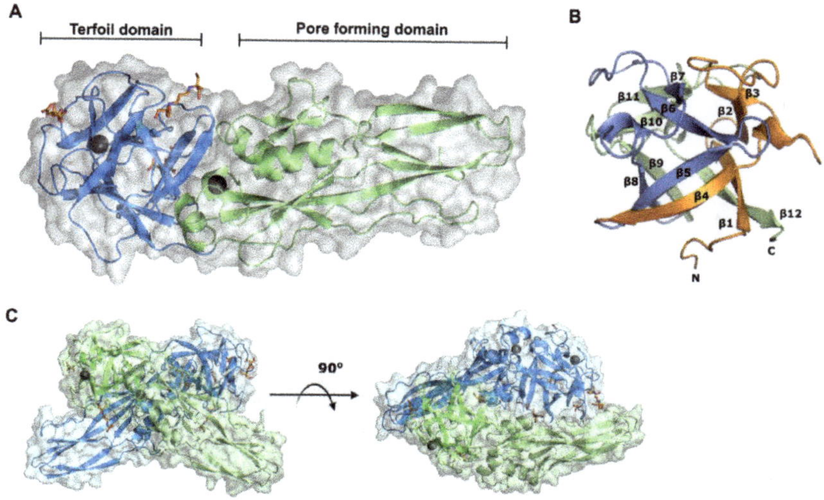

Figure 2. Tpp80Aa1 structure and homodimer packing. (**A**) Cartoon representation of Tpp80Aa1 shows an N-terminal Ricin-B like-lectin domain (blue) and a C-terminal putative pore-forming domain (green). Two bound calcium ions are represented by the black spheres and orange sticks represent buffer molecules (2 bis-tris propane, 1 citrate). (**B**) The carbohydrate binding domain is composed of three pseudo symmetric sections of β-trefoil fold corresponding with residues 9–57 (orange), 58–105 (blue), and 106–155 (green). (**C**) Tpp80Aa1 is present as a homodimer with a large molecular interface between monomer 1 (blue) and monomer 2 (green).

The N-terminal domain is composed of the well-described ricin B type β-trefoil lectin fold [27]. The β-trefoil consists of three subdomains α (β1–β4), β (β5–β8), and γ (β9–β12) assembling around a pseudo three-fold axis. The first and fourth β strand of each repeat form together a β-barrel, whereas the second and third form a β-hairpin (Figures 2B and 3C). The C-terminal domain of Tpp80Aa1 is rich in β-sheet topology characteristic of other Tpp proteins (Figure 3A) and the founding member of β-pore-forming toxins (β-PFTs), aerolysin [28]. Aerolysin and aerolysin-like proteins have a structurally conserved PFD, generally consisting of five β-strands with an insertion loop between strands β2 and β3 [29]. A structure reminiscent of an aerolysin insertion loop is present in Tpp80Aa1 between residues 259–272 (SWSIGADMGFS), as a short β-hairpin—with predominantly amphipathic structure—tucked under a loop. Similar structures are present in other Tpp proteins and are hypothesised to unfold in pore formation (Figures 3A and S4).

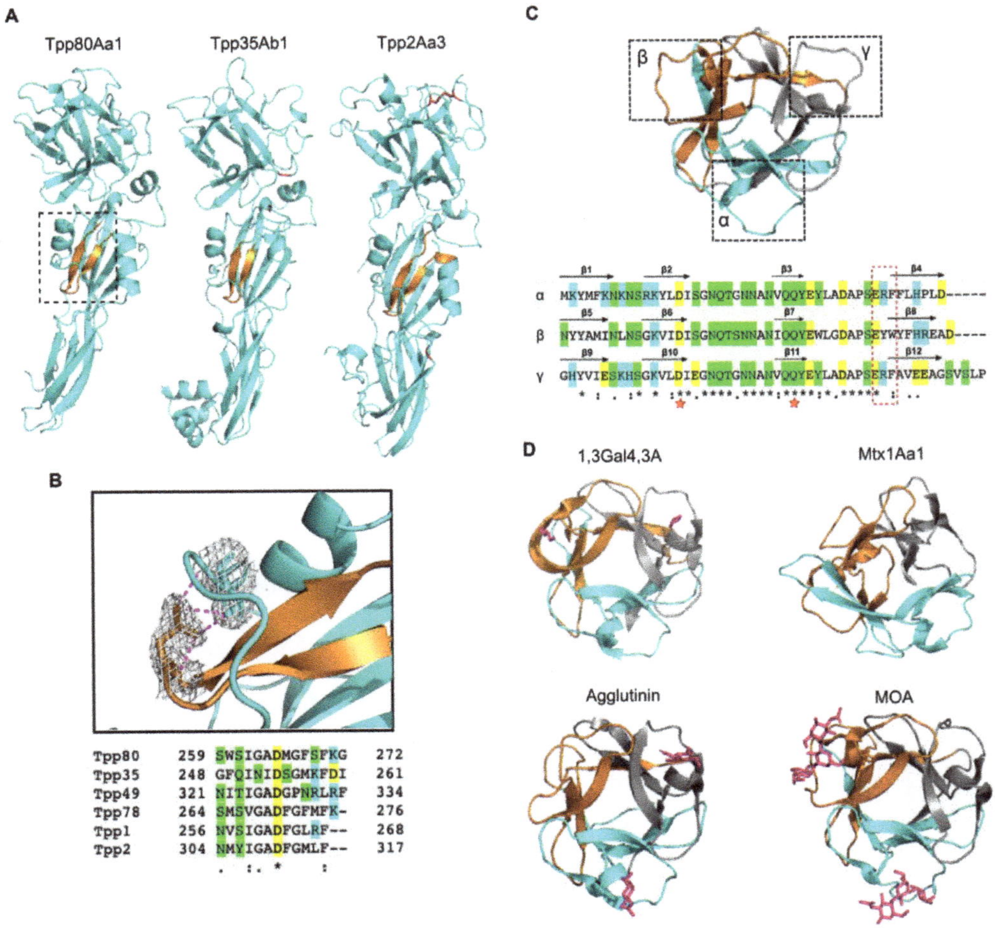

Figure 3. Tpp80Aa1 shows structural homology to other Tpp family members, and ricin B-lectin domains. (**A**) Cartoon depiction of Tpp80Aa1 and insecticidal Tpp family members with strong structural homology, Tpp35Ab1 and Tpp2Aa3. The putative insertion loop is depicted in orange, and disulphide bonds in Tpp35 and Tpp2 are shown in red (there are no Cys residues in Tpp80Aa1). (**B**) Cartoon depiction of the PFD putative insertion loop, boxed in (**A**). The PFD β-hairpin contains a conserved aspartic acid residue that forms polar contacts (magenta) with the backbone of histidine 291 and side chain of serine 290 on the adjacent loop. The conserved aspartic acid residue is marked with an (*) in the sequence alignment of all known Tpp structures, (:) indicates conservation between groups of strongly similar properties, and (.) indicates conservation between groups of weakly similar properties (determined by Clustal Omega). Residues are highlighted cyan = basic amino acids, yellow = acidic amino acids, green = polar uncharged side chains. (**C**) Lectin domain of Tpp80Aa1, highlighting the 3 subdomains α (cyan), β (orange) and γ (grey). A sequence alignment of the 3 domains indicating the regions of repeated β-sheet topology. The region where the 'QxW' motif is often found in ricin domains is boxed in red, and the regions of putative carbohydrate-binding residues in related structures are highlighted by a red star. (**D**) Cartoon depiction of lectin domains with strong structural homology include 1,3Gal4,3A bound to glycerol, MOA bound to Gal(1,3)Gal(1,4)GlcNAc, R. solani agglutinin bound to N-acetylgalactosamine, and Mtx1Aa1, with ligands are depicted in magenta. Sequence alignments were generated by Clustal Omega.

2.4. Interface Analysis

The final model shows the presence of a homodimer with a large molecular interface forming an 'X' structure (Figure 2C). Superposition of the Tpp80Aa1 monomers show the two copies to be highly similar with a root-mean-square deviation (RMSD) of 0.795Å estimated by PyMOL [30] (Figure S3A). This 'X' structure between Tpp monomers has also been observed in natural crystals of Tpp49Aa1 homodimers [31], and Tpp1Aa2/Tpp2Aa2 heterodimers [32], showing that similar dimer forms are produced in in vitro crystal screens. In the case of Tpp80Aa1, PDBePISA interface analysis identified 12 interfaces (Table S2); the largest interface (1010.9 Å^2) between the two monomers in the 'X' shape involves 34 residues of monomer 1 and 35 residues of monomer 2, with 13 hydrogen bonds (Figure S3B). Although the size of the Tpp80Aa1 molecular interface (1010.9 Å^2) exceeds the threshold estimated to discriminate between a biological and an artificial dimer (856 Å^2) [33], it is significantly smaller than that observed for the Tpp1/Tpp2 heterodimer (1833.1 Å^2) and has a much lower binding energy (Tpp80Aa1 $\Delta^{i}G$ of −7.1 kcal/mol, Tpp1/Tpp2 of −22.5 kcal/mol). Whereas the large Tpp1/Tpp2 interface may be preserved in solution, this is unlikely to be the case for the Tpp80Aa1 dimer—indeed SEC indicates Tpp80Aa1 is largely monomeric in solution (Figure S1A). Mechanistically this makes sense, given the Tpp1/Tpp2 1:1 molar ratio shown to be optimal for receptor binding and toxicity is 1:1. As Tpp80Aa1 does not need a partner to elicit toxicity, it is possible this 'X' shape is a requirement for packing and stability in the crystal structure.

Insecticidal proteins are usually produced in protoxin form and processed, often by trypsin-like enzymes, in the target insect gut [34,35]. N-terminal sequencing by Edman degradation shows the first 5 amino acid residues of trypsin-activated Tpp80Aa1 are XMTFX (where X is an amino acid that could not be assigned). Combined with LCMS analysis, which showed the molecular weight of activated Tpp80Aa1 to be 41,115.5 Da (Figure S1), this indicates that proteolytic activation removes the first three residues and the last eleven amino acids of Tpp80Aa1. In the Tpp1/Tpp2 heterodimer, the large 53-residue Tpp2 pro-region is hypothesised to maintain the heterodimer until receptor binding, where the slow release of the large pro-region signals pore formation [32]. Although the Tpp80Aa1 pro-region fragments are not visible in the electron density map, the pro-regions are located near crystal contact interfaces, and could play a role in stability prior to dissolution (Figure S3B–D).

2.5. Tpp80Aa1 Has Structural Similarity with Other Tpp Proteins and Ricin B-like Lectin Domain-Containing Proteins

Tpp80Aa1 structurally related proteins were identified by using the DALI server to search the Protein Data Bank (Figure 3, Table S3). The strongest matches were other Tpp insecticidal proteins: L. sphaericus Tpp2Aa2 (PDB 5FOY-B) and B. thuringiensis Tpp35Ab2 (PDB 4JP0-A). The related Tpp structures all share the N-terminal β-trefoil and C-terminal Toxin_10 family PFD (Figure 3A), and, within the PFD, a putative membrane insertion β-hairpin tucked under a loop. Multiple sequence alignment of the β-hairpin from all published Tpp structures indicated a conserved aspartic acid residue (Figure 3B), which forms polar contacts with the backbone of a residue in the overlying loop, and frequently with the sidechain of a semi-conserved serine/threonine residue preceding it in the loop (Figures 3B and S4).

In addition to the structural homology with other Tpp proteins, Tpp80Aa1 shows strong regional matches with the mosquitocidal holotoxin (Mtx1Aa1, PDB 2VSE-A), which belongs to a distinct structural class of toxin but contains 4 ricin-lectin repeats; an exo-beta-1,3-galactanase (Ct1,3Gal43A; PDB 3VSF-F) from the thermophilic bacterium *Clostridium thermocellum*; and agglutinins from both the fungi *Marasmius oreades* (MOA; PDB 5D63-A), and *Rhizoctonia solani* (PDB 4G9N). These homology matches are based on the structural similarity of the N-terminal ricin_B lectin-like domain (Figure 3D). Ricin B-lectins are frequently characterised by galactose binding, and this has been shown for the four lectin-domains with the greatest structural similarity. Mtx1Aa1 is a mosquitocidal protein comprising a catalytic domain with 4 ricin B-type lectin domains curled around it, containing 12 putative sugar binding sites [36]. These sites are structurally related to pierisin—a cytotoxin from *Pieris rapae* (cabbage white butterfly) that is reported to bind the glycolipids globotriaosylceramide (Gb3) and globotetraosylceramide (Gb4) which have a terminally linked galactose and N-acetylgalactosamine, respectively [37]. Furthermore, the structurally homologous proteins noted above—MOA [38], Ct1,3Gal43A protein [39], and *R. solani* agglutinin [40]—also display binding affinity towards galactose/GalNAc and galactose/GalNAc containing polysaccharides. A characteristic, although not completely conserved, sequence feature of ricin B lectin domains is the presence of a glutamine-any residue-tryptophan motif (QxW) as internal repeats near the origin of the fourth β-strand of each subdomain $(QxW)_3$ [41]. The tryptophan consistently forms part of the hydrophobic core [42], whereas the glutamine is hypothesised to be a putative carbohydrate binding residue [43]. Tpp80Aa1 does not have any fully conserved QxW motifs but does have conserved hydrophobic residues (either phenylalanine or tryptophan) in the 'W' position (Figure 3C and Figure S5). There is also a conserved aspartic acid in the second β-strand, and a 'QQY' repeat in the third β-strand of each domain, which have been proposed as putative carbohydrate binding sites in related structures (Figures 3C and S5).

2.6. Tpp80Aa1 Binds Galactose-Containing Glycolipids and Lipids from Target Species

Given the presence of the ricin_B lectin domain, we investigated the ability of trypsin-activated Tpp80Aa1 to bind carbohydrate residues. Based on the structural similarity between the N-terminus of Tpp80Aa1 and other galactose-binding lectins, we utilised glycolipids as a tool to investigate the ability of Tpp80Aa1 to bind galactose. Lipid binding blots (Figure 4A) show Tpp80Aa1 can bind mixed ganglioside extracts (which contain GM1, GD1a, GD1b, GT1b), purified GM1, and purified GM3, but does not interact with glucosylceramide, C20 ceramide, sphingomyelin, phosphatidylcholine, cholesterol, ceramide phosphoethanolamine, or lysosphingomyelin. This indicates specific binding activity of Tpp80Aa1 to glycolipid structures with terminal β-D-galactose residues (GM1, GM3, GD1a, GD1b, GT1b) and no interaction with a terminal glucose (GlcCer) or the lipid moieties. In the case of GM3 binding, this also indicates that Tpp80Aa1 can bind a terminal galactose residue conjugated to a sialic acid residue. Indeed, Tpp80Aa1 can strongly bind the sugar headgroup of GM1 after it had been cleaved from the lipid fraction (Figure 4B) and adding galactose as a competitive inhibitor substantially reduced GM1 binding (Figure 4C). Collectively these experiments indicate the observed binding is via an interaction with the glycolipid headgroup.

To see if trypsin-activated Tpp80Aa1 was interacting with mosquito-derived lipids, we isolated lipids from the guts of *Culex* and *Aedes* mosquitoes into two chemical phases—the upper of which is hydrophilic and attracts glycolipids with more polar carbohydrate structures and the hydrophobic lower fraction which contains non-polar lipids which tend to have less complexity or no sugar structure [44,45]. We observed binding to both phases (Figure 4D), indicating Tpp80Aa1 can interact with mosquito-derived lipids. Binding to the upper phase suggests binding to polar glycolipids, presumably in part through galactose binding. Further work is required to validate if glycolipid interaction occurs in vivo: it is equally possible that Tpp80Aa1 is also capable of interacting with a galactose-containing glycoprotein.

2.7. Galactose Competition Reduces Tpp80Aa1 Toxicity in Mosquito Cell Lines

To investigate the biological relevance of galactose interaction, we performed sugar inhibition assays using the MRA-918 cell line from target species *C. quinquefasciatus* (Figure 4E). Addition of trypsin-activated Tpp80Aa1 rendered cells 28% viable compared to untreated controls, yet addition of Tpp80Aa1 alongside galactose or N-acetyl-galactosamine has a protective effect and resulted cells being 66% and 54% viable, respectively. Addition of mannose, fucose and glucose had no protective effect (Figure 4E). The addition of sugars alone had no effect on cell viability (Figure S6). These results show addition of external galactose or GalNAc significantly decreases Tpp80Aa1 cytotoxicity.

A structural alignment of the Tpp80Aa1 N-terminal lectin domain with the recently identified galactose-interacting Tpp78Aa1 (PDB: 7Y78) shows the trefoil domain adopts a highly similar conformation (RMSD 0.927 Å, Figure 4F), as was noted previously with an alignment between Tpp78Aa1 and ricin B [15]. Tpp78Aa1 has four conserved residues (D86, Y100, N107, and Q108) with ricin B (PDB: 3VT1) that are attributed to binding and recognition of galactose (D416, Y431, N438, and Q439). In Tpp80Aa1 we see a similar conservation (D73, Y88, S96, and E97) where the sidechains superimpose with those present in Tpp78Aa1, and we propose this as the putative galactose binding site of Tpp80Aa1 (Figure 4F). These conserved residues are present in each of the 3 subdomains within the β-trefoil of Tpp80Aa1 (Figure 3C), indicating multiple putative sugar binding sites.

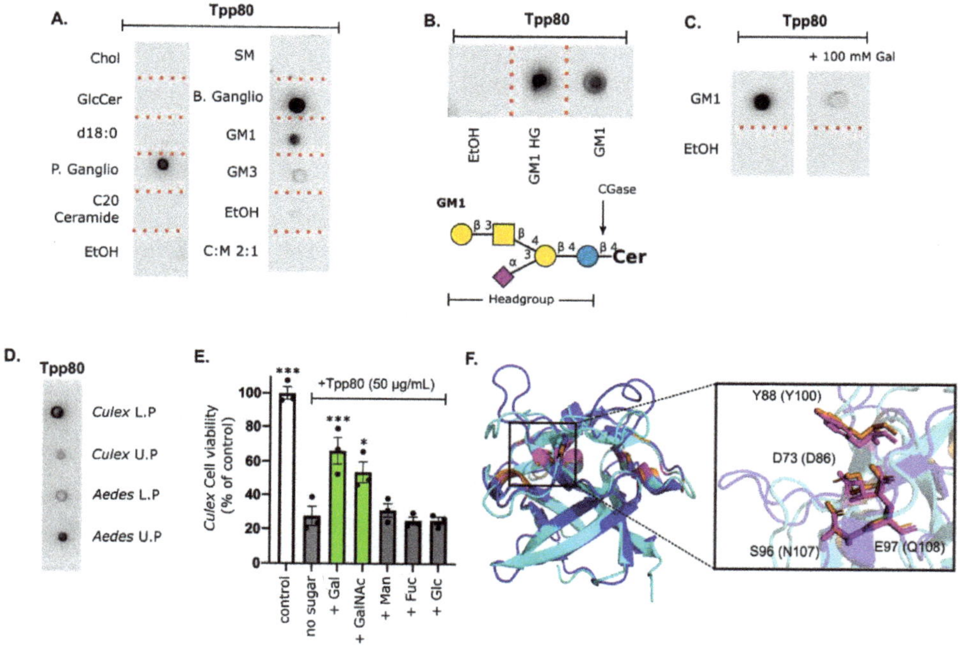

Figure 4. Tpp80Aa1 interacts with glycolipid moieties containing a terminal galactose residue. (**A**) Dot blots probed with biotinylated activated Tpp80Aa1 (20 μg/mL) show binding to porcine and bovine mixed gangliosides (P/B Ganglio), and ganglioside GM1 and GM3. No binding is observed to cholesterol (Chol), glucosylceramide (GlcCer), sphinganine (d18:0 Sph), C20 ceramide, or sphingomyelin (SM), and solvent only (C:M 2:1 or EtOH). (**B**) Tpp80Aa1 can bind the isolated sugar headgroup (HG) of GM1—CGase was utilised to cleave the sugar headgroup off GM1 as depicted in GM1 glycan (created using DrawGlycan 2.0 using standard sugar symbols). (**C**) Addition of galactose (100 mM) to the binding assay significantly reduces Tpp80Aa1 binding to GM1. (**D**) Tpp80Aa1 binds upper and lower phase lipid(s) isolated from *C. quinquefasciatus* and *Ae. aegypti* larvae. (**E**) Competition assay investigating the protective effects of sugars (15 mM) in the *C. quinquefasciatus* MRA-918 cell line. Galactose (Gal) and N-acetylgalactosamine (GalNAc) addition alongside Tpp80 reduces Tpp80 induced cytotoxicity (green bars). Mannose (Man), fucose (Fuc) and glucose (Glc) do not affect Tpp80-induced toxicity (grey bars). Statistical analysis was performed using GraphPad Prism (ver 8.2.0), using one-way ANOVAs comparing each group to the Tpp80 treated * $p < 0.05$, *** $p < 0.01$ (**F**) Alignment generated in PyMOL of Tpp80Aa1 (cyan) with Tpp78Aa1 (dark blue) with putative carbohydrate binding sites shown (Tpp78 = orange, Tpp80 = pink).

3. Discussion

Tpp80Aa1 is an interesting novel candidate for mosquito control, having recently been shown to cause mortality in *Culex pipiens pallens* larvae [3], and its demonstrated target range has been expanded in this work to include *C. quinquefasciatus*, *Ae. aegypti*, and *An. gambiae*. Tpp80Aa1 may also be an option for circumventing resistance (such as in Tpp1/Tpp2 resistant *Culex*) or combining with other mosquitocidal proteins to lower the chance of resistance development. Compared to the other Tpp family members, Tpp80Aa1 is the only one capable of exerting mosquitocidal activity alone, although, in contrast to the other mosquitocidal Tpp binary toxins (Tpp1/Tpp2, Tpp49/Cry48) the potency appears to be lower, with approximately a 1000-fold higher concentration required for activity—although the variable nature of bioassays make this difficult to compare directly. Conceivably, the binary nature of these proteins facilitates a higher potency, with other Tpp proteins reported to act alone against insect targets also showing a lower potency: Tpp36 against western corn rootworm (147.3 μg/well) [46]; Tpp78Aa1 and Tpp78Ba1 against their rice planthopper targets (between approximately 6 and 16 μg/mL) [47,48]. Here, we demonstrate Tpp80Aa1 is localised and internalised in the same regions of the mosquito larval midgut epithelium as the Tpp1/Tpp2 complex [7]. Coadministration of Cyt1A protein has been observed previously to facilitate Tpp1 internalisation in resistant mosquito populations where Tpp2 no longer binds to the Cqm1 receptor [49]. The ability of Cyt1A to act as a surrogate receptor for Tpp80Aa1 in the same manner remains to be investigated.

The fact that Tpp80Aa1 acts alone makes it particularly appealing for manipulation to understand the mechanisms underlying insecticidal activity, and for engineering mutants to increase/alter toxicity—for which our resolved structure of Tpp80Aa1 provides a template. Tpp80Aa1 consists of a ricin B-type N-terminal trefoil domain and a C-terminal putative pore forming domain. Lectin domains are commonplace in domain 1 of Tpp proteins, and domain 3 of Cry toxins, suggesting a wider role for carbohydrate binding in pesticidal activity. This is indeed the case with some pesticidal proteins, as illustrated by glycosphingolipid receptors mediating Cry5B and Cry14A toxicity in nematodes [45], and N-acetylgalactosamine (GalNAc) forming a component of the Cry1Ac receptor(s) in some lepidopteran species [50,51]. In terms of the Tpp family, Tpp78 has recently been identified to interact with galactose, GalNAc, and lactose, and several sugars—including chitobiose, chitotriose, N-acetylmuramic acid and N-acetylneuraminic acid—can reduce Tpp1/Tpp2 activity in *Culex* cells [35], with arabinose and fucose also shown to reduce Tpp1 toxicity towards *Culex* larvae [52]. Within the Tpp80Aa1 β-trefoil, we recognise three putative carbohydrate binding domains—one in each of the α, β & γ subdomains, indicating the potential to interact with several–sugar molecules simultaneously. Further investigation will be required to confirm if these putative sites are facilitating the Tpp80Aa1 galactose interaction, and whether other carbohydrates—such as GalNAc—can also interact with Tpp80Aa1.

Lipids are known to play crucial roles in the mode of action of most protein toxins through promoting binding, endocytosis, and/or cytoplasmic translocation [53]. Examples include cholera toxin binding to GM1 [54], anthrax toxin to lipid microdomains [55], Shiga toxin to Gb3 [56], and lysenin binding to sphingomyelin [57]. Initial binding on the cell surface is hypothesised to initiate toxin oligomerisation—a critical step for facilitating conformational change, receptor recognition, and pore formation with β-PFTs. We suggest Tpp80Aa1 oligomerisation may occur through an interaction with galactose or GalNAc present on proteins or lipids at the cell surface. Indeed, addition of galactose/GalNAc to our cell bioassays has a protective effect on the cells, suggesting Tpp80Aa1 binding to free galactose/GalNAc is preventing it from interacting with its putative receptor(s). We observed binding to lipids isolated from target species, indicating that lipid binding occurs—although whether this facilitates toxin action is still to be investigated. As terminal galactose residues are not specific to glycolipids or glycoproteins of the mosquito midgut, it is highly likely other receptor(s) are present in the mosquito midgut to confer target species/tissue specificity. However, a change in glycan binding profiles might be an indication of resistance development, as is exampled in previous studies investigating Cry5B toxicity in nematodes [45].

A two-stage binding process is hypothesised for many insecticidal toxins, with an initial low affinity interaction allowing the flexibility required for reorganisation into an oligomeric state prior to receptor interaction [58,59]. A well characterised example is the initial low-affinity interaction of Cry1Ac to GalNAc followed by secondary high-affinity binding to a glycoprotein receptor [60,61]. For the Tpp1/Tpp2 binary complex, receptor-mediated endocytosis appears to be a key component of pore formation, as demonstrated by the formation of cationic ion channels in—normally nontarget—MDCK cell lines engineered to express the relevant Cqm1 receptor [17]. Furthermore, the Cqm1 receptor is localised to lipid microdomains enriched in glycosphingolipids which could be playing an important role in initial oligomerization. Precisely how the pore-forming domain in Tpp proteins inserts into the membrane is unknown.

Future experiments to confirm pore forming activity and discover the Tpp80Aa1 receptor will be key to understanding its mechanism of action. Here, we pinpoint both putative carbohydrate-binding residues and residues hypothesised to initiate membrane-insertion and pore formation. This work can facilitate future studies exploring the mechanism of action, enhancing Tpp80Aa1 activity, and developing Tpp80Aa1 for potential use in the field.

4. Materials and Methods

4.1. Tpp80Aa1 Expression and Purification

A synthetic clone of the *tpp80Aa1* gene was produced in the pET30a plasmid to express a Tpp80Aa1 protein with a short N-terminal extension including a hexa-histidine tag (by inserting the entire *tpp80Aa1* reading frame, downstream of the *Bam*HI site in the vector). This plasmid was introduced into BL21 *E. coli* cells and cultured in 2× YT medium containing kanamycin. Once the OD_{600} reached ~0.6, protein expression was induced with 0.5 mM IPTG for 18 h with shaking at 25 °C (200 rpm). Bacterial cultures were collected (7000× *g*, 4 °C, 15 min) and lysed via two freeze–thaw cycles (−80 °C/37 °C) and sonication (10 × 10 s, with 20 s intervals, on ice). The lysate was clarified by centrifugation (23,000× *g*, 4 °C, 30 min) and then filtered through a 0.45 µm filter and proteins were purified using standard immobilised metal affinity chromatography (Protino Ni-TED, Macherey-Nagel, Düren, Germany). Samples were concentrated and imidazole removed via buffer exchanging the sample by 4 rounds of dilution/concentration in a 10 kDa cut-off centrifugal filter unit (Amicon Ultra-15, MilliporeSigma, Burlington, USA using 50 mM TrisHCl pH 7.4 supplemented with a protease inhibitor cocktail (cOmplete Protease Inhibitor Cocktail, Roche, Basel, Switzerland). Purified proteins were examined by SDS-PAGE. For ultra-pure samples, eluted samples were subjected to size-exclusion chromatography (SEC; Figure S1). Where trypsinised protein was required (all presented dot blots and cell assays), immobilised TPCK treated trypsin resin (20233, Thermo Scientific, Waltham, MA, USA) was added to the sample and incubated in a shaker at 37 °C for 16 h, followed by centrifugation to remove the trypsin resin prior to use. Liquid chromatography mass spectrometry (LCMS) was used to quantify molecular weight, and N-terminal sequencing by Edman degradation was performed by Alta Bioscience (Birmingham, UK).

4.2. Bioassays (Insects & Cells)

Bioassays were carried out against a range of mosquito larvae (*C. quinquefasciatus*, *Ae. aegypti*, and *An. gambiae*) and insect cell lines derived from *C. quinquefasciatus* (MRA-918), *Ae. aegypti* (C6/36) and *An. gambiae* (Ag55). MRA-918 cells were kindly gifted by Dr. Mario Soberón (Mexico City, Mexico) and c6/36 and Ag55 cells from Dr. Claire Donald (Glasgow, UK). For insect larvae, 10–15 third-instar larvae were placed in 5 mL of dH_2O and maintained in a humidified room at 24 °C. Mortality was assessed by counting live larvae at 24 h after the addition of purified toxin, or the equivalent amount of the relevant buffer (50 mM TrisHCl, pH 7.5). For all larval bioassays, non-trypsinised protein was used.

The concentration giving 50% mortality (LC_{50}) was calculated using GraphPad prism for Mac OS (Ver 8.2.0) plotting log(concentration of toxin) against % survival rate.

Insect cell lines were maintained at 27 °C in Schneider's Insect Medium supplemented with 10% FBS. For cellular bioassays, cells were plated at 10,000/well of a 96-well plate in 150 µL of medium until ~70% confluent. Cell viability was investigated using resazurin, as described previously. Trypsinised Tpp80Aa1 was used in cellular bioassays. For the sugar competition assay, sugars (glucose, galactose, mannose, fucose, N-acetylgalactosamine) were dissolved into the cell culture medium at a final concentration of 15 mM, alongside activated Tpp80 and sugar-only controls. GraphPad Prism for Mac OS (Ver 8.2.0), using one-way ANOVAs followed by Dunnett's multiple comparisons test was used to compare individual treatment groups back to the control. Data are presented as mean ± standard deviation.

4.3. *Tpp80Aa1 Labelling*

Fluorescent labelling of Tpp80Aa1 (non-activated) for midgut imaging was performed using the Alexa Fluor™ 488/555 Protein Labelling Kit (A10235, Invitrogen, Waltham, MA, USA), as per the manufacturer's instructions. Briefly, Tpp80Aa1 was diluted to 2 mg/mL in dPBS in a final volume of 0.5 mL, to this 50 µL of 1 M sodium bicarbonate buffer (pH 8.3) was added. Tpp80Aa1 solution was added to the dye-containing vial and stirred at RT for 1 h. Purification of labelled protein was achieved using the Zeba Dye Spin Columns provided. Labelled protein (Tpp80Aa1-555) was stored at 4 °C protected from light.

4.4. *In Vivo Midgut Imaging*

Labelled Tpp80Aa1-555 protein was added to 1.5 mL water at a final concentration of 50 µg/mL containing 4 mosquito larvae (third instar). After 45 min, larvae were put in fresh water (containing no Tpp80Aa1-555) and left for a further 30 min before gut dissection in PBS. To label cell nuclei, extracted guts were added to PBS containing 1 µg/mL Hoechst 33342 and gently rocked at RT for 30 min. Samples were mounted in 1 mm glass capillary tubes (10490413, Fisher Scientific, Waltham, MA, USA) with 1% low melting point agarose (16520050, ThermoFisher, Waltham, MA, USA). For imaging, we used a Zeiss Lightsheet Z.1 (Oberkochen, Germany) with a 405 nm (20 mW) and 561 nm laser (20 mW) with either a 10×/0.5 W Plan Apo or a 20×/1.0 Plan Apo (water immersion) objective.

4.5. *Crystallisation*

For crystallisation trials, Tpp80Aa1 was concentrated to 8 mg/mL in 50 mM TrisHCl pH 8.0. Crystallisation screening was performed in 96-well plates (3 Lens Crystallisation plate, SWISSCI, Zug, Switzerland) using a commercially available crystal screen, PACT Premier HT-96 screen (MD1-36, Molecular Dimensions, Rotherham, UK). Plates were set up using a Mosquito Crystallisation robot (SPT Labteck, Melbourn, UK), with 200 nL Tpp80Aa1 added to 200 nL screen solution. Very small crystals grew in several wells, the most promising single crystals appeared in F11 condition (0.2 M sodium citrate, 0.1M Bis Tris propane, 20% *w/v* PEG 3350, pH 6.5) and were used for seeding preparation. For making seed stock, 20 µL of F11 reservoir buffer was added to crystal drops, crystals were then crushed with a glass crystal crusher and transferred to an Eppendorf containing a PTFE seed bead for brief centrifugation. Crystal seeds were diluted 1:10 with well solution and added to the same 96-well screen (200 nL seed dilution, 200 nL protein, 200 nL buffer). Seeding produced multiple hits, which were harvested 2-weeks post seeding and flash-frozen in liquid nitrogen. H11 condition (0.2 M sodium citrate, 0.1 M Bis Tris propane, 20% *w/v* PEG3350, pH 8.5) produced the 1.8 Å Tpp80Aa1 dataset.

4.6. *Data Collection and Structure Determination*

Data were collected at Diamond Light Source (Harwell, UK) at beamline I03. Images were processed with the DIALS package and amplitudes estimated with TRUNCATE, in the CCP4 package [62]. The structure of Tpp80Aa1 was determined using molecular replacement in PHASER using the CCP4i2 software (ver 7.1.012). The starting search model was a synthetic construct, ThreeFoil (PDB entry 3PG0), and the C-terminus (aa 159–366) of Tpp1 (formerly BinA, PDB entry 5FOY). This was followed by successful model building of a partial model with Buccaneer. The resulting model and maps were inspected manually via Coot [63], followed by iterative rounds of real-space refinement and model building cycles using Coot and REFMAC5 [64], respectively. Data collection and processing statistics, and refinement statistics are summarised in Table S1.

4.7. Structural Analysis

Comparing Tpp80Aa1 structural similarity to other proteins in the Protein Data Bank was performed using the DALI server and a heuristic PDB search [65]. Interface analysis was performed using the PDBe PISA web server [66].

4.8. Lipid Extractions

Lipids were purified from larvae into two chemical phases using Svennerholm partitioning. Larvae were starved for 24 h prior to lipid extraction. Second/third instar larvae (0.5 g total mass) were rinsed three times in dH_2O, then flash frozen in liquid nitrogen and thawed at RT three times. Pellets were sonicated and lipid extraction was performed at 37 °C for 2 h with agitation in a mixture of chloroform, methanol, and water with a final ratio of 4:8:3. Samples were centrifuged at 1400× g for 5 min to split into upper (hydrophilic, attracts more polar lipids) and lower (hydrophobic, attracts generally simpler nonpolar lipids) phases. Silica-based hydrophobic cartridges (WAT036810, Sep-Pak tC18, Wilmslow, UK) were used to purify and concentrate upper phase glycolipids. All samples were dried under N_2 at 40 °C and resuspended in 50 µL methanol (upper phase) and 200 µL 1:1 chloroform to methanol (lower phase) for use in dot blots. Thin layer chromatography was used to check successful lipid extraction (not shown).

4.9. Lipid Dot Blots

All dot blots were performed using a PVDF 0.2 µm pore membrane (ISEQ00010 Immobilon®-PSQ PVDF, Millipore, Burlington, MA, USA). Lipid standards were added to the blot at a concentration of 2 µg in a volume no larger than 4 µL, or for larvae-extracted lipids, 2 µL of the final suspension. Blots were left to dry and then blocked with tris buffered saline (TBS) containing 5% BSA for 1 h at room temperature (RT). Trypsin-activated Tpp80Aa1 was added at 20 µg/mL in TBS containing 1% BSA (TBS-1%) and left agitating overnight at 4 °C. Blots were washed in TBS-1% for 10 min at RT, and then probed with an anti-polyHistidine-Peroxidase antibody (A7058, Sigma-Aldrich, St. Louis, MO, USA) for 1 h at RT with agitation. Blots were washed in TBS-1% and visualised using a LI-COR C-Digit chemiluminescence Western blot scanner and a WESTAR ECL-Sun HRP detection kit, (K1-0052, geneflow, Lichfield, UK). For activated toxin, the same process was used but with biotinylated Tpp80Aa1 and an ABC-HRP kit (PK-6100, vector laboratories, Burlingame, CA, USA). The lipids used in this manuscript were; porcine brain total ganglioside extract (860053P Avanti® Polar Lipids, 2 mg/mL in chloroform:ethanol 2:1), ganglioside GM3 (GM3, 860058P Avanti® Polar Lipids, 2 mg/mL in ethanol), glucosylceramide (GlcCer, 131304P Avanti® Polar Lipids, 2 mg/mL in ethanol), ganglioside GM1 (GM1, 860065P Avanti® Polar Lipids, 2 mg/mL in ethanol), mixed bovine gangliosides (1065 Matreya LLC, 10 mg/mL in chloroform:ethanol 2:1), C20 ceramide (860520P Avanti® Polar Lipids, 10 mg/mL in ethanol), sphingomyelin (860062 Avanti® Polar Lipids, 10 mg/mL in chloroform), cholesterol (700000P Avanti® Polar Lipids, 2 mg/mL in ethanol), and, sphinganine d18:0 (860498P Avanti® Polar Lipids, 2 mg/mL in ethanol). For the galactose competition assay, GM1 was added to the blot as described above, and 100 mM galactose was added into the Tpp80Aa1-containing TBS solution. Ceramide glycanase (LZ-CER-HM-KIT, LudgerZyme, Oxfordshire, UK) was used to remove the sugar headgroup from GM1, as per instructions from the supplier. Briefly, 10 µL of enzyme was added per 2 nmol of GM1 alongside 10 µL of reaction buffer and 16 µL of dH_2O, and the resulting solution was incubated at

37 °C for 24 h. To purify the glycan from the mixture, LudgerClean S cartridges (LC-S-A6, LudgerZyme, Oxfordshire, UK) were used as per the supplier's instructions. Eluted glycans were dried and resuspended in 50 μL ethanol prior to dot blot assay.

Supplementary Materials: The following supporting information can be downloaded at: https://www.mdpi.com/article/10.3390/toxins14120863/s1, Figure S1: Tpp80Aa1 expression, crystallisation, and N-terminal sequencing; Figure S2: Electron density map and model of the Tpp80Aa1 structure; Figure S3: Superposition and interfaces of the Tpp80Aa1 monomers; Figure S4: Conserved putative insertion loop contacts in Tpp family members; Figure S5: Multiple sequence alignment of QxW motifs in Tpp80Aa1 and structurally similar lectin domains; Figure S6. Sugar addition had no effect on MRA-918 cell viability. Galactose (Gal), N-acetylgalactose (GalNAc), mannose (Man), fucose (Fuc) or glucose (Glu) were added to C. quinquefasciatus derived cells (MRA-918) at a final concentration of 15 mM. Twenty-four hours post addition, no significant impact was observed on cell viability, as quantified by resazurin assay ($p > 0.05$, one-way ANOVA compared to control); Table S1: Data collection and refinement statistics for Tpp80Aa1; Table S2: Interfaces in the Tpp80Aa1 crystal structure, as calculated by PDBePISA; Table S3: Top 20 proteins with structural similarity to Tpp80Aa1, as identified by the DALI server. References [67–77] are cited in the supplementary materials.

Author Contributions: Conceptualization, H.L.B. and C.B.; methodology, M.L.-L., P.J.R., E.L.-E. and C.B.; investigation, H.L.B.; formal analysis, H.L.B., L.J.W. and P.J.R.; writing—original draft preparation, H.L.B.; writing—review and editing, H.L.B., L.J.W., M.L.-L., H.W.-E., E.L.-E., P.J.R. and C.B.; supervision, E.L.-E., P.J.R. and C.B.; funding acquisition, C.B., H.W.-E. and E.L.-E. All authors have read and agreed to the published version of the manuscript.

Funding: This work was supported by the Biotechnology and Biological Sciences Research Council (BBSRC, grant reference BB/S002774/1) and a BBSRC-funded South West Biosciences Doctoral Training Partnership (training grant reference BB/M009122/1). This work was also supported by Corteva Agriscience.

Institutional Review Board Statement: Not applicable.

Informed Consent Statement: Not applicable.

Data Availability Statement: The data presented in this study are openly available in the Protein Data Bank at www.rcsb.org with accession code 8BAD.

Acknowledgments: The authors would like to thank Diamond Light Source for beamtime (proposal mx20147, and the staff of beamlines I03. The authors would like to thank Mario Soberón for the MRA-918 cells (Instituto de Biotecnología, Universidad Nacional Autónoma de México, Cuernavaca 62250, Morelos, México) and Claire Donald for the C6/36 and Ag55 cells (School of Infection and Immunity University of Glasgow, Glasgow, Scotland, UK).

Conflicts of Interest: The authors declare no conflict of interest. The funders had no role in the design of the study; in the collection, analyses, or interpretation of data; in the writing of the manuscript; or in the decision to publish the results.

References

1. Schnepf, E.; Crickmore, N.; Van Rie, J.; Lereclus, D.; Baum, J.; Feitelson, J.; Zeigler, D.R.; Dean, D.H. *Bacillus thuringiensis* and its pesticidal crystal proteins. *Microbiol. Mol. Biol. Rev.* **1998**, *62*, 775–806. [CrossRef] [PubMed]
2. Silva-Filha, M.; Romao, T.P.; Rezende, T.M.T.; Carvalho, K.D.S.; Gouveia de Menezes, H.S.; Alexandre do Nascimento, N.; Soberon, M.; Bravo, A. Bacterial Toxins Active against Mosquitoes: Mode of Action and Resistance. *Toxins* **2021**, *13*, 523. [CrossRef] [PubMed]
3. Zhou, Y.; Wu, Z.Q.; Zhang, J.; Wan, Y.S.; Jin, W.J.; Li, Y.Z.; Fang, X.J. Cry80Aa1, a novel *Bacillus thuringiensis* toxin with mosquitocidal activity to *Culex pipiens* pallens. *J. Invertebr. Pathol.* **2020**, *173*, 107386. [CrossRef]
4. Valtierra-de-Luis, D.; Villanueva, M.; Berry, C.; Caballero, P. Potential for *Bacillus thuringiensis* and Other Bacterial Toxins as Biological Control Agents to Combat Dipteran Pests of Medical and Agronomic Importance. *Toxins* **2020**, *12*, 773. [CrossRef] [PubMed]
5. Lacey, L.A. *Bacillus thuringiensis* serovariety israelensis and *Bacillus sphaericus* for mosquito control. *J. Am. Mosq. Control Assoc.* **2007**, *23*, 133–163. [CrossRef] [PubMed]
6. Crickmore, N.; Bone, E.J.; Williams, J.A.; Ellar, D.J. Contribution of the Individual Components of the Delta-Endotoxin Crystal to the Mosquitocidal Activity of *Bacillus thuringiensis* Subsp Israelensis. *Fems. Microbiol. Lett.* **1995**, *131*, 249–254.

7. Oei, C.; Hindley, J.; Berry, C. Binding of purified *Bacillus sphaericus* binary toxin and its deletion derivatives to *Culex quinquefasciatus* gut: Elucidation of functional binding domains. *J. Gen. Microbiol.* **1992**, *138*, 1515–1526. [CrossRef] [PubMed]
8. Wirth, M.C.; Jiannino, J.A.; Federici, B.A.; Walton, W.E. Synergy between toxins of *Bacillus thuringiensis* subsp. israelensis and *Bacillus sphaericus*. *J. Med. Entomol.* **2004**, *41*, 935–941. [CrossRef]
9. Wirth, M.C.; Federici, B.A.; Walton, W.E. Cyt1A from *Bacillus thuringiensis* synergizes activity of *Bacillus sphaericus* against Aedes aegypti (Diptera: Culicidae). *Appl. Envion. Microbiol.* **2000**, *66*, 1093–1097. [CrossRef]
10. Wirth, M.C.; Walton, W.E.; Federici, B.A. Cyt1A from *Bacillus thuringiensis* restores toxicity of *Bacillus sphaericus* against resistant *Culex quinquefasciatus* (Diptera: Culicidae). *J. Med. Entomol.* **2000**, *37*, 401–407. [CrossRef]
11. Mulla, M.S.; Thavara, U.; Tawatsin, A.; Chomposrf, J.; Su, T.Y. Emergence of resistance and resistance management in field populations of tropical *Culex quinquefasciatus* to the microbial control agent *Bacillus sphaericus*. *J. Am. Mosq. Contr.* **2003**, *19*, 39–46.
12. Yuan, Z.M.; Zhang, Y.M.; Cai, Q.X.; Liu, E.Y. High-level field resistance to *Bacillus sphaericus* C3-41 in *Culex quinquefasciatus* from southern China. *Biocontrol. Sci. Technol.* **2000**, *10*, 41–49. [CrossRef]
13. Crickmore, N.; Berry, C.; Panneerselvam, S.; Mishra, R.; Connor, T.R.; Bonning, B.C. A structure-based nomenclature for *Bacillus thuringiensis* and other bacteria-derived pesticidal proteins. *J. Invertebr. Pathol.* **2020**, *186*, 107438. [CrossRef] [PubMed]
14. Berry, C.; Crickmore, N. Structural classification of insecticidal proteins—Towards an in silico characterisation of novel toxins. *J. Invertebr. Pathol.* **2017**, *142*, 16–22. [CrossRef]
15. Cao, B.; Nie, Y.; Guan, Z.; Chen, C.; Wang, N.; Wang, Z.; Shu, C.; Zhang, J.; Zhang, D. The crystal structure of Cry78Aa from *Bacillus thuringiensis* provides insights into its insecticidal activity. *Commun. Biol.* **2022**, *5*, 801. [CrossRef] [PubMed]
16. De Melo, J.V.; Jones, G.W.; Berry, C.; Vasconcelos, R.H.T.; de Oliveira, C.M.F.; Furtado, A.F.; Peixoto, C.A.; Silva, M.H.N.L. Cytopathological Effects of *Bacillus sphaericus* Cry48Aa/Cry49Aa Toxin on Binary Toxin-Susceptible and -Resistant *Culex quinquefasciatus* Larvae. *Appl. Envion. Microb.* **2009**, *75*, 4782–4789. [CrossRef]
17. Opota, O.; Gauthier, N.C.; Doye, A.; Berry, C.; Gounon, P.; Lemichez, E.; Pauron, D. *Bacillus sphaericus* binary toxin elicits host cell autophagy as a response to intoxication. *PLoS ONE* **2011**, *6*, e14682. [CrossRef] [PubMed]
18. Singh, G.J.; Gill, S.S. An electron microscope study of the toxic action of *Bacillus sphaericus* in *Culex quinquefasciatus* larvae. *J. Invertebr. Pathol.* **1988**, *52*, 237–247. [CrossRef]
19. De Melo, J.V.; Vasconcelos, R.H.; Furtado, A.F.; Peixoto, C.A.; Silva-Filha, M.H. Ultrastructural analysis of midgut cells from *Culex quinquefasciatus* (Diptera: Culicidae) larvae resistant to *Bacillus sphaericus*. *Micron* **2008**, *39*, 1342–1350. [CrossRef]
20. Davidson, E.W. Variation in binding of *Bacillus sphaericus* toxin and wheat germ agglutinin to larval midgut cells of six species of mosquitoes. *J. Invertebr. Pathol.* **1989**, *53*, 251–259. [CrossRef]
21. Lekakarn, H.; Promdonkoy, B.; Boonserm, P. Interaction of *Lysinibacillus sphaericus* binary toxin with mosquito larval gut cells: Binding and internalization. *J. Invertebr. Pathol.* **2015**, *132*, 125–131. [CrossRef]
22. Dadd, R.H. Alkalinity within the midgut of mosquito larvae with alkaline-active digestive enzymes. *J. Insect Physiol.* **1975**, *21*, 1847–1853. [CrossRef] [PubMed]
23. Boudko, D.Y.; Moroz, L.L.; Linser, P.J.; Trimarchi, J.R.; Smith, P.J.; Harvey, W.R. In situ analysis of pH gradients in mosquito larvae using non-invasive, self-referencing, pH-sensitive microelectrodes. *J. Exp. Biol.* **2001**, *204*, 691–699. [CrossRef] [PubMed]
24. Jones, G.W.; Nielsen-Leroux, C.; Yang, Y.; Yuan, Z.; Dumas, V.F.; Monnerat, R.G.; Berry, C. A new Cry toxin with a unique two-component dependency from *Bacillus sphaericus*. *Faseb. J.* **2007**, *21*, 4112–4120. [CrossRef] [PubMed]
25. Berry, C.; Hindley, J.; Ehrhardt, A.F.; Grounds, T.; de Souza, I.; Davidson, E.W. Genetic determinants of host ranges of *Bacillus sphaericus* mosquito larvicidal toxins. *J. Bacteriol.* **1993**, *175*, 510–518. [CrossRef] [PubMed]
26. Park, H.W.; Bideshi, D.K.; Federici, B.A. Recombinant strain of *Bacillus thuringiensis* producing Cyt1A, Cry11B, and the *Bacillus sphaericus* binary toxin. *Appl. Envion. Microbiol.* **2003**, *69*, 1331–1334. [CrossRef]
27. Cummings, R.D.; Etzler, M.E. R-type Lectins. In *Essentials of Glycobiology*, 2nd ed.; Varki, A., Cummings, R.D., Esko, J.D., Freeze, H.H., Stanley, P., Bertozzi, C.R., Hart, G.W., Etzler, M.E., Eds.; Cold Spring Harbor: New York, NY, USA, 2009.
28. Cirauqui, N.; Abriata, L.A.; van der Goot, F.G.; Dal Peraro, M. Structural, physicochemical and dynamic features conserved within the aerolysin pore-forming toxin family. *Sci. Rep.* **2017**, *7*, 13932. [CrossRef]
29. Szczesny, P.; Iacovache, I.; Muszewska, A.; Ginalski, K.; van der Goot, F.G.; Grynberg, M. Extending the aerolysin family: From bacteria to vertebrates. *PLoS ONE* **2011**, *6*, e20349. [CrossRef]
30. Schrödinger, L.; DeLano, W. PyMOL. 2020. Available online: http://www.pymol.org/pymol (accessed on 1 November 2022).
31. Williamson, L.J.; Galchenkova, M.; Best, H.L.; Bean, R.J.; Munke, A.; Awel, S.; Pena, G.; Knoska, J.; Schubert, R.; Doerner, K.; et al. Structure of the *Lysinibacillus sphaericus* Tpp49Aa1 pesticidal protein elucidated from natural crystals using MHz-SFX. *bioRxiv* **2022**. [CrossRef]
32. Colletier, J.P.; Sawaya, M.R.; Gingery, M.; Rodriguez, J.A.; Cascio, D.; Brewster, A.S.; Michels-Clark, T.; Hice, R.H.; Coquelle, N.; Boutet, S.; et al. De novo phasing with X-ray laser reveals mosquito larvicide BinAB structure. *Nature* **2016**, *539*, 43–47. [CrossRef]
33. Ponstingl, H.; Henrick, K.; Thornton, J.M. Discriminating between homodimeric and monomeric proteins in the crystalline state. *Proteins* **2000**, *41*, 47–57. [CrossRef] [PubMed]
34. Nicolas, L.; Lecroisey, A.; Charles, J.F. Role of the gut proteinases from mosquito larvae in the mechanism of action and the specificity of the *Bacillus sphaericus* toxin. *Can. J. Microbiol.* **1990**, *36*, 804–807. [CrossRef] [PubMed]
35. Broadwell, A.H.; Baumann, P. Proteolysis in the gut of mosquito larvae results in further activation of the *Bacillus sphaericus* toxin. *Appl. Envion. Microbiol.* **1987**, *53*, 1333–1337. [CrossRef]

36. Treiber, N.; Reinert, D.J.; Carpusca, I.; Aktories, K.; Schulz, G.E. Structure and mode of action of a mosquitocidal holotoxin. *J. Mol. Biol.* **2008**, *381*, 150–159. [CrossRef]
37. Matsushima-Hibiya, Y.; Watanabe, M.; Hidari, K.I.P.J.; Miyamoto, D.; Suzuki, Y.; Kasama, T.; Kanazawa, T.; Koyama, K.; Sugimura, T.; Wakabayashi, K. Identification of glycosphingolipid receptors for pierisin-1, a guanine-specific ADP-ribosylating toxin from the cabbage butterfly. *J. Biol. Chem.* **2003**, *278*, 9972–9978. [CrossRef] [PubMed]
38. Cordara, G.; van Eerde, A.; Grahn, E.M.; Winter, H.C.; Goldstein, I.J.; Krengel, U. An Unusual Member of the Papain Superfamily: Mapping the Catalytic Cleft of the Marasmius oreades agglutinin (MOA) with a Caspase Inhibitor. *PLoS ONE* **2016**, *11*, e0149407. [CrossRef]
39. Ichinose, H.; Kuno, A.; Kotake, T.; Yoshida, M.; Sakka, K.; Hirabayashi, J.; Tsumuraya, Y.; Kaneko, S. Characterization of an exo-beta-1,3-galactanase from Clostridium thermocellum. *Appl. Env. Microb.* **2006**, *72*, 3515–3523. [CrossRef]
40. Candy, L.; Peumans, W.J.; Menu-Bouaouiche, L.; Astoul, C.H.; Van Damme, J.; Van Damme, E.J.; Erard, M.; Rouge, P. The Gal/GalNAc-specific lectin from the plant pathogenic basidiomycete Rhizoctonia solani is a member of the ricin-B family. *Biochem. Biophys. Res. Commun.* **2001**, *282*, 655–661. [CrossRef]
41. Hazes, B. The (QxW)(3) domain: A flexible lectin scaffold. *Protein Sci.* **1996**, *5*, 1490–1501. [CrossRef]
42. Rutenber, E.; Ready, M.; Robertus, J.D. Structure and evolution of ricin B chain. *Nature* **1987**, *326*, 624–626. [CrossRef]
43. Hazes, B.; Read, R.J. A mosquitocidal toxin with a ricin-like cell-binding domain. *Nat. Struct. Biol.* **1995**, *2*, 358–359. [CrossRef] [PubMed]
44. Folch, J.; Lees, M.; Stanley, G.H.S. A Simple Method for the Isolation and Purification of Total Lipides from Animal Tissues. *J. Biol. Chem.* **1957**, *226*, 497–509. [CrossRef]
45. Hui, F.; Scheib, U.; Hu, Y.; Sommer, R.J.; Aroian, R.V.; Ghosh, P. Structure and glycolipid binding properties of the nematicidal protein Cry5B. *Biochemistry* **2012**, *51*, 9911–9921. [CrossRef]
46. Rupar, M.J.; Donovan, W.P.; Groat, R.G.; Slaney, A.C.; Mattison, J.W.; Johnson, T.B.; Charles, J.F.; Dumanoir, V.C.; de Barjac, H. Two novel strains of *Bacillus thuringiensis* toxic to coleopterans. *Appl. Envion. Microbiol.* **1991**, *57*, 3337–3344. [CrossRef] [PubMed]
47. Wang, Y.; Liu, Y.; Zhang, J.; Crickmore, N.; Song, F.; Gao, J.; Shu, C. Cry78Aa, a novel *Bacillus thuringiensis* insecticidal protein with activity against *Laodelphax striatellus* and *Nilaparvata lugens*. *J. Invertebr. Pathol.* **2018**, *158*, 1–5. [CrossRef]
48. Cao, B.; Shu, C.; Geng, L.; Song, F.; Zhang, J. Cry78Ba1, One Novel Crystal Protein from *Bacillus thuringiensis* with High Insecticidal Activity against Rice Planthopper. *J. Agric. Food Chem.* **2020**, *68*, 2539–2546. [CrossRef]
49. Nascimento, N.A.; Torres-Quintero, M.C.; Molina, S.L.; Pacheco, S.; Romao, T.P.; Pereira-Neves, A.; Soberon, M.; Bravo, A.; Silva-Filha, M. Functional *Bacillus thuringiensis* Cyt1Aa Is Necessary To Synergize *Lysinibacillus sphaericus* Binary Toxin (Bin) against Bin-Resistant and -Refractory Mosquito Species. *Appl. Envion. Microbiol.* **2020**, *86*, e02770-19. [CrossRef]
50. Burton, S.L.; Ellar, D.J.; Li, J.; Derbyshire, D.J. N-acetylgalactosamine on the putative insect receptor aminopeptidase N is recognised by a site on the domain III lectin-like fold of a *Bacillus thuringiensis* insecticidal toxin. *J. Mol. Biol.* **1999**, *287*, 1011–1022 [CrossRef]
51. Derbyshire, D.J.; Ellar, D.J.; Li, J. Crystallization of the *Bacillus thuringiensis* toxin Cry1Ac and its complex with the receptor ligand N-acetyl-D-galactosamine. *Acta Cryst. D Biol. Cryst.* **2001**, *57*, 1938–1944. [CrossRef]
52. Sharma, M.; Gupta, G.D.; Kumar, V. Mosquito-larvicidal BinA toxin displays affinity for glycoconjugates: Proposal for BinA mediated cytotoxicity. *J. Invertebr. Pathol.* **2018**, *156*, 29–40. [CrossRef] [PubMed]
53. Helms, B. Host-pathogen interactions: Lipids grease the way. *Eur. J. Lipid Sci. Technol.* **2006**, *108*, 895–897. [CrossRef]
54. Holmgren, J.; Lonnroth, I.; Mansson, J.; Svennerholm, L. Interaction of cholera toxin and membrane GM1 ganglioside of small intestine. *Proc. Natl. Acad. Sci. USA* **1975**, *72*, 2520–2524. [CrossRef] [PubMed]
55. Scobie, H.M.; Rainey, G.J.; Bradley, K.A.; Young, J.A. Human capillary morphogenesis protein 2 functions as an anthrax toxin receptor. *Proc. Natl. Acad. Sci. USA* **2003**, *100*, 5170–5174. [CrossRef] [PubMed]
56. Shin, I.S.; Ishii, S.; Shin, J.S.; Sung, K.I.; Park, B.S.; Jang, H.Y.; Kim, B.W. Globotriaosylceramide (Gb3) content in HeLa cells is correlated to Shiga toxin-induced cytotoxicity and Gb3 synthase expression. *BMB Rep.* **2009**, *42*, 310–314. [CrossRef]
57. Yamaji, A.; Sekizawa, Y.; Emoto, K.; Sakuraba, H.; Inoue, K.; Kobayashi, H.; Umeda, M. Lysenin, a novel sphingomyelin-specific binding protein. *J. Biol. Chem.* **1998**, *273*, 5300–5306. [CrossRef] [PubMed]
58. Bravo, A.; Gill, S.S.; Soberon, M. Mode of action of *Bacillus thuringiensis* Cry and Cyt toxins and their potential for insect control. *Toxicon* **2007**, *49*, 423–435. [CrossRef]
59. Bravo, A.; Gomez, I.; Conde, J.; Munoz-Garay, C.; Sanchez, J.; Miranda, R.; Zhuang, M.; Gill, S.S.; Soberon, M. Oligomerization triggers binding of a *Bacillus thuringiensis* Cry1Ab pore-forming toxin to aminopeptidase N receptor leading to insertion into membrane microdomains. *Biochim. Biophys. Acta* **2004**, *1667*, 38–46. [CrossRef]
60. Gomez, I.; Sanchez, J.; Munoz-Garay, C.; Matus, V.; Gill, S.S.; Soberon, M.; Bravo, A. *Bacillus thuringiensis* Cry1A toxins are versatile proteins with multiple modes of action: Two distinct pre-pores are involved in toxicity. *Biochem. J.* **2014**, *459*, 383–396. [CrossRef]
61. Jenkins, J.L.; Lee, M.K.; Valaitis, A.P.; Curtiss, A.; Dean, D.H. Bivalent sequential binding model of a *Bacillus thuringiensis* toxin to gypsy moth aminopeptidase N receptor. *J. Biol. Chem.* **2000**, *275*, 14423–14431. [CrossRef]
62. Winn, M.D.; Ballard, C.C.; Cowtan, K.D.; Dodson, E.J.; Emsley, P.; Evans, P.R.; Keegan, R.M.; Krissinel, E.B.; Leslie, A.G.; McCoy, A.; et al. Overview of the CCP4 suite and current developments. *Acta Cryst. D Biol. Cryst.* **2011**, *67*, 235–242. [CrossRef]

63. Emsley, P.; Lohkamp, B.; Scott, W.G.; Cowtan, K. Features and development of Coot. *Acta Cryst. D Biol. Cryst.* **2010**, *66*, 486–501. [CrossRef]
64. Murshudov, G.N.; Skubak, P.; Lebedev, A.A.; Pannu, N.S.; Steiner, R.A.; Nicholls, R.A.; Winn, M.D.; Long, F.; Vagin, A.A. REFMAC5 for the refinement of macromolecular crystal structures. *Acta Cryst. D Biol. Cryst.* **2011**, *67*, 355–367. [CrossRef]
65. Holm, L. Using Dali for Protein Structure Comparison. *Methods Mol. Biol.* **2020**, *2112*, 29–42. [CrossRef]
66. Krissinel, E.; Henrick, K. Inference of macromolecular assemblies from crystalline state. *J. Mol. Biol.* **2007**, *372*, 774–797. [CrossRef]
67. Grahn, E.; Askarieh, G.; Holmner, A.; Tateno, H.; Winter, H.C.; Goldstein, I.J.; Krengel, U. Crystal structure of the Marasmius oreades mushroom lectin in complex with a xenotransplantation epitope. *J. Mol. Biol.* **2007**, *369*, 710–721. [CrossRef]
68. Skamnaki, V.T.; Peumans, W.J.; Kantsadi, A.L.; Cubeta, M.A.; Plas, K.; Pakala, S.; Zographos, S.E.; Smagghe, G.; Nierman, W.C.; Van Damme, E.J.; et al. Structural analysis of the Rhizoctonia solani agglutinin reveals a domain-swapping dimeric assembly. *FEBS J.* **2013**, *280*, 1750–1763. [CrossRef]
69. Fujimoto, Z.; Kuno, A.; Kaneko, S.; Kobayashi, H.; Kusakabe, I.; Mizuno, H. Crystal structures of the sugar complexes of Streptomyces olivaceoviridis E-86 xylanase: Sugar binding structure of the family 13 carbohydrate binding module. *J. Mol. Biol.* **2002**, *316*, 65–78. [CrossRef]
70. Ichinose, H.; Fujimoto, Z.; Honda, M.; Harazono, K.; Nishimoto, Y.; Uzura, A.; Kaneko, S. A beta-l-Arabinopyranosidase from Streptomyces avermitilis is a novel member of glycoside hydrolase family 27. *J. Biol. Chem.* **2009**, *284*, 25097–25106. [CrossRef] [PubMed]
71. Sulzenbacher, G.; Roig-Zamboni, V.; Peumans, W.J.; Rouge, P.; Van Damme, E.J.; Bourne, Y. Crystal structure of the GalNAc/Gal-specific agglutinin from the phytopathogenic ascomycete Sclerotinia sclerotiorum reveals novel adaptation of a beta-trefoil domain. *J. Mol. Biol.* **2010**, *400*, 715–723. [CrossRef] [PubMed]
72. Khan, F.; Kurre, D.; Suguna, K. Crystal structures of a beta-trefoil lectin from Entamoeba histolytica in monomeric and a novel disulfide bond-mediated dimeric forms. *Glycobiology* **2020**, *30*, 474–488. [CrossRef] [PubMed]
73. Pohleven, J.; Renko, M.; Magister, S.; Smith, D.F.; Kunzler, M.; Strukelj, B.; Turk, D.; Kos, J.; Sabotic, J. Bivalent carbohydrate binding is required for biological activity of Clitocybe nebularis lectin (CNL), the N,N′-diacetyllactosediamine (GalNAcbeta1-4GlcNAc, LacdiNAc)-specific lectin from basidiomycete *C. nebularis*. *J. Biol. Chem.* **2012**, *287*, 10602–10612. [CrossRef] [PubMed]
74. Terada, D.; Kawai, F.; Noguchi, H.; Unzai, S.; Hasan, I.; Fujii, Y.; Park, S.Y.; Ozeki, Y.; Tame, J.R. Crystal structure of MytiLec, a galactose-binding lectin from the mussel Mytilus galloprovincialis with cytotoxicity against certain cancer cell types. *Sci. Rep.* **2016**, *6*, 28344. [CrossRef] [PubMed]
75. Uchida, T.; Yamasaki, T.; Eto, S.; Sugawara, H.; Kurisu, G.; Nakagawa, A.; Kusunoki, M.; Hatakeyama, T. Crystal structure of the hemolytic lectin CEL-III isolated from the marine invertebrate Cucumaria echinata: Implications of domain structure for its membrane pore-formation mechanism. *J. Biol. Chem.* **2004**, *279*, 37133–37141. [CrossRef] [PubMed]
76. Claesson, M.; Lindqvist, Y.; Madrid, S.; Sandalova, T.; Fiskesund, R.; Yu, S.; Schneider, G. Crystal structure of bifunctional aldos-2-ulose dehydratase/isomerase from Phanerochaete chrysosporium with the reaction intermediate ascopyrone M. *J. Mol. Biol.* **2012**, *417*, 279–293. [CrossRef] [PubMed]
77. Kern, O.; Leon, P.C.V.; Gittis, A.G.; Bonilla, B.; Cruz, P.; Chagas, A.C.; Ganesan, S.; Ribeiro, J.M.C.; Garboczi, D.N.; Martin-Martin, I.; et al. The structures of two salivary proteins from the West Nile vector *Culex quinquefasciatus* reveal a beta-trefoil fold with putative sugar binding properties. *Curr. Res. Struct. Biol.* **2021**, *3*, 95–105. [CrossRef] [PubMed]

toxins

MDPI

Article

Insights from the Structure of an Active Form of *Bacillus thuringiensis* Cry5B

Jiaxin Li [1,†], Lin Wang [1,†], Masayo Kotaka [1,2], Marianne M. Lee [1] and Michael K. Chan [1,*]

1 School of Life Sciences and Center of Novel Biomaterials, The Chinese University of Hong Kong, Shatin, Hong Kong SAR 999077, China
2 School of Biomedical Sciences, LKS Faculty of Medicine, The University of Hong Kong, Pokfulam, Hong Kong SAR 999077, China
* Correspondence: michaelkchan88@cuhk.edu.hk; Tel.: +852-39431487
† These authors contributed equally to this work.

Abstract: The crystal protein Cry5B, a pore-forming protein produced by the soil bacterium *Bacillus thuringiensis*, has been demonstrated to have excellent anthelmintic activity. While a previous structure of the three-domain core region of Cry5B(112–698) had been reported, this structure lacked a key N-terminal extension critical to function. Here we report the structure of Cry5B(27–698) containing this N-terminal extension. This new structure adopts a distinct quaternary structure compared to the previous Cry5B(112–698) structure, and also exhibits a change in the conformation of residues 112–140 involved in linking the N-terminal extension to the three-domain core by forming a random coil and an extended α-helix. A role for the N-terminal extension is suggested based on a computational model of the tetramer with the conformation of residues 112–140 in its alternate α-helix conformation. Finally, based on the Cry5B(27–698) structure, site-directed mutagenesis studies were performed on Tyr495, which revealed that having an aromatic group or bulky group at this residue 495 is important for Cry5B toxicity.

Keywords: Cry5B; crystal structure; nematicidal activity; oligomerization

Key Contribution: This structure provides insight into the structure of the N-terminal extension of Cry5B that is required for function. Site-directed mutagenesis and *C. elegans* toxicity assays are used to show that Tyr495 is important for the toxicity of Cry5B.

Citation: Li, J.; Wang, L.; Kotaka, M.; Lee, M.M.; Chan, M.K. Insights from the Structure of an Active Form of *Bacillus thuringiensis* Cry5B. *Toxins* **2022**, *14*, 823. https://doi.org/10.3390/toxins14120823

Received: 31 October 2022
Accepted: 20 November 2022
Published: 23 November 2022

1. Introduction

Crystal (Cry) proteins produced during sporulation by the spore-forming soil bacterium *Bacillus thuringiensis* (*Bt*) are best known for their ability to serve as natural insecticides. The specificity of Cry proteins to insects, and the absence of their toxicity towards mammals and other vertebrates have contributed to the incorporation of their genes into various transgenic crops for protection against insect infestations [1–3]. In recent years, a select group of Cry proteins has been identified and shown to be toxic to nematodes. One of the most extensively characterized of these is Cry5B, which has been demonstrated to be highly active in vitro and in vivo against multiple nematodes including free-living and parasitic species [4]. Plant-parasitic nematodes, such as root-knot nematodes, cyst nematodes and lesion nematodes, annually cause a hundred billion US dollars in damage to crops worldwide, and thus there is interest in combining the use of Cry5B together with other insecticidal Cry proteins in order to protect the agricultural crop production to provide broader protection against crop pests.

Parasitic nematode infections also occur in humans and animals, and thus there has been interest in exploring their use as a therapy to treat intestinal nematode infections with some promising results having been obtained in early in vivo studies [5–8]. Oral administration of purified Cry5B crystals to *Ancylostoma ceylanicum*-infected hamsters elicited a comparable effect to the standard anthelminthic drug mebendazole [5], while *Bt* lysate with cytosolic Cry5B treatment of pigs infected with *Ascaris suum* showed 97% elimination of L4 larvae [6]. These findings support the use of Cry5B as a potential drug for treating livestock and human nematode infections.

The nematicidal activity of Cry5B has been proposed to follow a sequence of events similar to those of the insecticidal Cry toxins binding to a host receptor, followed by membrane insertion that causes intestinal damage and ultimately the death of the host [9]. Previous investigations on the toxicity of Cry5B against *Caenorhabditis elegans* have shown that Cry5B binds to a glycolipid comprised of an invertebrate-specific tetrasaccharide core *N*-acetylgalactosamine (GalNAc) β1–4 *N*-acetylglucosamine (GlcNAc) β1–3 mannose (Man) β1–4 glucose (Glc) attached to the lipids on the surface of *C. elegans'* intestinal epithelial cells [10,11]. How the binding affects the Cry5B conformation and/or triggers the oligomerization is currently unclear.

A structure of an elastase-treated Cry5B encompassing residues 112–698 has been reported [11]. The structure revealed that Cry5B(112–698) possesses a three-domain core similar to that found in the structures of other Cry proteins (e.g., Cry1Ac, Cry3Aa, Cry4Ba). In these structures, domain I consists of a helix bundle, domain II adopts a β-prism motif, and domain III has a β-sandwich motif. Notably, the fold of domain II in Cry5B exhibits similarity to banana lectin (BanLec), a mannose-specific jacalin-related lectin whose two separate glycan-binding sites are located in the loop regions on the top of the β-prism [12]. The glycan-binding sites in BanLec are made up of loops with a GG or GXXXD motif, while in Cry5B domain II they are conserved as GG and GXXXE, respectively.

Given that proper proteolytic activation is required to convert Cry protoxin to its toxic counterpart, and the fact that previous studies found that while C-terminal truncation of Cry5B at residue 698 does not affect its toxicity to *C. elegans*, truncation of the N-terminus before nucleotide 63 (residue 21) weakens the toxicity dramatically [13], a key question with respect to the Cry5B(112–698) protein was whether this elastase-treated Cry5B construct retained its toxicity to *C. elegans*. We subsequently confirmed that Cry5B(1–698) was active as reported [13] and exhibited similar nematicidal activity as Cry5B(1–772), while Cry5B(112–698) was inactive (Figure S1). The significance of these findings was the implication that the N-terminus of Cry5B contains a feature critical to function. We thus sought to determine the structure of Cry5B containing its N-terminal region and herein report such a structure of Cry5B(27–698). In addition to the N-terminal extension, the newly determined crystal structure reveals a network of inter-residue hydrogen bonding mediated by Tyr495; we thus evaluated the functional significance of Tyr495 by site-directed mutagenesis and showed that mutation of Tyr495 can significantly impact the toxicity of Cry5B against *C. elegans*.

2. Results

2.1. Structure of Cry5B(27–698) with Its Functionally Required N-Terminal Extension

Recombinant Cry5B(1–772) with an N-terminal His-tag was overexpressed in *E. coli* and the soluble protein was purified by Ni-affinity and gel-filtration chromatography (Figure S2). Screening of crystallization conditions led to the identification of protein crystals that diffracted to 4.5 Å resolution. The structure was determined by molecular replacement using the Cry5B(112–698) (PDB ID: 4D8M) structure as the initial model and found to contain two molecules in the asymmetric unit (Table 1).

The dominant feature of the new Cry5B(27–698) structure is the three-domain core (residues 141–698) observed in the previous elastase-treated Cry5B(112–698) structure. A new region does not present in previously determined Cry5B structures; however, it is an N-terminal extension attached to the three-domain core consisting of five short helices buttressed against a pocket formed by domains I and II. While an initial trace could be generated, the assignment of residues based on the electron density map was hindered by the low resolution. Thus to assist in the residue assignments of this N-terminal extension, the amino acid sequence of residues 1–134 was submitted to Robetta [14] to generate a predicted model. Notably, the resulting Robetta model fitted the polyalanine model built based on the electron density quite well for at least the first four helices. Using this Robetta model to guide the side chain assignments, a final model was obtained spanning residues 27–698 (Figure 1A). N-terminal additional helices were assigned as α1(29–39), α2(50–61), α3(71–78), α4(91–97) and α5(115–122). Residues 1–24 were predicted by Robetta to be flexible loops with two very short helices, and thus we presume residues 1–26, which could not be modeled, are disordered. No density for amino acids beyond residue 698 was observed, suggesting either their disorder or cleavage during crystallization. The presence of aromatic amino acids (Phe, Trp and Tyr) with strong electron density, and the location of Pro and Gly residues at the turns were also used to guide the sequence alignment and side chains were fitted to the electron density map to the best extent possible. The positions of residues 141–698 were based on the existing Cry5B(112–698) structure used for molecular replacement.

Table 1. Data Collection and Refinement Statistic.

	Cry5B(27–698)
Data collection	
Wavelength (Å)	0.99984
Space group	P 4 21 2
Cell dimensions: *a*, *b*, *c* (Å)	114.4 114.4 263.4
α, β, γ (°)	90.0 90.0 90.0
Resolution (Å)	20 (4.5)
R_{merge}	0.274 (0.559)
$I/\sigma I$	2.2 (1.3)
Completeness (%)	99.7 (100)
Redundancy	5.3 (5.4)
Refinement	
No. reflections	10,836
R_{work}/R_{free}	23.76/28.85
No. atoms: Protein	10,254
Ligand/ion	0
Water	0
B-factors(Å^2): Protein	107.08
R.m.s. deviations: Bond lengths (Å)	0.002
Bond angles (°)	0.54
Range of residues	27–83, 88–107, 113–165, 173–214, 224–698

Outer shell statistics are in parentheses.

In addition to the presence of the N-terminal extension, another significant difference between our Cry5B(27–698) and the elastase-treated Cry5B(112–698) structures is in the orientation and secondary structure conformations of residues 112–140. In the previous Cry5B(112–698) structure, these residues form part of an extended α-helix (residues 112–162) that helps to tighten the homotrimer (Figure 1D,E). However, in our new structure, residues 113–127 form an α-helix that lies adjacent to the short helices of the N-terminal extension, while residues 128–140 unwind to adopt a random coil (Figure 1A). Presumably this unwinding is needed to accommodate the N-terminal extension and remain connected to the three-domain core.

Arguably the biggest difference in the two Cry5B structures, however, is the oligomerization state of the protein. While the elastase-treated Cry5B(112–698) structure forms a trimer with no apparent channel (Figure 1E,F), which is similar to that observed for Cry4Ba [15–17], our Cry5B(27–698) structure appears to form a weakly associated tetramer with a wide opening that narrows into a 10 Å pore (Figure 1B,C). Considering the interaction of Cry5B with cadherin sequences has been shown to lead to a tetrameric species [18], we also verified that Cry5B(1–772) can oligomerize when co-expressing with cadherin repeats 7 and 8 (CD7/8) from nematode-specific cadherin CDH-8 (Figure 2). It is possible that the cadherin peptides can strengthen the interaction between Cry5B subunits and stabilize a tetramer similar to that observed in the structure.

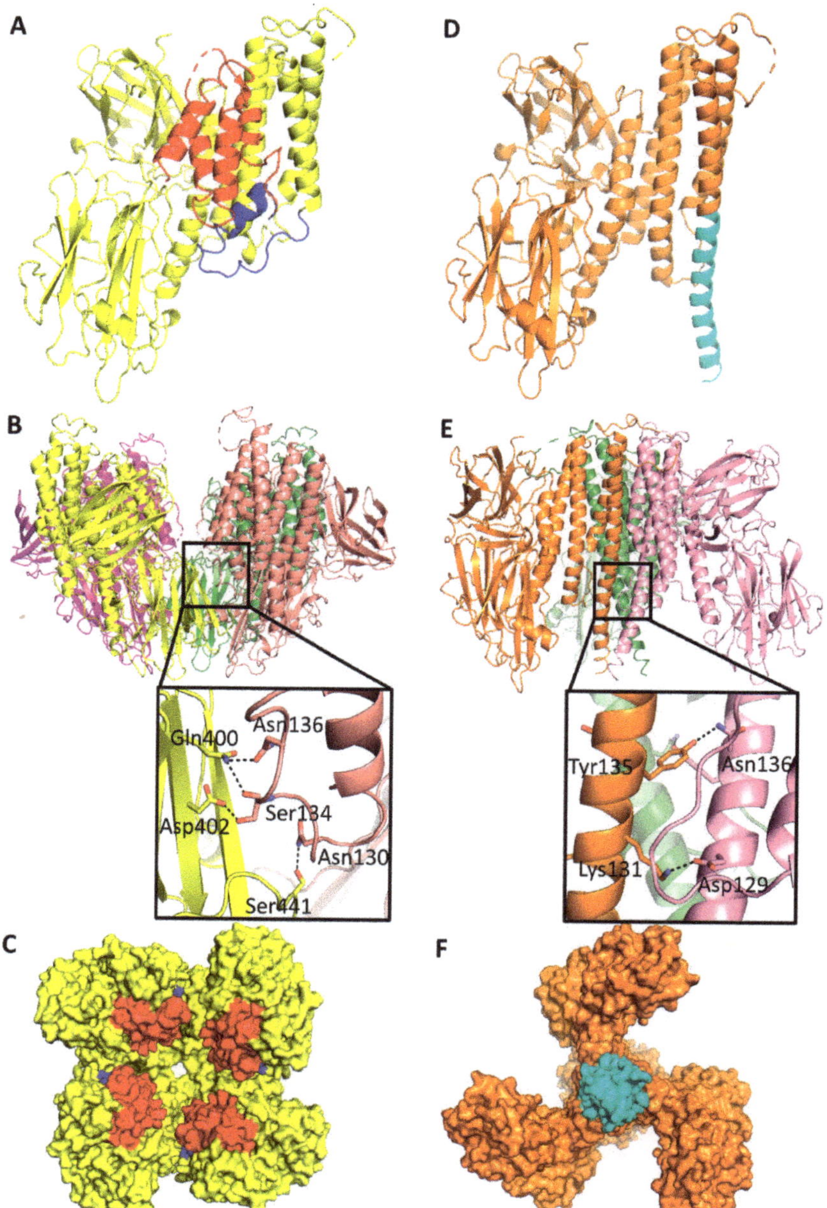

Figure 1. Structures of Cry5B(27–698) and Cry5B(112–698). Structure of (**A**) single subunit and tetramer in (**B**) side view and (**C**) top view of Cry5B(27–698). N-terminal functional extension (residues 27–107) in red, residues 112–140 in blue and the three-domain core (residues 141–698) colored in yellow. Structure of (**D**) single subunit and trimer in (**E**) side view and (**F**) bottom view of Cry5B(112–698) (PDB ID: 4D8M). Residues 112–140 in cyan and the three-domain core (residues 141–698) colored in orange. The black boxes in (**B**,**E**) showing the interaction between neighboring subunits. Hydrogen bond shown as black dashed lines.

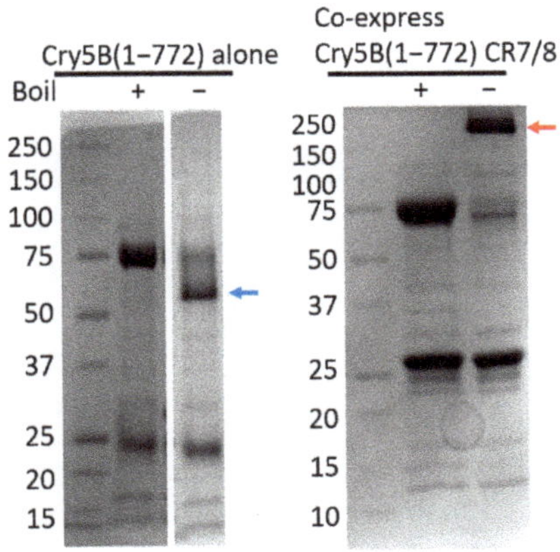

Figure 2. CDH-8 CR7/8 can trigger the in vitro oligomerization of Cry5B(1–772). The blue and red arrows indicate the Cry5B(1–772) monomer and oligomer in native condition, respectively.

2.2. *Cry5B (27–698) Is an Active Toxin*

As mentioned above, the previous structure of Cry5B(112–698) was of an inactive form of the toxin. Based on the structure, the shortest N-terminal truncation of the protein forming the crystal used for the Cry5B(27–698) would be at residue 26, though additional residues could be present but not observed due to disorder. To determine whether Cry5B(27–698) represents an active form, *E. coli* expressing Cry5B with different truncations, Cry5B(1–698), Cry5B(12–698), Cry5B(21–698), Cry5B(27–698) and Cry5B(112–698), were fed *C. elegans* to see which were toxic (Figures 3 and S3A). These results confirmed that Cry5B(112–698) is non-toxic, while N-terminal truncations before residue 21 showed no effect. Importantly, while the Cry5B(27–698) construct showed a statistically significant reduction in the toxicity compared to constructs with the shorter N-terminal truncation, it still remained highly active and showed much higher toxicity than Cry5B(112–698). These data indicate that the Cry5B(27–698) structure represents an active form of the toxin.

2.3. *Identification of Tyr495 as a Residue with a Key Role in the Nematicidal Activity of Cry5B*

In the course of analyzing the putative tetramer of our Cry5B structure, we noticed that Tyr495 formed a hydrogen bond to a neighboring Leu carbonyl, which together with the 4-fold symmetry, formed a ring of hydrogen bonds at the apex of the tetramer (Figure 4). We thus decided to evaluate whether this residue and the relevant inter-residue interactions were important. Tyr495 was therefore substituted with Ala and the toxicity of the corresponding protein was investigated (Figures 5 and S3B). It was found that a mutation of Tyr495 to Ala in the Cry5B(1–772) construct resulted in the loss of nematicidal activity, demonstrating the importance of this residue. Previous studies of Cry1Ab identified a Phe residue involved in membrane insertion, which lost activity when mutated to Ala but retained activity when mutated to other aromatic residues [19,20]. We thus prepared Phe and Trp mutants of Tyr495 and found that these mutations resulted in retention in activity as well (Figure 5). Replacing Tyr495 with an Arg appeared to retain some activity, though there was a statistically significant decrease compared to the wild type (Figure 5). Collectively, these data suggested that the hydrogen bond in this region was not critical for activity, and instead that having an amino acid with an aromatic or possibly bulky group at residue 495 is important for function.

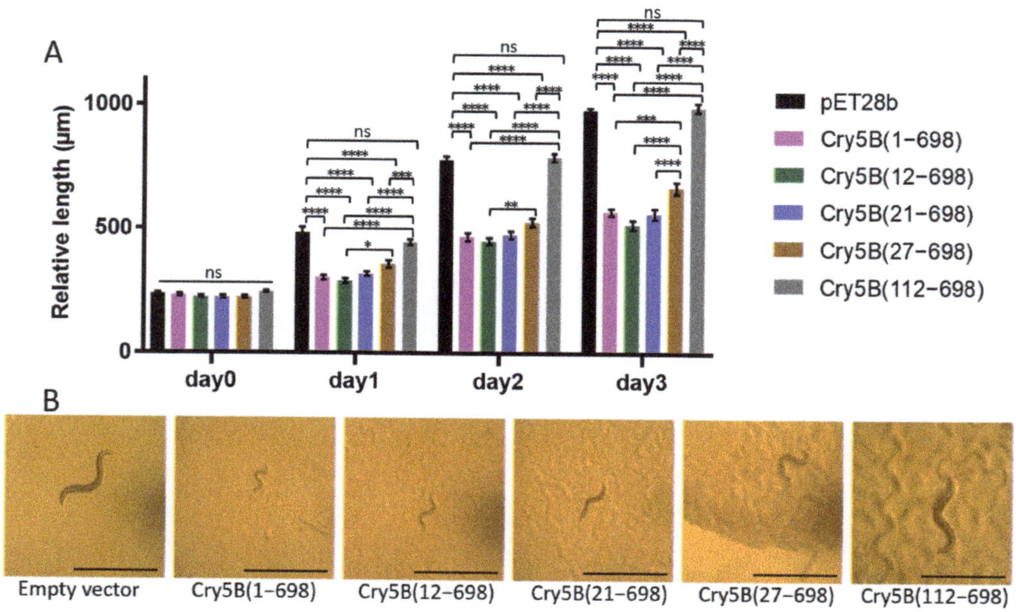

Figure 3. Toxicity test of Cry5B N-terminal truncations against *C. elegans*. (**A**) Relative length of worms indicated the toxicity of Cry5B constructs. (**B**) Each panel shows a typical *C. elegans* grown on *E. coli* BL21 expressing different constructs after three days. Scale bar represents 1 mm. Worms grown on empty vector and Cry5B(112–698) are obviously large and healthy, while those grown on Cry5B(1–698), Cry5B(12–698), Cry5B(21–698) and Cry5B(27–698) are much smaller, indicating intoxication. Two-way ANOVA, N = 15–25, **** $p < 0.0001$, *** $p < 0.001$, ** $p < 0.01$, * $p < 0.05$; ns, not significant.

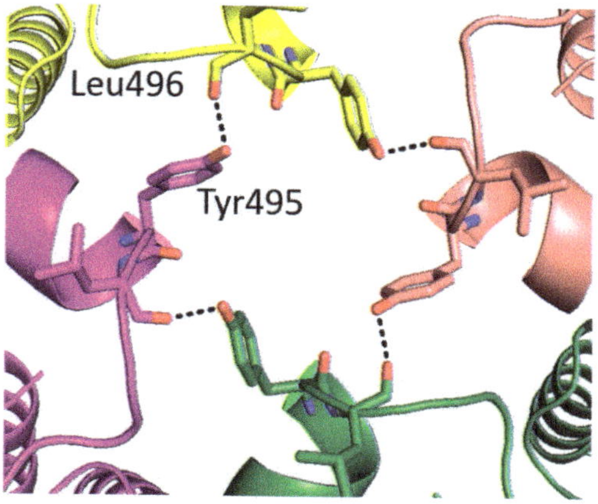

Figure 4. A ring of hydrogen bonds at the apex of the tetramer Cry5B(27–698). Four subunits colored as in Figure 1B. Hydrogen bonds shown as black dashed lines.

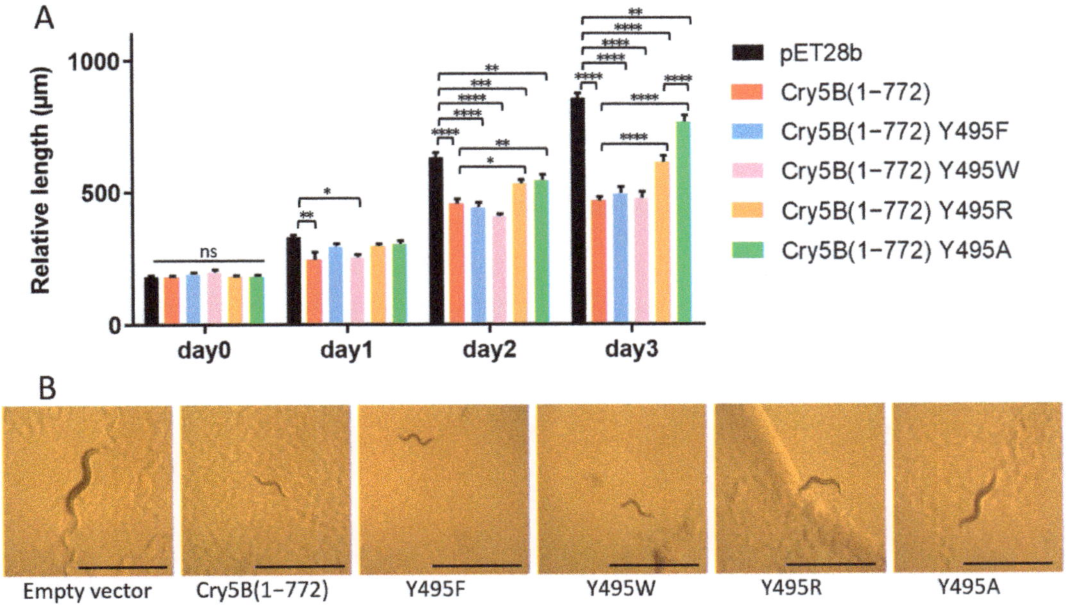

Figure 5. Toxicity test of Cry5B(1–772) Y495 mutants against *C. elegans*. (**A**) Relative length of worms indicated the toxicity of Cry5B constructs. (**B**) Each panel shows a typical *C. elegans* grown on *E. coli* BL21 expressing different constructs after three days. Scale bar represents 1 mm. Worms grown on empty vector and Cry5B(1–772) Y495A are obviously larger, while those grown on wild-type Cry5B(1–772), Cry5B(1–772) Y495F, Cry5B(1–772) Y495W and Cry5B(1–772) Y495R are much smaller. Two-way ANOVA, N = 10, **** $p < 0.0001$, *** $p < 0.001$, ** $p < 0.01$, * $p < 0.05$; ns, not significant.

3. Discussion

3.1. Computational Model for the N-Terminal Extension in Cry5B Tetramer

As mentioned previously, one surprising difference in our Cry5B(27–698) structure and the previous elastase-treated Cry5B(112–698) structure is the location and conformation of residues 112–140 (Figure 1A,D). In the Cry5B(112–698) structure, this region forms an extended helix (residues 112–162) belonging to the five-helix bundle. In our Cry5B(27–698) structure, however, it appears to be partially unwound, connecting the N-terminal extension (residues 27–107), which was not present in Cry5B(112–698), to domain I of the three-domain core.

Given that forming the extended α-helix(112–140) observed in the Cry5B(112–698) should be enthalpically favorable, we decided to produce a computational model of the tetrameric structure of Cry5B(27–698) with an extended helix (residues 112–162) to see where the N-terminal extension (residue 27–107) would be located. The model shown in Figure 6 reveals that in the tetramer, the rearranged N-terminal region would be positioned near (~7 Å) the proposed glycan-binding motif in domain II [11] of a neighboring subunit. Given that the N-terminal extension is important for eliciting the nematicidal activity of Cry5B, we wondered whether the possible rearrangement could provide a mechanism for the N-terminal extension to impact glycan binding, and decided to test this hypothesis by further mutagenesis and functional analyses.

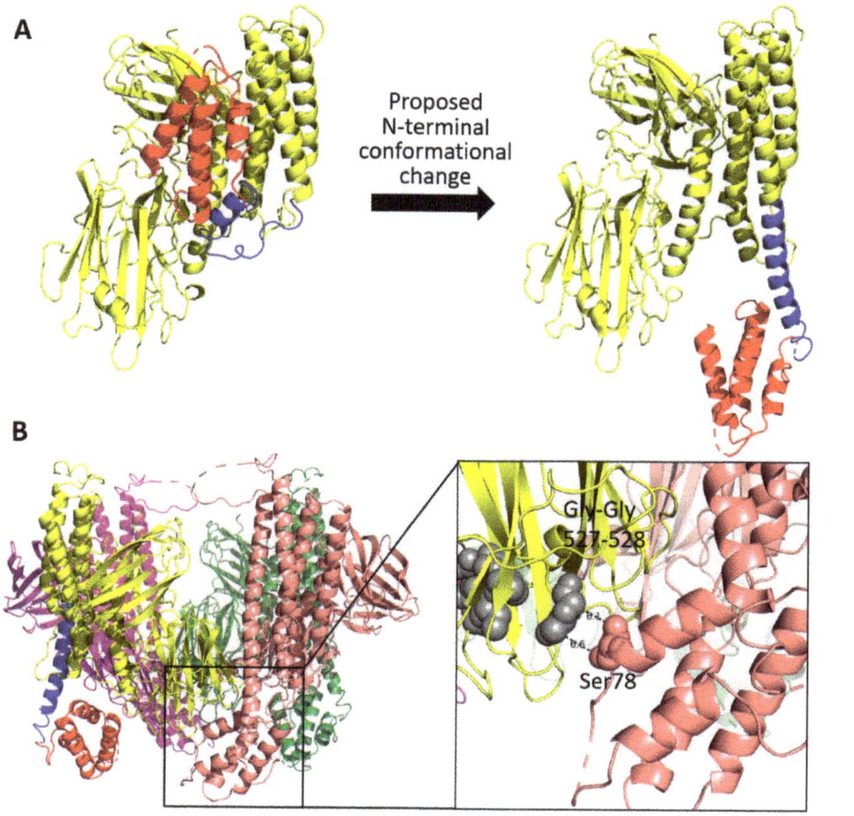

Figure 6. Computational model of the Cry5B(27–698) with extended helix. (**A**) Proposed reorientation of N-terminal region. (**B**) Modeled Cry5B(27–698) tetramer structure when N-terminal region rearranged to form an extended helix (residues 112–162). Colored as in Figure 1A. Black boxes showing the close distance (~7 Å) between the proposed glycan-binding motif (GG and GPIEE shown as grey spheres) and the possible position of N-terminal region (red).

3.2. Cry5B N-Terminal Region (Residues 27–111) Contributes to Binding with Galactose

Considering the galactose-dependent binding mechanism between Cry5B and glycolipid in *C. elegans*, and the fact that galactose can inhibit their binding [10], the binding affinity between the Cry5B constructs (residues 27–698 and 112–698) with galactose was determined by microscale thermophoresis (MST). The data show that the dissociation constant (Kd) between Cry5B(27–698) and galactose was 1.64 ± 1.04 μM, while that for Cry5B(112–698) was 21.5 ± 10.3 μM, suggesting a 10-fold weaker binding (Figure S4). These data support the notion that the N-terminal region (residues 27–111) of Cry5B is likely involved in its binding with the glycolipid receptor with one possibility being the rearrangement of the N-terminal extension predicted by the computational model. Further insights could be obtained from the complex structure of Cry5B(27–698) and galactose.

3.3. Comparison of the Cry5B(27–698) with Other Three-Domain Cry Proteins

The toxicity studies of different N-terminal truncations of Cry5B (Figure 3) showed that N-terminal truncations before residue 21 had no impact on its toxicity, and that Cry5B(27–698) was still active though less toxic, while Cry5B(112–698), which lacks the N-terminal extension, showed complete loss of activity, thus highlighting the importance of the N-terminal extension on the function of Cry5B. Notably, similar N-terminal helical extensions are present in other Cry proteins. For Cry1A proteins, a widely used toxin against lepidopteran insects, the presence of these N-terminal regions was also shown to be functionally important. For Cry1Ai, a construct consisting of residues (36–605) (corresponding to residues 36–606 in Cry1Ac) was found to be its minimal active fragment. Loss of one N- or C-terminal residue of this construct abolished its toxicity against *P. xylostella* larvae [21]. In another Cry1A protein, Cry1Ah(50–639) was toxic, but losing residues 50–108 abrogated its toxicity against *P. xylostella* larvae [22]. Significantly, sequence alignment of these Cry1A proteins to the structurally determined Cry1Ac proteins (Figure S5) shows the presence of an N-terminal extension whose truncation suggests a similar requirement for activity. Our findings with the Cry5B(27–698) structure corroborate these previous results and reaffirm the vital role the N-terminal regions that extend from the Domain I of three-domain Cry proteins play in the activity of Cry proteins against both insects and nematodes. Notably, many other structures of Cry proteins have been shown to contain a similar N-terminal extension (Figure S6), which could be similarly important for their function.

4. Conclusions

In summary, we determined the structure of an active form of Cry5B that contains an N-terminal extension encompassing residues 27–107 that is absent in the previously determined Cry5B structures. The Cry5B(27–698) structure allowed us to elucidate the structural features of this N-terminal extension, and suggests a possible role in eliciting its toxicity toward nematodes. This structure led to the identification of Tyr495 as a residue critical for Cry5B nematicidal function. These studies could help to improve the use of Cry5B as a treatment for gastrointestinal nematode infections and for preventing nematode damage on crops.

5. Materials and Methods

5.1. Cloning of Cry5B Fragments and Mutants

Cry5B(1–772) gene was amplified from *Bt* strain YBT-1518 genomic DNA by PCR using KAPA HiFi polymerase using primers Cry5B 1-EcoRI-F and Cry5B 772-XhoI-R in Table S1. The PCR products were then ligated into the EcoRI and XhoI sites of the double digested pET28b vector using Gibson Assembly. Other fragments of Cry5B such as Cry5B(1–698), Cry5B(12–698), Cry5B(21–698), Cry5B(27–698) and Cry5B(112–698) were cloned using similar strategy with corresponding primers listed in Table S1.

Using the wild-type pET28b-Cry5B(1–772) plasmid as a template, different mutations at Tyr495 were introduced using site-directed mutagenesis. Primers for each construct are listed in Table S1.

5.2. Expression and Purification of Cry5B(1–772)

A pET28b plasmid containing the *cry5B(1–772)* gene was transformed into *E. coli* BL21(DE3) and the recombinant protein was overexpressed in Luria–Bertani (LB) media supplemented with 50 μg/mL kanamycin and induced with 0.1 mM isopropyl β-D-1-thiogalactopyranoside (IPTG) at 20 °C and grown overnight. The harvested cells were lysed by sonication in buffer A (20 mM Tris pH 8.0, 500 mM NaCl, 0.1% phenylmethylsulfonyl fluoride (PMSF) and 0.1% benzydamine) and loaded on to a 5 mL Ni-NTA column (GE Healthcare), which was then washed with buffer containing a 0 to 500 mM imidazole gradient. The recombinant Cry5B(1–772) protein was eluted at ~150 mM imidazole and further purified by size-exclusion chromatography (GE Healthcare) with buffer B (20 mM HEPES pH 8.0, 50 mM NaCl). The fractions were analyzed by SDS-PAGE and checked by mass spectrometry-based protein identification.

5.3. Crystallization and Data Collection

Cry5B(1–772) was crystallized using the sitting-drop vapor-diffusion method from plates initially incubated at 18 °C, and then transferred to 32 °C after 2 weeks. Several crystals were obtained from similar conditions of the Cubic Screen (Hampton) containing 2500 mM NaCl as precipitant. The crystal used for data collection was obtained from condition with 2500 mM NaCl, 100 mM sodium cacodylate pH 6.5 and 200 mM Li_2SO_4. The diffraction dataset was collected on beamline 13B at the NSRRC in Hsinchu, Taiwan. Data processing and scaling were performed with iMosflm [23] and SCALA [24] from the CCP4i suite [25].

5.4. Phase Determination and Refinement

The initial structure was generated using PHASER [26] using the structure of Cry5B(112–698) (PDB ID: 4D8M) as searching model. The extra N-terminal region beyond this model was manually added and adjusted to fit the electron density using the program Coot [27]. To assist in the side chain identification, the N-terminal sequence (residues 1–134) was submitted to Robetta [12], which was used to generate a predicted model. In addition to guiding the side chain assignments, some density at the ends of helices that we had initially fitted as random coils were recognized as being helical when Robetta model was used for comparison. Iterative cycles of model building and refinement with the programs Coot and Phenix [28], respectively, were carried out to improve the model. The quality of the final model was evaluated using the program MolProbity [29] and summarized in Table 1. The coordinates and structure factors have been deposited in the Protein Data Bank (PDB ID: 8HHE).

5.5. Cry5B Toxicity Study on C. elegans

5.5.1. General Preparation

The *C. elegans* strain Bristol N2 and *E. coli* OP50 used in this study were a gift from Professor King-Lau Chow of HKUST. Nematode growth media (NGM) was prepared according to published protocol [30], and the *C. elegans* worms were grown on NGM plates inoculated with 100 μL *E. coli* OP50 overnight culture. After 3–5 days' growth and reproduction at 20 °C, the worms were synchronized based on the protocol [30], and the larvae harvested at the L1–L2 stage were used in toxicity test.

5.5.2. Toxicity Test

E. coli BL21 transformed with pET28b vectors harboring either Cry5B(1–772) or one of its Y495 mutants or Cry5B(1–698) or one of its N-terminal truncations were induced with 0.1 mM IPTG and grown at 20 °C in 5 mL LB culture. The overnight culture supplemented with an additional 5 mM IPTG was used to inoculate NGM plates. The plates seeded with *E. coli* were incubated at 20 °C overnight before use. For the control group, empty pET28b vector was used.

Five synchronized L1–L2 larvae were transferred to each NGM plate. The appearance of worms was monitored for three days. As the toxin inhibits the growth of worms, the relative length of *C. elegans* reflects the toxicity of the Cry5B constructs. Photographs of every worm were taken every 24 h. The length was determined using the software ImageJ [31], and data were analyzed using software GraphPad Prism 6.

5.5.3. Microscale Thermophoresis (MST)

Purified His-Cry5B(27–698) and His-Cry5B(112–698) as protein targets were labelled using Alexa Fluor 647 NHS ester dye according to the instruction. Unreacted dye was removed with a desalting column (30K MWCO, Bio-Rad). The labelled targets were adjusted to appropriate concentrations for detection, and the galactose as ligand was freshly solubilized in the same buffer—aforementioned buffer B (20 mM HEPES pH 8.0, 50 mM NaCl).

For each assay, 16 different serially-diluted concentrations of ligand were firstly prepared, and then mixed with equal volume of labelled protein target at room temperature. The reaction mixtures were loaded into standard Monolith NT.115 capillaries and measured using a Monolith NT.115 instrument (NanoTemper Technologies). Instrument parameters were adjusted to 40% MST power and 2% or 6% excitation power (2% for Cry5B(27–698) and 6% for Cry5B(112–698). The Kd values were calculated using MO.Affinity Analysis v.2.2.4

software (NanoTemper Technologies) as mean ± SEM from at least three independent experiments with a single site-specific binding model.

Supplementary Materials: The following supporting information can be downloaded at: https://www.mdpi.com/article/10.3390/toxins14120823/s1, Figure S1: Toxicity test of Cry5B(1–698), Cry5B(1–772) and Cry5B(112–698) against *C. elegans*. Figure S2: Purified Cry5B(1–772). Figure S3: Western blot detects His-tag indicating the expression level of His-Cry5B constructs. Figure S4: Binding curves from MST assays for the binding of Cry5B constructs and galactose. Figure S5: Sequence alignment of Cry1A proteins. Figure S6: Structures of three-domain Cry proteins possessed with the extra amino acid beyond five-helix bundle. Table S1: Primers used for making Cry5B constructs.

Author Contributions: Conceptualization, M.K.C. and M.M.L.; methodology, M.K.C., J.L., L.W. and M.K.; validation, M.K.C. and J.L.; formal analysis, M.K.C., L.W. and J.L.; investigation, M.K.C., J.L. and L.W.; resources, M.K.C.; writing—original draft preparation, M.K.C. and J.L.; writing—review and editing, M.K.C., J.L. and M.M.L.; supervision, M.K.C. and M.M.L.; project administration, M.K.C. and M.M.L.; funding acquisition, M.K.C. and M.M.L. All authors have read and agreed to the published version of the manuscript.

Funding: The work described in this paper was funded by the University Grants Committee (UGC) of Hong Kong, GRF 14176017, the CUHK Center of Novel Biomaterials, and a CUHK Direct Grant.

Institutional Review Board Statement: Not applicable.

Informed Consent Statement: Not applicable.

Data Availability Statement: The atomic coordinates and structure factors of Cry5B(27–698) have been deposited in the protein data bank under the accession codes 8HHE.

Acknowledgments: We thank the experimental facility and the technical services provided by the Synchrotron Radiation Protein Crystallography Facility of the National Core Facility Program for Biotechnology, Ministry of Science and Technology, and the National Synchrotron Radiation Research Center, a national user facility supported by the Ministry of Science and Technology, Taiwan, ROC. We thank the Taiwan Protein Project (AS-KPQ-109-TPP2) for their assistance on this project.

Conflicts of Interest: The authors declare no conflict of interest.

References

1. Tabashnik, B.E. Communal benefits of transgenic corn. *Science* **2010**, *330*, 189–190. [CrossRef] [PubMed]
2. Rahman, M.; Zaman, M.; Shaheen, T.; Irem, S.; Zafar, Y. Safe use of Cry genes in genetically modified crops. *Environ. Chem. Lett.* **2015**, *13*, 239–249. [CrossRef]
3. Briefs, I. Global status of commercialized biotech/GM crops in 2017: Biotech crop adoption surges as economic benefits accumulate in 22 years. *ISAAA Brief* **2017**, *53*, 25–26.
4. Wei, J.-Z.; Hale, K.; Carta, L.; Platzer, E.; Wong, C.; Fang, S.-C.; Aroian, R.V. *Bacillus thuringiensis* crystal proteins that target nematodes. *Proc. Natl. Acad. Sci. USA* **2003**, *100*, 2760–2765. [CrossRef] [PubMed]
5. Cappello, M.; Bungiro, R.D.; Harrison, L.M.; Bischof, L.J.; Griffitts, J.S.; Barrows, B.D.; Aroian, R.V. A purified *Bacillus thuringiensis* crystal protein with therapeutic activity against the hookworm parasite *Ancylostoma ceylanicum*. *Proc. Natl. Acad. Sci. USA* **2006**, *103*, 15154–15159. [CrossRef]
6. Urban Jr, J.F.; Hu, Y.; Miller, M.M.; Scheib, U.; Yiu, Y.Y.; Aroian, R.V. *Bacillus thuringiensis*-derived Cry5B has potent anthelmintic activity against *Ascaris suum*. *PLoS Negl. Trop. Dis.* **2013**, *7*, e2263. [CrossRef]
7. Li, H.; Abraham, A.; Gazzola, D.; Hu, Y.; Beamer, G.; Flanagan, K.; Soto, E.; Rus, F.; Mirza, Z.; Draper, A. Recombinant paraprobiotics as a new paradigm for treating gastrointestinal nematode parasites of humans. *Antimicrob. Agents Chemother.* **2021**, *65*, e01469-20. [CrossRef]
8. Urban, J.F., Jr.; Nielsen, M.K.; Gazzola, D.; Xie, Y.; Beshah, E.; Hu, Y.; Li, H.; Rus, F.; Flanagan, K.; Draper, A. An inactivated bacterium (paraprobiotic) expressing *Bacillus thuringiensis* Cry5B as a therapeutic for *Ascaris* and *Parascaris* spp. infections in large animals. *One Health* **2021**, *12*, 100241. [CrossRef]
9. Bravo, A.; Gill, S.S.; Soberon, M. Mode of action of *Bacillus thuringiensis* Cry and Cyt toxins and their potential for insect control. *Toxicon* **2007**, *49*, 423–435. [CrossRef]
10. Griffitts, J.S.; Haslam, S.M.; Yang, T.; Garczynski, S.F.; Mulloy, B.; Morris, H.; Cremer, P.S.; Dell, A.; Adang, M.J.; Aroian, R.V. Glycolipids as receptors for *Bacillus thuringiensis* crystal toxin. *Science* **2005**, *307*, 922–925. [CrossRef]
11. Hui, F.; Scheib, U.; Hu, Y.; Sommer, R.J.; Aroian, R.V.; Ghosh, P. Structure and glycolipid binding properties of the nematicidal protein Cry5B. *Biochemistry* **2012**, *51*, 9911–9921. [CrossRef] [PubMed]

12. Meagher, J.L.; Winter, H.C.; Ezell, P.; Goldstein, I.J.; Stuckey, J.A. Crystal structure of banana lectin reveals a novel second sugar binding site. *Glycobiology* **2005**, *15*, 1033–1042. [CrossRef] [PubMed]
13. Li, X.-Q.; Tan, A.; Voegtline, M.; Bekele, S.; Chen, C.-S.; Aroian, R.V. Expression of Cry5B protein from *Bacillus thuringiensis* in plant roots confers resistance to root-knot nematode. *Biol. Control.* **2008**, *47*, 97–102. [CrossRef]
14. Baek, M.; DiMaio, F.; Anishchenko, I.; Dauparas, J.; Ovchinnikov, S.; Lee, G.R.; Wang, J.; Cong, Q.; Kinch, L.N.; Schaeffer, R.D. Accurate prediction of protein structures and interactions using a three-track neural network. *Science* **2021**, *373*, 871–876. [CrossRef] [PubMed]
15. Buzdin, A.; Revina, L.; Kostina, L.; Zalunin, I.; Chestukhina, G. Interaction of 65-and 62-kD proteins from the apical membranes of the *Aedes aegypti* larvae midgut epithelium with Cry4B and Cry11A endotoxins of *Bacillus thuringiensis*. *Biochemistry (Moscow)* **2002**, *67*, 540–546. [CrossRef] [PubMed]
16. Boonserm, P.; Davis, P.; Ellar, D.J.; Li, J. Crystal structure of the mosquito-larvicidal toxin Cry4Ba and its biological implications. *J. Mol. Biol.* **2005**, *348*, 363–382. [CrossRef] [PubMed]
17. Thamwiriyasati, N.; Sakdee, S.; Chuankhayan, P.; Katzenmeier, G.; Chen, C.-J.; Angsuthanasombat, C. Crystallization and preliminary X-ray crystallographic analysis of a full-length active form of the Cry4Ba toxin from *Bacillus thuringiensis*. *Acta Crystallogr. F* **2010**, *66*, 721–724. [CrossRef] [PubMed]
18. Peng, D.; Wan, D.; Cheng, C.; Ye, X.; Sun, M. Nematode-specific cadherin CDH-8 acts as a receptor for Cry5B toxin in *Caenorhabditis elegans*. *Appl. Microbiol. Biotechnol.* **2018**, *102*, 3663–3673. [CrossRef] [PubMed]
19. Rajamohan, F.; Cotrill, J.A.; Gould, F.; Dean, D.H. Role of domain II, loop 2 residues of *Bacillus thuringiensis* CryIAb δ-endotoxin in reversible and irreversible binding to *Manduca sexta* and *Heliothis virescens*. *J. Biol. Chem.* **1996**, *271*, 2390–2396. [CrossRef] [PubMed]
20. Nair, M.S.; Liu, X.S.; Dean, D.H. Membrane insertion of the *Bacillus thuringiensis* Cry1Ab toxin: Single mutation in domain II block partitioning of the toxin into the brush border membrane. *Biochemistry* **2008**, *47*, 5814–5822. [CrossRef] [PubMed]
21. Zhou, Z.-S.; Lin, H.-Y.; Ying, L.; Shu, C.-L.; Song, F.-P.; Zhang, J. The minimal active fragment of the Cry1Ai toxin is located between 36I and 605I. *J. Integr. Agric.* **2014**, *13*, 1036–1042. [CrossRef]
22. Xue, J.; Zhou, Z.; Song, F.; Shu, C.; Huang, D.; Zhang, J. Identification of the minimal active fragment of the Cry1Ah toxin. *Biotechnol. Lett.* **2011**, *33*, 531–537. [CrossRef] [PubMed]
23. Battye, T.G.G.; Kontogiannis, L.; Johnson, O.; Powell, H.R.; Leslie, A.G. *iMOSFLM*: A new graphical interface for diffraction-image processing with MOSFLM. *Acta Crystallogr. D* **2011**, *67*, 271–281. [CrossRef]
24. Evans, P. Scaling and assessment of data quality. *Acta Crystallogr. D* **2006**, *62*, 72–82. [CrossRef] [PubMed]
25. Project, C.C. The CCP4 suite: Programs for protein crystallography. *Acta Crystallogr. D* **1994**, *50*, 760–763.
26. McCoy, A.J.; Grosse-Kunstleve, R.W.; Adams, P.D.; Winn, M.D.; Storoni, L.C.; Read, R.J. Phaser crystallographic software. *J. Appl. Crystallogr.* **2007**, *40*, 658–674. [CrossRef] [PubMed]
27. Emsley, P.; Cowtan, K. *Coot*: Model-building tools for molecular graphics. *Acta Crystallogr. D* **2004**, *60*, 2126–2132. [CrossRef]
28. Liebschner, D.; Afonine, P.V.; Baker, M.L.; Bunkóczi, G.; Chen, V.B.; Croll, T.I.; Hintze, B.; Hung, L.-W.; Jain, S.; McCoy, A.J. Macromolecular structure determination using X-rays, neutrons and electrons: Recent developments in Phenix. *Acta Crystallogr. D* **2019**, *75*, 861–877. [CrossRef]
29. Williams, C.J.; Headd, J.J.; Moriarty, N.W.; Prisant, M.G.; Videau, L.L.; Deis, L.N.; Verma, V.; Keedy, D.A.; Hintze, B.J.; Chen, V.B. MolProbity: More and better reference data for improved all-atom structure validation. *Protein Sci.* **2018**, *27*, 293–315. [CrossRef]
30. Stiernagle, T. Maintenance of *C. elegans* (February 11, 2006), The *C. elegans* Research Community. In *WormBook*; Available online: http://www.wormbook.org/chapters/www_strainmaintain/strainmaintain.html (accessed on 30 October 2022). [CrossRef]
31. Schneider, C.A.; Rasband, W.S.; Eliceiri, K.W. NIH Image to ImageJ: 25 years of image analysis. *Nat. Methods* **2012**, *9*, 671–675. [CrossRef]

toxins

MDPI

Article

Multifunctional Properties of a *Bacillus thuringiensis* Strain (BST-122): Beyond the Parasporal Crystal

Argine Unzue [1,2], Carlos J. Caballero [2], Maite Villanueva [2], Ana Beatriz Fernández [2] and Primitivo Caballero [1,*]

[1] Institute of Multidisciplinary Research in Applied Biology-IMAB, Universidad Pública de Navarra, 31192 Mutilva, Spain

[2] Departamento de Investigación y Desarrollo, Bioinsectis SL, Plaza Cein 5, Nave A14, 31110 Noáin, Spain

* Correspondence: pcm92@unavarra.es

Abstract: Chemical products still represent the most common form of controlling crop pests and diseases. However, their extensive use has led to the selection of resistances. This makes the finding of new solutions paramount to countering the economic losses that pests and diseases represent in modern agriculture. *Bacillus thuringiensis* (Bt) is one of the most reliable alternatives to chemical-based solutions. In this study, we aimed to further expand the global applicability of Bt strains beyond their spores and crystals. To this end, we selected a new Bt strain (BST-122) with relevant toxicity factors and tested its activity against species belonging to different phyla. The spore and crystal mixture showed toxicity to coleopterans. Additionally, a novel Cry5-like protein proved active against the two-spotted spider mite. In vivo and plant assays revealed significant control of the parasitic nematode, *Meloidogyne incognita*. Surprisingly, our data indicated that the nematocidal determinants may be secreted. When evaluated against phytopathogenic fungi, the strain seemed to decelerate their growth. Overall, our research has highlighted the potential of Bt strains, expanding their use beyond the confinements of spores and crystals. However, further studies are required to pinpoint the factors responsible for the wide host range properties of the BST-122 strain.

Keywords: *Bacillus thuringiensis*; bioassays; toxicity; multifunctional; Coleoptera; plant-parasitic nematode; mite; phytopathogenic fungi

Key Contribution: The potential of Bt strains beyond the confinements of spores and crystals. Extend the use of Bt products to pests and diseases that are conventionally controlled with synthetic products.

Citation: Unzue, A.; Caballero, C.J.; Villanueva, M.; Fernández, A.B.; Caballero, P. Multifunctional Properties of a *Bacillus thuringiensis* Strain (BST-122): Beyond the Parasporal Crystal. *Toxins* **2022**, *14*, 768. https://doi.org/10.3390/toxins14110768

Received: 10 October 2022
Accepted: 3 November 2022
Published: 7 November 2022

1. Introduction

Bacillus thuringiensis (Bt) is a Gram-positive, ubiquitous, spore-forming bacterium that produces a parasporal inclusion body (crystal) during its stationary phase of growth. The crystal is primarily composed of insecticidal proteins, which have been effectively used to control lepidopteran and coleopteran crop pests in agriculture, as well as mosquito disease vectors [1–3]. Bt-based solutions currently represent 1% of the agrochemical market, including insecticides, fungicides, and herbicides [4]. Within the microbial pesticide segment, Bt solutions lead the bacterial product market, with close to 70% of the total share [5].

Bt crystals are mainly composed of Cry and Cyt proteins which are toxic to species of insects of different orders such as Lepidoptera, Diptera, Coleoptera, Hymenoptera, etc. Additionally, Bt synthesizes other water-soluble components and insecticidal toxins during the growth phase, named Vip (vegetative insecticidal protein) and Sip (secreted insecticidal protein), among others [1,3,6]. In the industry, there are some examples of Bt-based products that are manufactured by concentrating the fermentation broth (such as heat spray drying), hence they retain water-soluble components [7]. However, many of the most relevant products today do not incorporate these pesticidal factors since the method of production involves retrieving spores and crystals through centrifugation, leaving behind non-crystal toxins and other factors.

Today, Bt represents a popular biological resource when controlling populations of lepidopteran, dipteran, and coleopteran pests. One of the reasons for this is the specificity of its host spectrum, making it an alternative to the wide spectrum of chemical synthesized pesticides [1]. For instance, the Cry1, Cry2, and Cry9 groups are mainly toxic to lepidopterans [8–10]. Analogously, Cry3, Cry7, Cry8, and Cry1Ia (a subgroup of Cry1) have increased activity against coleopterans [11–14]. For dipterans, Cry4, Cry10, Cry11, and Cyt have been proven as the most relevant so far [15–21]. Some exceptions to these are the Cry1B, Cry2A, App6, Xpp22, and Mpp51 proteins, among others, with proven activity across various insect orders [8,22–24]. However, its potential for becoming a realistic solution against other challenging pest organisms, such as mites, parasitic nematodes, or phytopathogenic fungi, is still being addressed and studies to demonstrate this are still being conducted [25–30]. In the case of phytophagous mites, a number of species from the Tetranychidae (Arachnida) family constitute relevant agricultural pests. The most widely distributed species of these types of spider mites is the highly polyphagous and ubiquitous *Tetranychus urticae* (two-spotted spider mite) [31]. It is relevant in the most extensive crops, but their damage is more severe in vegetable and ornamental crops, which are cultivated in greenhouse conditions [32]. Although detailed information on the damage caused by PPNs (plant parasitic nematodes) is not easily accessible, the current estimations place the world's economic losses due to these types of pathogens at USD 75–125 billion per year [33]. One of the most relevant PPNs are the root-knot nematodes (RKN) (*Meloidogyne* spp.), which are widespread geographically and have a large host range. For instance, *M. incognita* can parasite more than 2000 species of plants [34,35]. Another significant source of agriculture losses in crops, such as wheat, rice, maize, potato, and soybean, are phytopathogenic fungi [36]. Phytopathogenic fungi can damage the plant (aerial or root part) or affect post-harvesting products. The main fungi species that cause disease in agriculture are *Alternaria* spp., *Fusarium* spp., *Rhizoctonia solani*, *Puccinia* spp., *Sclerotinia* spp., *Phytophthora infestans*, and *Verticillium dahliae*, among others [36,37]. The particularity of some of these organisms is that their control relies on the use of synthetic chemical products [34,38].

Here, we focused on finding and characterizing a wild-type Bt strain as a source of toxicity factors to confer protection and control against diverse pests and pathogens that produce significant economic losses in agriculture. Specifically, its microbial pesticide properties against the Colorado potato beetle (*Leptinotarsa decemlineata*), the two-spotted spider mite (*Tetranychus urticae*), the root-knot nematode *Meloidogyne incognita*, and the *Verticillium dahliae* and *Fusarium oxysporum* species of phytopathogenic fungi were evaluated. Our study showed that the spores and crystals contain toxicity factors with significant activity against coleopterans and mites in vivo. However, the selected strain showed nematocidal properties in vivo and in cucumber plants in the pot assay, which were found to reside within the supernatant after the fermentation process. Additionally, it decelerated the growth of diverse phytopathogenic fungal species in plate assays. Although these preliminary results, under controlled conditions, could represent a good starting point in understanding the potential of strain BST-122, further studies would be required to address the precise factors, mechanisms of action, and overall biosafety of the microorganism. All in all, the results suggest that the potential of Bt strains for controlling different pests may be overlooked by the currently available commercial solutions, which are usually based on mixtures of spores and crystals. Uncovering additional secreted factors and understanding their mechanism of action may prove a reliable source for expanding their uses in the crop protection sector.

2. Results

2.1. Selection of a Bt Strain Harboring Previously Described Toxicity Factors against Pest Organisms Belonging to Different Phyla

The currently available Bt-based solutions often focus on an insect order at once. Meaning that a Bt solution aimed at controlling dipterans may not be effective for the control of lepidopterans and vice versa. One of the objectives of this study was to find a Bt strain with multifunctional properties in order to control species of phytophagous or plant parasitic pests belonging to different orders, phyla, or even kingdoms of life. In order to search for the right candidate, we screened our library of sequenced Bt strains using the crystal genes *cry5*, *app6*, *cry12*, *cry13*, *cry14*, *cry21*, and *xpp55*, and secreted factors—such as chitinases and metalloproteases—as the main criterium. The reasoning behind this selection was as follows: Cry5, App6, and Cry12, had previously been described as nematocidal/acaricidal proteins [25–28,39,40], Cry14 and Xpp55 as active against nematocidal and coleopteran [14,25,39], and Cry13 and Cry21 as nematocidal proteins [39–41]. Additionally, chitinases and metalloproteases had been reported by other research groups to enhance their activity against nematodes and to possibly affect fungi [42–44].

Out of the screened Bt strains from our laboratory library, BST-122 (isolated from a dust sample) was selected for this study since it presented an interesting gene content, with the potential for controlling pests out of the scope of most currently available Bt-based products (Table 1). Specifically, we found the sequence of a cry5-like gene (96% identical to Cry5Ad1 (ABQ82087.1) using local alignment (BLASTP), but with global alignment the identity dropped sharply to 36% pairwise identity and endotoxin_N and delta_endotoxin_C domain-containing protein), which could potentially represent a novel protein for the control of mites and nematodes. Moreover, sequences that codified for the coleopteran- and hemipteran-specific toxins—Mpp51 (ADK94873.1) and ORF2 of cry65Aa (Orf2_cry65Aa; AEB52308.1)—and secreted factors, such as an exochitinase (AIE34993.1) and metalloproteases (Bmp1 (AFZ77001.1) and ColB (ACZ37253.1)), were also present (Table 1). To further characterize the strain, we grew it in CCY medium at 28 °C and 200 rpm, for 72 h. The isolate produced parasporal crystals consisting of two bodies that were attached to the spores. One appeared as a dark, round shape and the other as a dark, bar-shaped crystal (Figure 1A). To further analyze which of the biocidal genes (Table 1) were expressed in the tested conditions and, as a result, integrated into the crystals, the mixture of spores and crystals was run in an SDS-PAGE. As Figure 1B shows, the main three characteristic bands observed in the protein profile of strain BST-122 were approximately 120, 58, and 34 kDa, which correlated with the predicted size of the individual Cry5-like, Orf2_cry65A, and Mpp51Aa crystal proteins, respectively.

Table 1. Insecticidal protein content of the BST-122 Bt strain.

Target Database	Pairwise Identity (%)	MW (kDa)	Length (No. Amino Acid Residues)	Amino Acids Overlap in Global Alignment	Accession Number of Reference	Accession Number
Crystal proteins						
Mpp51Aa2	98	34.5	312	312	ADK94873.1	OP696897
Cry5Ad1	36	119.2	1092	1148	ABQ82087.1	OP604599
Orf2_cry65Aa	96	58.4	512	490	AEB52308.1	OP722690
Mpp2Aa6	30	46.2	412	136	U41822.1	OP722691
Secreted factors						
Exochitinase	99	39.4	360	360	AIE34993.1	OP722692
ColB (metalloprotease)	77	48.3	426	428	ACZ37253.1	OP722693
Bmp1 (metalloprotease)	34	65.3	592	367	AFZ77001.1	OP722694

In addition to the morphology and crystal protein content and their expression, we tested the BST-122 Bt strain for the production of type I β-exotoxin, which is nonspecific and toxic to vertebrates. HPLC analyses confirmed the lack of this determinant in the BST-122 strain (Figure S1).

To understand the potential of strain BST-122 as a broad microbial control agent we decided to evaluate the activity of its spores and crystals against a model species for each of the potential target orders of organisms.

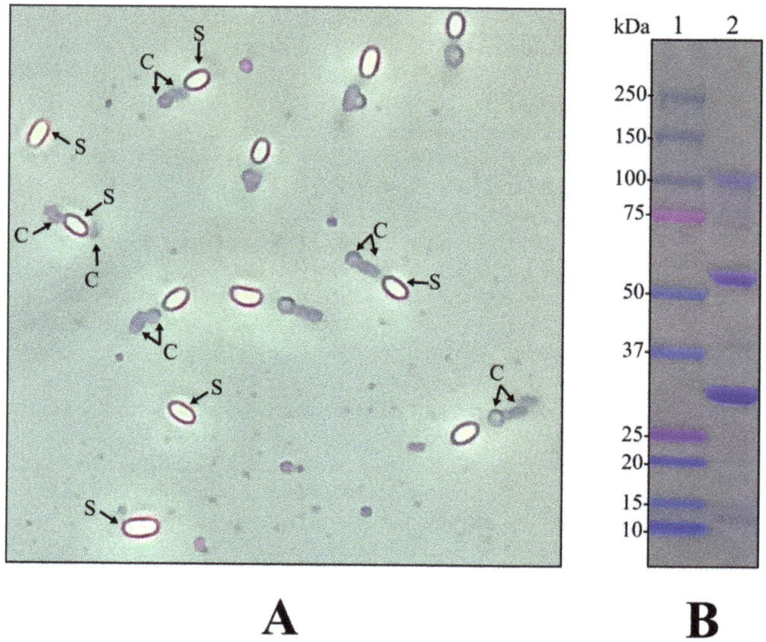

Figure 1. (**A**): picture of the BST-122 Bt strain, as seen under the optical microscope (1000× magnification) after growth in CCY medium for 72 h, at 28 °C and 200 rpm. S—spore; C—crystal. (**B**): SDS-PAGE of BST-122 mixture of spores and crystals. The strain was grown as described in A. In total, 15 µL of the mixture of spores and crystals were denaturalized, mixed with 15 µL of loading buffer for 5 min, at 95 °C. In total, 20 µL of the mix was loaded into the 4–20% polyacrylamide gel and Coomassie staining was applied to reveal discrete bands, corresponding to crystal and spore proteins: line 1—protein marker; line 2—BST-122 wild-type strain.

2.2. The BST-122 Bt Strain as a Biological Control Agent against Newly Hatched Larva of Leptinotarsa decemlineata

After analyzing the BST-122 crystals in SDS-PAGE (Figure 1B), we found bands corresponding to the predicted Mpp51Aa protein, indicating that it would be actively expressed by the Bt strain. Mpp51Aa was previously reported as active against coleopterans, among other insect orders [25,45]; therefore, we decided to test the biological activity of the BST-122 wild-type strain against a well-known representative of this order, namely the Colorado potato beetle (*L. decemlineata*). For this purpose, we carried out bioassays in which we addressed the mortality of first-instar larvae, which had fed from superficially contaminated potato leaves, at different concentrations of the BST-122 spore and crystal mixture (s + c). Mortality was registered 4 days post-treatment and the mean lethal concentration (LC_{50}) value was calculated from three independent biological replicates. Table 2 shows the effectiveness of the treatments, with a calculated LC_{50} of 10.5 µg/mL.

Table 2. Mean lethal concentration (LC_{50}) value of the BST-122 Bt strain, calculated for newly hatched larva of *L. decemlineata*.

LC	Concentration (µg/mL)	Lower Limits	Upper Limits	χ^2	df	Slope	SE Slope	Intercept
LC_{50}	10.5	6.99	15.0	1.96	5	0.947	0.111	−0.969

LC—lethal concentration; χ^2—chi-square; df—degree of freedom; SE—standard error.

2.3. Activity of BST-122 Bt Strain against *Tetranychus urticae* Protonymphs

The other major toxicity factor that was found within the BST-122 crystals was a Cry5-like protein. Sharing only 36% of identity with protein Cry5Ad1, this would constitute an entirely new protein, with the potential to control mites and nematodes [26,28,39]. To test the biological relevance of the BST-122 crystals against both types of potential hosts, we first conducted in vivo assays against *T. urticae*. A single concentration of a spores and crystals mixture (100 µg/mL) was fed to protonymphs of *T. urticae* on a Petri dish. The treatment was mixed with a saccharose and blue food dye (fluorella blue) and placed between two layers of Parafilm®, which covered the plate. In this manner, mites ingested the treatment by piercing through the parafilm with their sucking/feeding system. After 16 h of contact with the treatment, blue-colored mites were selected and placed on a bean leaf disc. Mortality was recorded after 3 days of treatment ingestion. However, the BST-122 did not present significant mortality compared to the control (data not shown).

Since the mixture of spores and crystals was ineffective against *T. urticae*, we speculated that purifying the Cry5-like protein would help increase its relative concentration and address its true potential in controlling mites. The *cry5*-like gene is found in an operon architecture, alongside the *orf2cry65Aa* gene in BST-122. Therefore, we constructed plasmid pSTAB-Cry5- orf65, which expressed the proteins Cry5-like and Orf2_Cry65Aa under the *cyt1A* promoter (P*cyt1A*) and used it to transform the BMB171 acrystalliferous strain. The resulting strain, BMB171 pSTAB-Cry5-orf65, was grown until the stationary phase, and the mixture of spores and crystals was retrieved by centrifugation. In order to verify the correct expression of both proteins, we ran an SDS-PAGE of the newly generated crystals and found that their predicted molecular masses—119.5 kDa and 58.4 kDa—correlated with the observed bands (Figure 2).

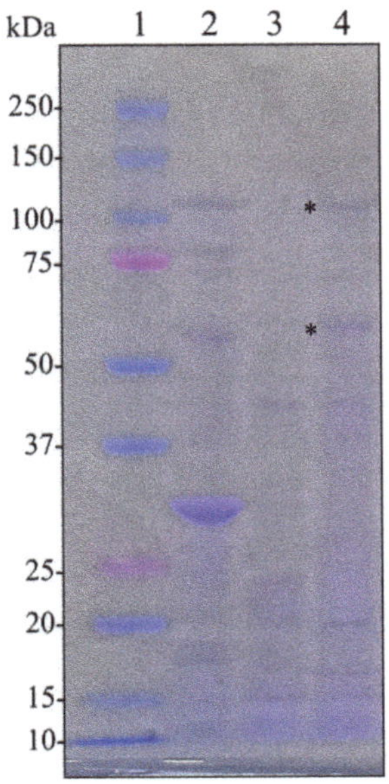

Figure 2. SDS-PAGE of the BST-122 recombinant protein (Cry5-orf65): 1—protein marker; 2—BST-122 wild-type strain; 3—BMB171-empty; 4—BMB171-Cry5-orf65 (* marked bands 119.5 and 58.4 kDa).

We then carried out in vivo bioassays following the same methodology to test the efficacy of the purified Cry5-like and Orf2_Cry65 crystals. In this case, applying a high concentration—as in the previous experiment (100 µg/mL)—resulted in 53.3% mortality against the two-spotted spider mite nymphs, 72 h after the treatment (Table 3). Strain BMB171, carrying an empty pSTAB vector, was used as a negative control.

Table 3. Acaricidal activity of the BST-122 recombinant protein (Cry5-orf65) at a single concentration (100 μg/mL) on protonymph of *Tetranychus urticae*.

Treatment	Mortality (%)
BMB171-Cry5-orf65	53.3 ± 16.3 a
BMB171-pSTAB	11.7 ± 3.4 b

Percentage of mortality (mean ± standard error) recorded 72 h after treatment. Different letters were used to denote statistical significance between values, Welch's *t*-test ($F_{1, 3.224}$ = 10.105, *p*-value = 0.04538).

2.4. The BST-122 Bt Strain as a Biological Control Agent against Meloidogyne incognita J2 Juveniles

To evaluate the potential nematocidal activity of the BST-122 strain on plant parasitic nematodes, we chose *M. incognita*, the most economically important PPN species in the tropical and subtropical regions. For this purpose, we performed bioassays on J2 individuals and evaluated its activity seven days post-treatment. We tested solubilized proteins of the s + c mixture against J2, at different concentrations. The mortality of juveniles registered a tendency dependent on the protein concentration (Table S1 and Figure S2). Since the solubilized protein of the wild-type strain performed below our expectations against *M. incognita*, we decided to test the activity of the Cry5-like protein, following the same reasoning that we applied for *T. urticae*. For this purpose, the BMB171-Cry5-orf65 was tested at two concentrations (50 and 150 μg/mL). Strain BMB171, carrying the pSTAB-empty vector, was used as a control. The results showed relatively high mortality rates for both, the treatment and the control, suggesting that there might be factors that are not present in the crystals that are responsible for exerting toxicity on the nematodes (Table S2). When comparing the secreted factors of both strains, we found chitinases and metalloproteases Bmp1 and ColB (Tables S3 and S4)—with the previously described nematocidal activities—to be present in both strains [42–44]. To address the potential toxicity of these factors, we conducted tests in which the supernatant (SN) and the whole culture (WC) were used as treatments. The results showed 61.1% and 52.5% mortality rates, respectively (ANOVA $F_{3,8}$ = 5.685; *p* = 0.0221 and post-hoc Tukey at *p*-value < 0.05) (Table 4 and Figure 3B). These findings indicated that toxicity factors within the SN may provide BST-122 with the potential to control *M. incognita* populations. Analogously, we tested the effect of the same culture fractions in nematode eggs but found no activity (data not shown).

Table 4. Mortality of *M. incognita* J2, treated with different fractions of a BST-122 fermented culture.

Treatment	J2 Mortality (%) Mean ± SE
Control	15.5 ± 4.3 a
BST-122 s + c	24.7 ± 7.9 ab
BST-122 SN	61.1 ± 14.7 b
BST-122 WC	52.5 ± 6.3 ab

SE—standard error. Different letters were used to denote statistical significance between values. ANOVA test ($F_{3,8}$ = 5.685; *p* = 0.0221) and post-hoc Tukey test at *p*-value < 0.05.

Our in vivo results suggested that an effective control of *M. incognita* populations may be achieved by using the SN of BST-122 cultures. Since *M. incognita* is an obligate parasite, we decided to corroborate such an observation by performing assays to test the effect of an infestation on cucumber plants. Here, we evaluated the nematocidal activity of s + c, SN, and the WC of our Bt strain when the cultures reached 10^8 spores/mL (50 μg/mL). As a negative control, water was used. All treatment plants were used 14 days after seeding and infested with 1000 freshly retrieved eggs of *M. incognita*. The experiment was evaluated 28 days after the Bt treatments. A total of three treatments were applied, with a week-long gap between them. The first treatment was implemented a day before the infestation (Figure 4A).

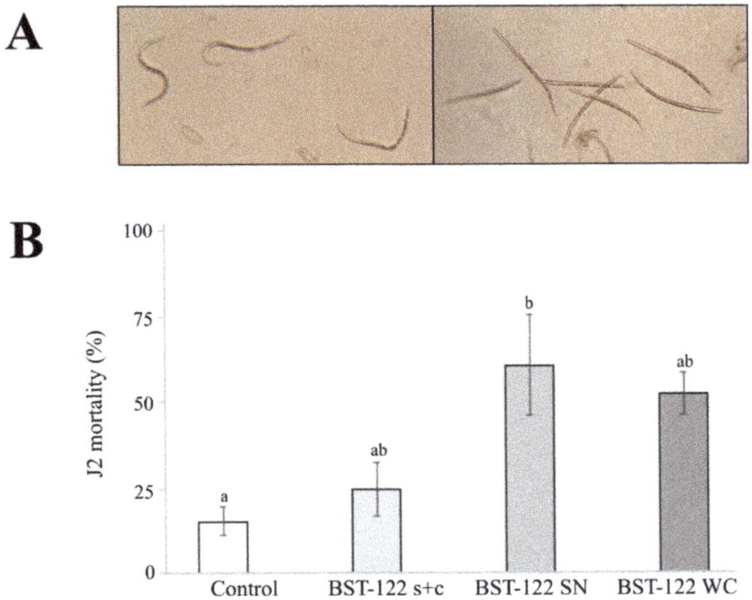

Figure 3. (**A**): pictures of in vivo bioassays of J2 *M. incognita,* as seen under an inverted microscope (100× magnification). Left—alive J2; right—dead J2. (**B**): results of in vivo bioassays of J2 mortality in *M. incognita,* after 7 days treated with the BST-122 Bt strain, at a 50 μg/mL concentration. Control—20 mM HEPES (pH 8.0); BST-122 s + c—mixture of spores and crystals of Bt BST-122 strain; BST-122 SN—supernatant of the fermentation of Bt BST-122 strain; and BST-122 WC—the whole culture of the fermentation of Bt BST-122 strain. Different letters were used to denote statistical significance between values.

The nematode infection was evaluated by counting the total number of galls in the root parts per plant. The results of the analysis were statistically analyzed using the ANOVA ($F_{3,78} = 26.28$; $p = 7.54 \times 10^{-12}$) and the Tukey post-hoc (p-value < 0.05) tests. As shown in Table 5, the plants treated with the s + c mixture did not differ significantly, compared to the control (209.2 and 257.6 galls per plant, respectively). However, plants treated with the fermentation supernatant (SN) and with the whole fermentation culture (WC) had a significant drop in the number of recorded galls (112.4 and 76 galls per plant, respectively). Overall, the gall reduction was up to 56.4% in the plants treated with SN and 70.5% in the plants treated with the WC, when compared to the control. This indicated that Bt- secreted factors may play an important role in achieving an effective control of PPNs in plant experiments.

2.5. *Effect of BST-122 in the Growth of Phytopathogenic Fungi: Verticillium dahliae and Fusarium oxysporum*

The above-mentioned results revealed that secreted factors of strain BST-122 may prove useful for the growth inhibition of parasitic nematodes. Phytopathogenic fungi represent another source of organisms that could potentially be controlled by said strain. To address this, we conducted experiments that consisted of comparing the biomass surface of the root phytopathogenic fungi species, *Verticillium dahliae,* and variants of *Fusarium oxysporum* that were grown in LB plates, in the presence or absence of BST-122. Pictures of the growing plates were taken on different days, depending on the fungal species. The results obtained from these experiments are shown in Table 6 and Figure 4. The growth of *Verticillium dahliae* was negatively affected by the presence of the Bt strain, BST-122 (Figure 5A). After 6 days of incubation, a significant slowdown in growth of 24.76% was observed (t-test: $t = 3.464$, $df = 19$, $p < 0.01$). At 12 days of incubation, this effect was further increased, scoring 63.90% (t-test: $t = 7.3254$, $df = 19$, $p < 0.001$). Additionally, two different varieties of *Fusarium oxysporum*—*F. oxysporum lycopersici,* which affects tomato plants, and *F. oxysporum melonis*, which infects melon plants—were tested. In the former, the deceleration in growth was significant at 4, 5, 6, and 7 days after the inoculation of the plates, reaching a total biomass surface reduction of 33.72% (*t*-test: t = 6.0417, df = 16, $p < 0.001$) (Figure 5B). In the latter, comparable reduction levels were achieved (30.33%) (*t*-test: t = 4.9758, df = 16, $p < 0.001$) (Figure 5C).

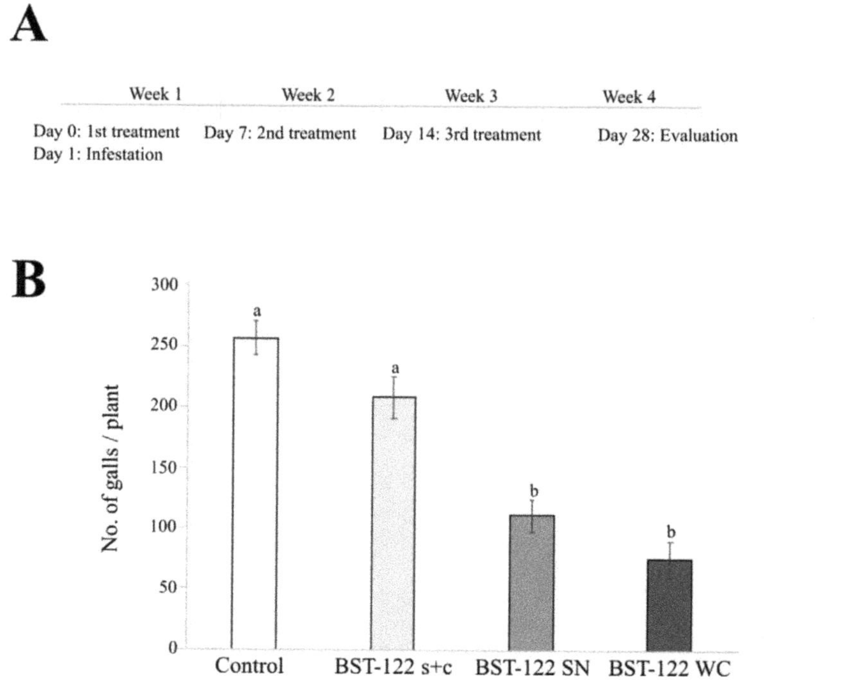

Figure 4. Activity of the BST-122 Bt strain against *M. incognita* eggs, tested in cucumber plants. (**A**): scheme of the methodology established to perform the experiments of BST-122 activity against *M. incognita* in cucumber plants. (**B**): Activity of BST-122 Bt strain against *M. incognita* eggs, tested in cucumber plants. It represents the no. of galls per plant in different treatments. Plants were infested with 1000 freshly laid eggs. Control—infested plants treated with H_2O; s + c—infested plants treated with the fermentation centrifugation or spores and crystal mixture; SN—infested plants treated with the fermentation supernatant; WC—infested plants treated with the whole fermentation culture. Different letters were used to denote statistical significance between values (ANOVA, followed by post-hoc Tukey test at $p < 0.05$).

Table 5. Activity of the BST-122 strain against *M. incognita* in cucumber plants. The effectiveness was addressed by evaluating the number of galls per plant after the different treatments.

Treatment	No. of Galls/ Plant [1]	% Reduction of Galls/Plant
Control	257.6 ± 13.9 a	-
BST-122 s + c	209.2 ± 17.1 a	18.8%
BST-122 SN	112.4 ± 14.1 b	56.4%
BST-122 WC	76.0 ± 14.8 b	70.5%

[1]—mean ± standard error. Different letters were used to denote statistical significance between values. ANOVA ($F_{3,78} = 26.28$; $p = 7.54 \times 10^{-12}$) and the post-hoc Tukey (*p*-value < 0.05).

Table 6. Effect of the BST-122 Bt strain on biomass production of different plant pathogenic fungi species.

Fungal Pathogen	Incubation Time (Days)	Biomass Surface		% Growth Inhibition	Significative Differences
		Control	Bt Treated		
Verticillium dahliae	6	5.17 ± 0.26	3.89 ± 0.24	24.76%	**
	12	10.46 ± 0.70	3.92 ± 0.42	63.90%	***

Table 6. *Cont.*

Fungal Pathogen	Incubation Time (Days)	Biomass Surface		% Growth Inhibition	Significative Differences
		Control	Bt Treated		
Fusarium oxysporum lycopersici	4	17.69 ± 0.89	14.96 ± 0.50	15.43%	*
	5	24.77 ± 1.07	18.48 ± 0.51	25.39%	***
	6	35.82 ± 2.43	24.08 ± 1.18	32.77%	**
	7	39.2 ± 1.96	25.98 ± 0.97	33.72%	***
Fusarium oxysporum melonis	4	19.32 ± 0.51	15.88 ± 0.29	17.80%	***
	5	27.30 ± 0.78	20.61 ± 0.46	24.50%	***
	6	37.34 ± 2.15	25.89 ± 1.34	30.66%	***
	7	39.70 ± 1.92	27.66 ± 1.47	30.33%	***

* *p*-value > 0.01–0.05; ** *p*-value > 0.001–0.01; *** *p*-value < 0.001.

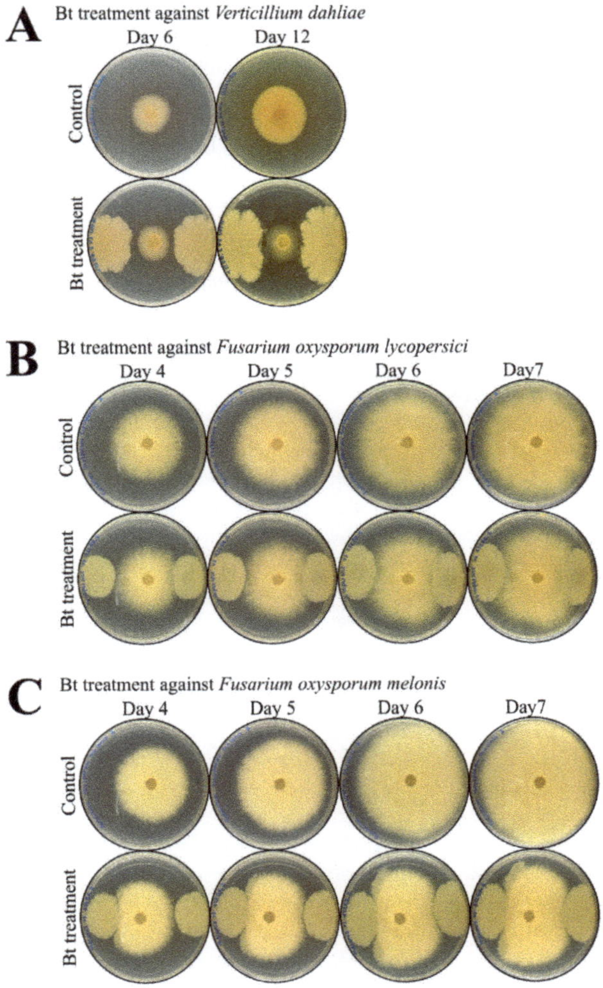

Figure 5. In vitro assays testing BST-122 Bt strain activity against diverse phytopathogenic fungi. The fungi species tested were *Verticillium dahliae* (**A**), *Fusarium oxysporum lycopersici* (**B**), and *Fusarium oxysporum melonis* (**C**). Briefly, a PDA agar disc with the fungus was inoculated in the center of an LB plate, and 5 μL of treatment was inoculated on each side of the fungus. A physiological solution (0.85% NaCl) was used as control, and an overnight culture in LB of the BST-122 strain, adjusted to 10^6 spores/mL, was used as the tested treatment. Pictures of fungal growth were taken at different development moments and the growth inhibition was measured with ImageJ (v.1.53a) software.

3. Discussion

In this study, we selected the strain BST-122, based on its gene content and associated potential as an insecticidal, nematocidal, fungicidal, and acaricidal agent. Some of the most interesting toxins harbored by this strain were a new *mpp51Aa* and a *cry5*-like gene. In a previous study, an Mpp51Aa protein was depicted as active against different insect orders, such as the hemipteran species (*Lygus hesperus* and *L. lineolaris*) and *L. decemlineata*—a coleopteran pest of economic relevance in agriculture [22,46]. In the case of Cry5 proteins, US patents, 5,211,946 and 5,350,576, had attributed Cry5Aa1 and Cry5Ab1 a potential activity against mites belonging to the *Tetranychus* and *Dermatophagoides* genus [26,28] and nematodes (*Caenorhabditis elegans* and the phytoparasitic nematodes, *Pratylenchus* spp. and *M. incognita*) [39,45,47]. Other proteins—such as secreted chitinases and metalloproteinases, namely *colB* and *bmp1* (a type of collagenase)—were also of interest due to their associated nematocidal and fungicidal properties [29,42–44,48,49]. Altogether, these characteristics made the selected strain a strong candidate for evaluating its potential multitarget properties.

The next step was testing its overall toxicity against representatives of the different orders and phyla of potential target organisms. One of BST-122's most relevant crystal proteins was a new Mpp51Aa protein, which could be active against coleopterans [46]. Insects belonging to this order have been traditionally controlled with Cry3 protein variants in formulated Bt-based products, or through their expression in transgenic plants [14]. The lack of diversity in Bt-based biopesticides, as well as the extensive use of engineered plants, have contributed to the emergence of resistant populations [50,51]. For this reason, promoting the use of Bt strains with novel or less commonly used insecticidal toxins is an interesting strategy towards diminishing this problem. Here, we propose strain BST-122 as an alternative to Bt strains harboring Cry3 for the control of *L. decemlineata*. Although we did not determine the activity of the new Mpp51Aa protein, Mpp51Aa1 is known for being active against the Colorado potato beetle (LC_{50} = 19.5 μg/mL) [22,46]. Based on this previous information, we evaluated the activity of a mixture of spores and crystals from BST-122 (LC_{50} of 10.5 μg/mL) (Table 2). Since Mpp51Aa seems to be present in the BST-122 crystals, one would assume that this protein could contribute to its overall activity against *L. decemlineata* (Figure 2). However, other proteins present in the crystal, such as Cry5-like, Orf2_cry65A, or Mpp2-like, could represent an additional source of toxicity or interact synergistically with the Mpp51Aa protein.

Another crystal protein of interest that is expressed by BST-122 was a Cry5-like candidate. Although there is little information available on the activity of this protein against mites, two patent publications on its potential acaricidal activity made us consider whether it might be possible to use BST-122 to control *T. urticae* populations [26,28]. Phytophagous mites have a particular feeding system that differs from the coleopteran or lepidopteran insect larvae. They typically feed on leaf tissue by inserting a stylet into it and sucking the epidermic and parenchymatic cell content. As a result, the plant cells collapse and die, causing chlorotic spots on the leaves and reducing the rate of transpiration and photosynthesis in the plant [32,52]. In this study, the characterization of Bt BST-122 as a potential acaricidal source was evaluated under in vivo conditions. For this purpose, we developed a new assay method that consisted of placing *T. urticae* nymphs in a Petri dish and providing spore and crystal mixtures of the selected strain in a drop containing a sucrose solution and food dye. After 16 h, colored nymphs were selected and placed on a bean leaf. The mortality of individuals was registered 72 h after treatment ingestion. Although the BST-122 strain did not reveal toxicity in mites, when testing the Cry5-like protein alone, at the same rate, 53.3% of mortality was recorded (Table 3). Most available studies on Bt acting as a control agent against mites have mainly focused on the activity of wild-type strains. For example, the Bt strains EA3, EA11.3, and EA26.1 were reported as being considerably active (LC_{50} of 7.111, 12.839, and 1.509 μg/mL, respectively) against the honeybee ectoparasite, *Varroa destructor* [53]. Other studies reported that strain GP532 was effective against the rabbit ear mite ectoparasite, *Psoroptes cuniculi* [54,55]. Additionally, strain

BPU5 has been described as displaying acaricidal properties against the phytophagous mite, *Tetranychus marcfarlanei*, at relatively high concentrations (LC_{50} = 8.024 mg/mL) [56]. However, few publications have described the factor responsible for the acaricidal properties of Bt. Some research studies have described β-exotoxin as acaricidal against the following phytophagous mites: *Panonynchus citri*, *P. ulmi*, *Tetranychus telarius*, *T. urticae*, *T. pacificus*, and the predator, *Metaseiulus occidentalis* [57–61]. Nonetheless, the β-exotoxin is toxic to mammals and, therefore, strains expressing such a determinant are not eligible for product development. Regarding delta-endotoxins, only US patents 5,211,946; 5,350,576; and 5,262,158 mention Cry5Aa1, Cry5Ab1, App6Aa1, App6Ba1, Cry8Aa1, and Cry12Aa1 as potentially active against the two-spotted spider mite (*T. urticae*) and the house dust mite, *Dermatophagoides pteronyssinus* [26–28]. Regardless, experiments have been conducted using the wild-type strains, expressing crystals with the said combinations of toxins, and significant results have only been obtained when applying high concentrations of the active ingredients (5 mg/mL). As opposed to the publicly available data, our results showed significant mortality when we used a novel protein against *T. urticae* nymphs, at a relatively low dose (100 μg/mL). However, the tested concentration was still too high from a cost-effective standpoint. Possibly, the most interesting contribution of these results is the finding of a new protein with acaricidal properties that could serve as a basis for searching new Cry5-like proteins with improved activity. Additionally, a foliar application would not be effective for the control of phytophagous mites due to their feeding behavior. Therefore, opting for the expression of Cry5-like candidates in the vascular system of plants may represent a more realistic approach for the control of these pests in agriculture.

Furthermore, Cry5 proteins had previously been described as potentially nematocidal [25,39]. This made us consider whether it would be effective against nematodes of agronomic importance. To test the potential of the strain, we performed in vivo assays using second-stage juveniles of the phytoparasitic nematode, *M. incognita*. The physiology of plant parasitic nematodes should be considered, whose feeding systems consist of a stylet, with a diameter of only 28 kDa for the cyst nematode, *Heterodera schachtii* [62]. However, it is known that solubilized proteins of up to 140 kDa can be incorporated into the gut of *Meloidogyne hapla* [63]. For this reason, we solubilized BST-122 crystals to test their activity against *M. incognita* J2 juveniles. After exposing individuals to increasing concentrations of BST-122 solubilized crystals, we found a correlation between the concentration and mortality (Table S1). However, significant mortality was only reached at a relatively high concentration. Following the same strategy as with *T. urticae*, we decided to test whether the purified Cry5-like protein had increased activity against J2 individuals. The results showed that treatments with the Cry5-orf65 crystal expressed from the BMB171 strain were undifferentiated from the negative control, the BMB171 strain carried an empty plasmid (Table S2). This suggested that the acrystalliferous BMB171 strain may harbor virulence factors that are not related to crystal pesticidal proteins (Table S3). This was in agreement with previous reports, in which BMB171 was attributed nematocidal properties [44,64]. Additionally, other studies addressed how the activity of purified crystals containing App6Aa and Cry5Ba in *C. elegans* was improved (4.73-fold and 3.59-fold, respectively) in the presence of the spores. It was suggested that the ColB metalloproteinase could be involved in nematocidal pathogenesis [44]. In a similar manner, a previous study demonstrated the synergistic activity of the Bmp1 metalloproteinase when mixed with Cry5Ba in *C. elegans* [43]. These studies indicated that Bt may produce factors that are not present in the crystal and that may represent an alternative to δ-endotoxins in the field of microbial pesticides. Therefore, since BST-122 shared many of the BMB171-secreted factors (Table S4), we considered whether these would be enriched in the supernatant and, perhaps, exert a greater nematocidal activity than the proteins of the crystal alone. To address this idea, we conducted treatments with the supernatant and the whole fermentation culture of the BST-122 strain against *M. incognita* juveniles. The results showed increased activity for both treatments when compared to the spore and crystal mixture. This indicated that the supernatant contained one or more toxicity factors that were responsible for the observed mortality (Table 4 and Figure 3).

Since *M. incognita* is an obligate parasite, we performed experiments treating the radicular system of nematode-infested cucumber plants to validate the in vivo results. When applying similar concentrations of the aforementioned treatments to cucumber plant pots, a reduction of 56.4% and 70.5% of galls per plant was observed for the fermentation supernatant and the whole culture, respectively (Table 5 and Figure 4). This was consistent with our in vivo observations and with studies from other research groups, in which the activity of Bt strains, ToIr65 and ToIr67, against *M. javanica* in tomato plants were more effective when the fermentation supernatant was applied [65]. Moreover, other studies, in which the whole culture, the fermentation supernatant, and the spores and crystals of diverse Bt strains were tested against *M. incognita*, showed how the application of the whole culture reduced the number of eggs by 84% [66]. These studies did not specify the main factors behind the observed toxicity. However, other research studies did indicate that the nematocidal properties of Bt supernatants may be due to the presence of β-exotoxin [67–69]. Although the toxicity of β-exotoxin in mammals is yet unclear, the public use of this active substance is forbidden by the recommendation of the World Health Organization [70]. In this study, the possibility of I β-exotoxin (thuringiensin) being responsible for the observed mortality and gall reduction in plant assays was excluded since its presence could not be detected by genome sequencing nor by HPLC analysis.

Some of the BST-122 secreted factors involved chitinases. These had been previously described as potential antifungal agents [29,49]. This made us consider whether BST-122 may also prove useful in controlling some of the most relevant phytopathogenic fungi in agriculture. To characterize the potential fungicidal activity of BST-122, we evaluated the growth inhibition of several fungal species that cause plant disease in the presence of the BST-122 strain. The experiment consisted of inoculating LB agar plates with an agar disc of each fungus flanked by two drops of a BST-122 culture (10^6 CFU/mL) to evaluate the effects. The tested species were *Verticillium dahliae* and two different serovars of *Fusarium oxysporum* (*lycopersici* and *meloni*). The fungal growth was measured at different moments of incubation, depending on the species. The in vivo growth inhibition was 60% in *V. dahliae* and approximately 30% in *F. oxysporum*, when compared to the control (Table 6 and Figure 5). These results suggest that the presence of the bacterium may decelerate the growth of the tested fungi species. Nonetheless, further experiments would be required to address the overall potential fungicidal activity of BST-122—for instance, by infecting plant tissue to evaluate the damage caused by the phytopathogenic fungi when treated with BST-122. Several compounds produced by the *Bacillus* spp. have been described as active against fungal phytopathogens [71]. In a previous study, isolates of *Bacillus haloterans* exhibited strong activity against diverse *Fusarium* sp.—*Botrytis cinerea*, *Alternaria alternata*, *Phytophthora infestans*, and *Rhizoctonia bataticola* [72]. In the particular case of *B. thuringiensis*, different compounds have been demonstrated as being beneficial for the control of phytopathogenic fungi. In another report, volatile organic compounds produced by the BCN10 strain resulted as antifungal in vitro and in vivo against five postharvest pathogens, such as *F. oxysporym* [73]. Other studies have revealed that chitinase extracted from Bt strains could significantly inhibit the mycelial growth of several pathogenic fungi [29,49]. Moreover, the fungicidal activity of diverse Bt strains has been attributed to the production of some peptides, for instance, the fengycin-like lipopeptides [74–76].

To summarize, we have characterized the potential of Bt strain BST-122 for the control of diverse plant pests and diseases, such as coleopteran pests, plant-parasitic nematodes, phytophagous mites, and plant pathogenic fungi—which are organisms that cause significant economical losses in agriculture worldwide. However, further studies on BST-122 are required to pinpoint the toxicological factors responsible for controlling each of the tested hosts; evaluate their activity against a broader host spectrum (for instance, hemipteran pests) analyze their impact on beneficial insects—such as pollinators or natural enemies—and verify soil ecosystem and agricultural compatibility. Overall, this work contributes to highlighting the importance of the multitarget potential of Bt strains, a notion that goes beyond the spores and crystals and that takes into consideration additional factors that could extend the use of Bt products to pests and diseases that are currently treated with synthetic chemicals.

4. Materials and Methods

4.1. Selection of Bacterial Strains

4.1.1. Bacterial Isolation

A laboratory library of Bt strains was built from diverse substrate samples, belonging to different geographical locations in Spain. Samples were mixed with sterile dH_2O, incubated at 72 °C for 20 min, and serial dilutions plated onto CCY agar plates [77,78]. Next, plates were incubated for 48 h, at 28 °C, and single colonies were inspected under an optical microscope to confirm the presence of parasporal crystals. Positive single colonies were cultured until cell lysis, and stored at −20 °C.

4.1.2. Total DNA Extraction and Genomic Sequencing

Total genomic DNA (chromosomal + plasmid) was extracted from the isolated strains, following the protocol for DNA isolation from Gram-positive bacteria supplied in the Wizard® Genomic DNA Purification Kit (Promega, Madison, WI, USA). The DNA library was prepared from total DNA and subsequently sequenced by Illumina NextSeq500 Sequencer (Genomics Research Hub Laboratory, School of Biosciences, Cardiff University, Cardiff, UK).

4.1.3. Identification of Potential Nematocidal/Insecticidal, Acaricidal, and Fungicidal Genes

The genomic raw data were processed and assembled using CLC Genomic Workbench v10.1.1 (Aarhus, Denmark). Reads were trimmed and filtered, and those shorter than 50 bp were removed. Processed reads were assembled de novo using a stringent criterion of overlap of at least 95 bp of the read and 95% identity, and reads were then mapped back to the contigs for assembly correction. Genes were predicted using GeneMark v2.5 (Atlanta, GA, USA) [79].

To assist the identification process of potential pesticidal toxin proteins, local BLASTP [74] was executed against a database built in our laboratory, including the amino acid sequences of known bacterial toxins from the bacterial pesticidal protein database (https://camtech-bpp.ifas.ufl.edu, accessed on 18 July 2022) [80,81], as well as other proteins of interest such as chitinase, enhancin-like metalloprotease, Bmp1, CalY, ColB, and InhA proteins. Proteins involved in the synthesis of zwittermycin A and β-exotoxin were also included [82]. Prediction of structurally conserved domains was carried out using CD-search [83]. Pairwise sequence alignment was carried out using GGSEARCH2SEQ v36.3.8h (Cambridgeshire, UK.) [84].

4.1.4. Production of Spores and Crystals from the Wild Bt Strain

For BST-122, single colonies from LB plates were inoculated in 50 mL of CCY sporulation culture medium [78], and grown under constant temperature (28 °C) and shaking (200 rpm). Crystal formation was observed daily under an optical microscope, at the magnification of ×1000 (Zeiss Axiolab 5, Carl Zeiss Microscopy GmbH, Jena, Germany). After 48 or 72 h, when approximately 95% of the cells had lysed, cultures were stored at 4 °C, until required. The number of spores/mL at the end of the fermentation process was addressed by plating serial dilutions in LB agar Petri dishes. The supernatant and the mixture of spores and crystals were collected by centrifugation at 9000× *g*, at 4 °C, for 10 min. The supernatant was stored at 4 °C. The spore and crystal mixture, after being washed with a saline solution (1 M NaCl, 10 mM EDTA), was resuspended in dH_2O, and kept at 4 °C. For protein quantification, the spore and crystal mixture was solubilized in carbonate buffer (50 mM Na_2CO_3, 100 mM NaCl, pH 11.3) and 10 mM DTT, for two hours, at 37 °C, and total protein concentration was quantified by Bradford assay [85], using bovine serum albumin (BSA) as a standard.

4.1.5. SDS-PAGE

Samples of spores and crystals, as well as solubilized proteins, were mixed with 2× sample buffer (Bio-Rad Laboratories Inc., Hercules, CA, USA), boiled at 100 °C for 5 min, and then subjected to electrophoresis, using Criterion TGX TM 4–20% Precast Gel (Bio-Rad Laboratories Inc., Hercules, CA, USA) [86]. Gels were stained with Coomassie Brilliant Blue R-250 (Bio-Rad Laboratories Inc., Hercules, CA, USA) and then distained with 30% ethanol and 10% acetic acid solution.

4.2. Detection of β-Exotoxin

The presence of type I β-exotoxin (thuringiensin) was evaluated in culture supernatants through HPLC analysis (Department de Gènetica, Universitat de València, Burjassot, Spain) [87]. The standard strain, HD-2 strain, was used as a positive control.

4.3. Coleopteran, Acaricidal, Nematocidal, and Fungicidal Activity of BST-122 Bt Strain

4.3.1. *Leptinotarsa decemlineata*, *Tetranychus urticae*, and *Meloidogyne incognita* Rearing

A laboratory colony of *L. decemlineata* was established from adults collected from organic potato fields near Pamplona (Spain) and maintained on potato plants (*Solanum tuberosum* L. cv. Jaerla). The population of *M. incognita* and *Tetranychus urticae* were obtained from the company Koppert-Spain (Almería, Spain). Nematodes were maintained on tomato plants (*Lycopersicum esculentum*, Mill. cv. Roma). Mites were maintained on bean plants (*Phaseolus vulgaris* L. cv. Garrafal Oro). All populations were kept in a phytotron, under controlled conditions of temperature, humidity, and photoperiod (25 ± 1 °C, 70 ± 5% RH, and L16:D8 h). Nematode eggs were extracted from the root-knot of at least 6-week-old infected roots [88]. The eggs were collected and rinsed with distilled water in sieves (25 and 20 μm pore). To collect the second-stage juveniles (J2), the eggs were placed in hatching dishes and incubated in moist chambers, in distilled water at 25 °C, in darkness. Freshly hatched J2 were collected every 48 h and used for experiments.

4.3.2. Phytopathogen Fungi Species Maintenance

Different phytopathogen fungi species were provided by the Regional Diagnostic Center of the Regional Government of Castilla y León (Salamanca, Spain). The tested fungi were *F. oxysporum* sv. *lycopersici* and *F. oxysporum* sv. *melonis* as root-pathogens, and *V. dahliae*, as an aerial plant pathogen. Fungi were grown on potato dextrose agar (PDA) medium (Sigma-Aldrich, St. Louis, MO, USA), and incubated at 28 °C.

4.3.3. *L. decemlineata* Bioassays In Vivo

The bioassays were performed by the leaf dip method on first instar larvae [89]. Five different spore and crystal mixture concentrations, ranging from 1.8 to 150 μg/mL, were prepared to determine the concentration–mortality responses and calculate the mean lethal concentration (LC_{50}). This bioassay was repeated at least three times. Total insect mortality was recorded after 4 days.

4.3.4. *T. urticae* Bioassays In Vivo

The bioassays were performed with first instar nymphs of *T. urticae*. For each treatment, 20–25 nymphs were placed in 35 mm diameter Petri plates, under a dissecting microscope (Zeiss Stemi 508, Carl Zeiss Microscopy GmbH, Jena, Germany) and the plates were covered with Parafilm®. A 200 μL drop of treatment (100 μL of Bt spore and crystal mixture, 80 μL of 79% of sucrose w/v, and 20 μL of food dye, diluted in sucrose solution) was placed between two layers of Parafilm®. The mites were incubated at 25 °C, under light conditions, during 16 h, and the colored nymphs were placed on a bean leaf disc (20 mm diameter, on a

2% agar plate). The mortality was scored 72 h after the ingestion. Bt treatments were tested at a single dose of 100 μg/mL. A mixture of sucrose and food dye was used as control.

4.3.5. *Meloidogyne incognita* Bioassays In Vivo

J2 Mortality Treated with Spores and Crystal Mixture, Supernatant, and the Whole Culture

The spore–crystal mixture (s + c), supernatant (SN), and the whole culture of the BST-122 Bt strain were tested against *M. incognita* J2 juveniles. All three different fractions came from a culture grown at 28 °C and 200 rpm, until a concentration of 50 μg/mL (10^8 spores/mL) was reached. This concentration of Bt strains was prepared by adding the appropriate volumes of 20 mM HEPES (pH 8.0) to the standard solution. A direct contact assay was carried out in a 96-well plate by modification of the standard method [67]. A total of 200 μL of each treatment was added to 5 μL of a J2 nematode suspension, containing a minimum of 20 individuals per well. Treatments were prepared in triplicate and incubated at room temperature. A solution containing 20 mM HEPES (pH 8.0) was used as a control. All treatments were replicated three times. The number of active and dead J2 were counted 7 days after the treatment, under an inverted microscope (Zeiss Primovert, Carl Zeiss Microscopy GmbH, Jena, Germany), and mortality percentages were calculated.

J2 Mortality Treated with Solubilized Protein

The spore and crystal mixtures were collected and washed, as described above. The spore and crystal mixture pellet was resuspended in solubilization buffer (0.05 M Na_2CO_3, 0.1 M NaCl, pH 11.3 + 0.01 M DL-Dithiothreitol) and incubated at 37 °C, for two hours. Insoluble debris were removed by centrifugation and the solubilized proteins from the supernatant run in SDS-PAGE. This solubilization was dialyzed in 20 mM HEPES (pH 8.0), with a 10 kDa pore Slide-A-Lyzer® Dialysis Cassette (Thermo-Scientific, Rockford, IL, USA) (adapted from [90]). The protein quantification was performed as described above. Different concentrations of the soluble crystal/spore protein of Bt strains, ranging from 50 to 400 μg/mL, were prepared by adding the appropriate volumes of 20 mM HEPES (pH 8.0) to the standard solution. The assay was conducted as described above. A solution containing 20 mM HEPES (pH 8.0) was used as a control treatment.

4.3.6. Assay of Control of *M. incognita* Infestation in Cucumber Plants

Surface-sterilized cucumber seeds (*Cucumis sativus* cv. Marketer) were germinated in 200 cc pots, in autoclaved vermiculite. Plants were fertilized with Osmocote® at the recommended doses and grown in a phytotron, under controlled conditions of temperature, humidity, and photoperiod (25 ± 1 °C, 70 ± 5% RH, and L16:D8 h). Fourteen days after seeding, when the 2nd real leaf had developed, 30 mL of the test solution was added around the root zone as a first treatment (adapted from [68]). 24 h later, plants were inoculated with a 1 mL suspension, containing approximately 1000 freshly extracted eggs of *M. incognita* (adapted from [91]). Treatments were repeated 7 and 14 days after the first application (Figure 4A). Bt strains were tested at a concentration of 10^8 spores/mL (adapted from [92,93]). The nematicide potential of each Bt strain was checked, testing the effectiveness from the whole culture (WC), the spore and crystal mixture (s + c), and the supernatant (SN). dH_2O was used as a negative control. Irrigation with dH_2O was conducted 3 times per week. Plants were harvested 28 days after the first treatment, and symptoms of nematode infection, namely gall formation on the roots, were recorded 28 days after the first treatment. The experiment was reproduced 4 times, with 5 technical replicates for each biological repetition.

4.3.7. Phytopathogen Fungi Species Bioassays In Vitro

The Bt strain was grown from a single colony, in LB medium, in an overnight culture, at 28 °C and at 200 rpm. Treatments were prepared at a final concentration of 10^6 CFU/mL, in a 0.85% NaCl solution, and two drops of 5 μL were placed flanking a 7 mm diameter fungal disc on an LB plate. As a negative control, LB plates inoculated with fungal disks were used. All plates were incubated at 28 °C until the negative control grew around the surface of the plate (methodology adapted from [72]). Fungal growth measures were analyzed using the ImageJ (v1.53a) software (https://imagej.nih.gov/ij/download.html, accessed on 15 March 2022). The percentage of growth inhibition was calculated using the following equation: $I = (C-T)/C \times 100$, where I = percent growth inhibition; C = growth in control; T = growth in treatment [94]. All treatments were tested per triplicate.

4.4. *Recombinant Protein Expression, Purification, and In Vivo Testing*

4.4.1. Amplification and Cloning of the cry5-orf2cry_65 Operon Genes

For the construction of plasmids expressing Bt toxins, the *cry5-orf2cry65* was first amplified by PCR from the BST-122 Bt strain genomic DNA, using the Phusion High-Fidelity DNA polymerase (NEB, Ipswich, MA, USA) and the corresponding primers harboring restriction enzyme recognition sequences in their extremes (Table S5). The resulting PCR products were purified by the NucleoSpin® Gel and PCR Clean Up kit (Macherey-Nagel Inc., Bethlehem, PA, USA), and ligated into the pJET plasmid (CloneJET PCR Cloning Kit, Thermo Scientific, Waltham, MA, USA). The ligation products were then electroporated into *E. coli* XL1 blue cells by using a standard protocol [95]. Colony-PCR was performed to check positive clones from which the plasmids were purified using the NucleoSpin® Plasmid Kit (Macherey-Nagel Inc., Bethlehem, PA, USA), following the manufacturer's instructions. Subsequently, the pJET plasmids were verified by sequencing (StabVida, Caparica, Portugal) and digested with the corresponding restriction enzymes to excise the fragment and interest. These were then purified from agarose gels and ligated into a pre-digested pSTABr vector, using the Rapid DNA ligation kit (ThermoScientific, Vilnius, Lithuania), to obtain the recombinant plasmid pSTABr-*cry5-orf2cry65*. The ligation products were then electroporated into *E. coli* XL1 blue cells by using standard protocols [95]. Positive clones were verified by colony-PCR and plasmids were purified and electroporated into the acrystalliferous BMB171 Bt strain.

4.4.2. Production, Purification, and Activity Assay against Diverse Target Species of the Recombinant Toxins

Single colonies from the BMB171-Cry5-orf65 recombinant strain were inoculated in 100 mL of LB culture medium, supplemented with erythromycin, and grown at 28 °C and 200 rpm, for 72 h. The whole was collected by centrifuging at 9000× *g*, at 4 °C, for 10 min, and the pellet was washed with dH_2O, twice. All protein quantifications were performed by Bradford and tests of the recombinant toxins against protonymphs of *T. urticae* and juveniles of *M. incognita* were carried out as described above. The BMB171-pSTAB-empty strain was used as a negative control in the bioassays.

4.4.3. Nucleotide Sequence Accession Numbers

The nucleotide sequence data reported in this paper have been deposited in the GeneBank database, under the following accession numbers: OP604599 for the *cry5*-like gene; OP696897 for the *mpp51Aa* gene; OP722690 for the *orf2_cry65A* gene; OP722691 for the *mpp2*-like gene; OP722692 for the *exochitinase* gene; OP722693 for the *colB* gene; and OP722694 for the *bmp1*-like gene.

4.5. *Statistical Analyses*

The mean lethal concentration (LC_{50}) of the spore and crystal mixture (*L. decemlineata*) and solubilized protein (*M. incognita*) of BST-122 Bt strain were determined based on the probit model [96]. The analysis of BST-122 activity against *M. incognita* J2 mortality (in vivo) and nematode infection in roots was performed by running the ANOVA (p-value < 0.05) and Tukey post-hoc tests. For the Cry5-like activity analysis, for *T. urticae* it was performed by Welch's *t*-test ($p < 0.05$), and for *M. incognita*, as the data were non-parametric, it was performed with the Kruskal–Wallis test ($p < 0.05$), and pairwise comparisons between group levels were performed with Bonferroni correction. The statistical analyses of fungal growth were performed by running Welch's t-test and the t-test for non-homoscedastic and homoscedastic data, respectively. All analyses were performed with the R (v.4.1.1.) software.

Supplementary Materials: The following supporting information can be downloaded at: https://www.mdpi.com/article/10.3390/toxins14110768/s1, Figure S1: results of HPLC analysis for detection of production of type I β-exotoxin (thuringiensin); Figure S2: SDS-PAGE from BST-122; Table S1: mean lethal concentration (LC50) value of the BST-122 solubilized protein crystals for J2 M. incognita individuals; Table S2: mortality percentages of J2 M. incognita individuals treated with Cry5_orf65 solubilized protein crystals; Table S3: insecticidal protein content of Bacillus thuringiensis BMB171; Table S4: comparative of strains BST-122 and BMB171 secreted factors; and Table S5: primers used in this study.

Author Contributions: Conceptualization, P.C. and C.J.C.; methodology, A.U., C.J.C. and M.V.; software, A.U., C.J.C., M.V. and A.B.F.; formal analysis, A.U. and C.J.C.; writing—original draft preparation, A.U., C.J.C., M.V., A.B.F. and P.C.; writing—review and editing, C.J.C., M.V., A.B.F. and P.C.; funding acquisition, P.C. All authors have read and agreed to the published version of the manuscript.

Funding: A. Unzue Doctorados Industriales predoctoral contract was cofounded by Gobierno de Navarra (0011-1408-2017-000027) and Bioinsectis SL. A. B. Fernández contract was cofounded by the Program Torres Quevedo (Ministerio de Ciencia, Innovación y Universidades, Gobierno de España) (PTQ-2018-010091) and Bioinsectis SL.

Institutional Review Board Statement: Not applicable.

Informed Consent Statement: Not applicable.

Data Availability Statement: Not applicable.

Acknowledgments: We are deeply grateful to Magda Galeano and Javier Calvo (Koppert, Spain) for providing nematode management protocols and mite populations. We thank Jorge Poveda for providing fungi species and experiment design assistance (Universidad Pública de Navarra) and Juan Ferré (Universitat de Valéncia) for the HPLC analysis.

Conflicts of Interest: The authors declare no conflict of interest.

References

1. Schnepf, E.; Crickmore, N.; Van Rie, J.; Lereclus, D.; Baum, J.; Feitelson, J.; Zeigler, D.R.; Dean, D.H. *Bacillus thuringiensis* and Its Pesticidal Crystal Proteins. *Microbiol. Mol. Biol. Rev.* **1998**, *62*, 775–806. [CrossRef] [PubMed]
2. Bravo, A.; Likitvivatanavong, S.; Gill, S.S.; Soberón, M. Bacillus Thuringensis: A Story of a Sucessful Bioinsecticide. *Insect Biochem. Mol. Biol.* **2011**, *41*, 423–431. [CrossRef] [PubMed]
3. Dulmage, H.T. Insecticidal Activity of Isolated of *Bacillus thuringiensis* and Their Potential for Pest Control. In *Microbial Control of Pests and Plant Diseases 1970–1980*; Burges, H.D., Ed.; Academic Press: London, UK, 1981; pp. 193–222.
4. Ortiz, A.; Sansinenea, E. Bacillus thuringiensis Based Biopesticides for Integrated Crop Management. In *Biopesticides*; Elsevier: Amsterdam, The Netherlands, 2022; pp. 1–6.
5. Dunham, W.; Trimmer, M. *Biological Products around the World, Bioproducts Industry Alliance Spring Meeting & International Symposium (BPIA.Org)*; 2BMonthly Newsletter: Nairobi, Kenya, 2018.
6. Palma, L.; Muñoz, D.; Berry, C.; Murillo, J.; Caballero, P. *Bacillus thuringiensis* Toxins: An Overview of Their Biocidal Activity. *Toxins* **2014**, *6*, 3296–3325. [CrossRef] [PubMed]
7. Jones, K.A.; Burges, H.D. Technology of Formulation and Application. In *Formulation of Microbial Biopesticides*; Burges, H.D., Ed.; Kluwer Academic Publishers: Dordrecht, The Netherlands, 1998; pp. 7–30.
8. Van Frankenhuyzen, K. Insecticidal Activity of *Bacillus thuringiensis* Crystal Proteins. *J. Invertebr. Pathol.* **2009**, *101*, 1–16. [CrossRef]
9. Hernández-Martínez, P.; Ferré, J.; Escriche, B. Susceptibility of *Spodoptera exigua* to 9 Toxins from *Bacillus thuringiensis*. *J. Invertebr. Pathol.* **2008**, *97*, 245–250. [CrossRef]
10. Ruiz De Escudero, I.; Estela, A.; Escriche, B.; Caballero, P. Potential of the *Bacillus thuringiensis* Toxin Reservoir for the Control of *Lobesia botrana* (Lepidoptera: Tortricidae), a Major Pest of Grape Plants. *Appl. Environ. Microbiol.* **2007**, *73*, 337–340. [CrossRef]
11. Haffani, Y.Z.; Cloutier, C.; Belzile, F.J. *Bacillus thuringiensis* Cry3Ca1 Protein Is Toxic to the Colorado Potato Beetle, *Leptinotarsa decemlineata* (Say). *Biotechnol. Prog.* **2001**, *17*, 211–216. [CrossRef]
12. Krieg, A.; Huger, A.M.; Langenbruch, G.A.; Schnetter, W. *Bacillus thuringiensis* Var. *tenebrionis*: Ein Neuer, Gegenüber Larven von Coleopteren Wirksamer Pathotyp. *Z. Angew. Entomol.* **1983**, *96*, 500–508. [CrossRef]

13. Lambert, B.; Hofte, H.; Annys, K.; Jansens, S.; Soetaert, P.; Peferoen, M. Novel *Bacillus thuringiensis* Insecticidal Crystal Protein with a Silent Activity against Coleopteran Larvae. *Appl. Environ. Microbiol.* **1992**, *58*, 2536–2542. [CrossRef]
14. Domínguez-Arrizabalaga, M.; Villanueva, M.; Escriche, B.; Ancín-Azpilicueta, C.; Caballero, P. Insecticidal Activity of *Bacillus thuringiensis* Proteins against Coleopteran Pests. *Toxins* **2020**, *12*, 430. [CrossRef]
15. Barloy, F.; Delécluse, A.; Nicolas, L.; Lecadet, M.M. Cloning and Expression of the First Anaerobic Toxin Gene from *Clostridium bifermentans* subsp. *malaysia*, Encoding a New Mosquitocidal Protein with Homologies to *Bacillus thuringiensis* Delta-Endotoxins. *J. Bacteriol.* **1996**, *178*, 3099–3105. [CrossRef] [PubMed]
16. Chilcott, C.N.; Ellar, D.J. Comparative Toxicity of *Bacillus thuringiensis* Var. *israelensis* Crystal Proteins *in Vivo* and *in Vitro*. *J. Gen. Microbiol.* **1988**, *134*, 2551–2558. [CrossRef] [PubMed]
17. Donovan, W.P.; Dankocsik, C.; Gilbert, M.P. Molecular Characterization of a Gene Encoding a 72-Kilodalton Mosquito-Toxic Crystal Protein from *Bacillus thuringiensis* subsp. *israelensis*. *J. Bacteriol.* **1988**, *170*, 4732–4738. [CrossRef] [PubMed]
18. Hernández-Soto, A.; Del Rincón-Castro, M.C.; Espinoza, A.M.; Ibarra, J.E. Parasporal Body Formation via Overexpression of the Cry10Aa Toxin of *Bacillus thuringiensis* subsp. *israelensis*, and Cry10Aa-Cyt1Aa Synergism. *Appl. Environ. Microbiol.* **2009**, *75*, 4661–4667. [CrossRef] [PubMed]
19. Rosso, M.L.; Delécluse, A. Contribution of the 65-Kilodalton Protein Encoded by the Cloned Gene *Cry19A* to the Mosquitocidal Activity of *Bacillus thuringiensis* subsp. *jegathesan*. *Appl. Environ. Microbiol.* **1997**, *63*, 4449–4455. [CrossRef]
20. Ward, E.S.; Ridley, A.R.; Ellar, D.J.; Todd, J.A. *Bacillus thuringiensis* Var. *israelensis* δ-Endotoxin. Cloning and Expression of the Toxin in Sporogenic and Asporogenic Strains of *Bacillus subtilis*. *J. Mol. Biol.* **1986**, *191*, 13–22. [CrossRef]
21. Valtierra-de-Luis, D.; Villanueva, M.; Berry, C.; Caballero, P. Potential for *Bacillus thuringiensis* and Other Bacterial Toxins as Biological Control Agents to Combat Dipteran Pests of Medical and Agronomic Importance. *Toxins* **2020**, *12*, 773. [CrossRef]
22. Baum, J.A.; Sukuru, U.R.; Penn, S.R.; Meyer, S.E.; Subbarao, S.; Shi, X.; Flasinski, S.; Heck, G.R.; Brown, R.S.; Clark, T.L. Cotton Plants Expressing a Hemipteran-Active *Bacillus thuringiensis* Crystal Protein Impact the Development and Survival of *Lygus hesperus* (Hemiptera: Miridae) Nymphs. *J. Econ. Entomol.* **2012**, *105*, 616–624. [CrossRef]
23. Thomson, M.; Knuth, M.; Cardineau, G. Bacillus thuringiensis Toxins with Improved Activity. U.S. Patent 5,874,288 A, 1999.
24. Li, H.; Olson, M.; Lin, G.; Hey, T.; Tan, S.Y.; Narva, K.E. *Bacillus thuringiensis* Cry34Ab1/Cry35Ab1 Interactions with Western Corn Rootworm Midgut Membrane Binding Sites. *PLoS ONE* **2013**, *8*, e53079. [CrossRef]
25. Guo, S.; Liu, M.; Peng, D.; Ji, S.; Wang, P.; Yu, Z.; Sun, M. New Strategy for Isolating Novel Nematicidal Crystal Protein Genes from *Bacillus thuringiensis* Strain YBT-1518. *Appl. Environ. Microbiol.* **2008**, *74*, 6997–7001. [CrossRef]
26. Payne, J.M.; Cannon, R.J.C.; Bagley, A.L. Bacillus thuringiensis Isolates for Controlling Acarides. U.S. Patent 5,211,946, 1993.
27. Payne, J.M.; Cannon, R.J.C.; Bagley, A.L. Bacillus thuringiensis Isolates for Controlling Acarida. U.S. Patent 5,262,158, 1993.
28. Payne, J.; Raymond, J.C.C.; Ralph, A.L. Bacillus thuringiensis Isolates for Controlling Acarides. U.S. Patent 5,350,576, 1994.
29. Reyes-Ramírez, A.; Escudero-Abarca, B.I.; Aguilar-Uscanga, G.; Hayward-Jones, P.M.; Eleazar Barboza-Corona, J. Antifungal Activity of *Bacillus thuringiensis* Chitinase and Its Potential for the Biocontrol of Phytopathogenic Fungi in Soybean Seeds. *J. Food Sci.* **2004**, *69*, 131–134. [CrossRef]
30. Zhou, Y.; Choi, Y.L.; Sun, M.; Yu, Z. Novel Roles of *Bacillus thuringiensis* to Control Plant Diseases. *Appl. Microbiol. Biotechnol.* **2008**, *80*, 563–572. [CrossRef] [PubMed]
31. Migeon, A.; Dorkeld, F. Spider Mites Web: A Comprehensive Database for the Tetranychidae. Available online: http://www.montpellier.inra.fr/CBGP/spmweb (accessed on 12 September 2022).
32. Helle, W.; Sabelis, M.W. *Spider Mites. Their Biology, Natural Enemies and Control*; Helle, W., Sabelis, M.W., Eds.; Elsevier Science Publisher B.V.: Amsterdam, The Netherlands, 1985.
33. Engelbrecht, G.; Horak, I.; Jansen van Rensburg, P.J.; Claassens, S. *Bacillus*-Based Bionematicides: Development, Modes of Action and Commercialisation. *Biocontrol Sci. Technol.* **2018**, *28*, 629–653. [CrossRef]
34. Ciancio, A.; Mukerji, K.G. *Integrated Management of Fruit Crops and Forest Nematodes*; Ciancio, A., Mukerji, K.G., Eds.; Springer Science & Business Media: Bari, Italy, 2009.
35. Jung, C.; Wyss, U. New Approaches to Control Plant Parasitic Nematodes. *Appl. Microbiol. Biotechnol.* **1999**, *51*, 439–446. [CrossRef]
36. Savary, S.; Willocquet, L.; Pethybridge, S.J.; Esker, P.; McRoberts, N.; Nelson, A. The Global Burden of Pathogens and Pests on Major Food Crops. *Nat. Ecol. Evol.* **2019**, *3*, 430–439. [CrossRef] [PubMed]
37. Mahmood, S.; Lakra, N.; Marwal, A.; Sudheep, N.M.; Anwar, K. Crop Genetic Engineering: An Approach to Improve Fungal Resistance in Plant System. In *Plant-Microbe Interactions in Agro-Ecological Perspectives*; Singh, D.P., Singh, H.B., Prabha, R., Eds.; Springer: Berlin, Germany, 2017; pp. 581–591. ISBN 9789811065934.
38. Nauen, R.; Stumpf, N.; Elbert, A.; Zebitz, C.P.W.; Kraus, W. Acaricide Toxicity and Resistance in Larvae of Different Strains of *Tetranychus urticae* and *Panonychus ulmi* (Acari: Tetranychidae). *Pest Manag. Sci.* **2001**, *57*, 253–261. [CrossRef]
39. Wei, J.-Z.; Hale, K.; Carta, L.; Platzer, E.; Wong, C.; Fang, S.-C.; Aroian, R. V *Bacillus Thuringiensis* Crystal Proteins That Target Nematodes. *Proc. Natl. Acad. Sci. USA* **2003**, *100*, 2760–2765. [CrossRef]
40. Guo, Y.; Weng, M.; Sun, Y.; Carballar-Lejarazú, R.; Wu, S.; Lian, C. *Bacillus thuringiensis* Toxins with Nematocidal Activity against the Pinewood Nematode *Bursaphelenchus xylophilus*. *J. Invertebr. Pathol.* **2022**, *189*, 107726. [CrossRef]
41. Iatsenko, I.; Boichenko, I.; Sommer, R.J. *Bacillus thuringiensis* DB27 Produces Two Novel Protoxins, Cry21Fa1 and Cry21Ha1, Which Act Synergistically against Nematodes. *Appl. Environ. Microbiol.* **2014**, *80*, 3266–3275. [CrossRef]

42. Zhang, L.; Yu, J.; Xie, Y.; Lin, H.; Huang, Z.; Xu, L.; Gelbic, I.; Guan, X. Biological Activity of *Bacillus thuringiensis* (Bacillales: Bacillaceae) Chitinase against *Caenorhabditis elegans* (Rhabditida: Rhabditidae). *J. Econ. Entomol.* **2014**, *107*, 551–558. [CrossRef]
43. Luo, X.; Chen, L.; Huang, Q.; Zheng, J.; Zhou, W.; Peng, D.; Ruan, L.; Sun, M. *Bacillus thuringiensis* Metalloproteinase Bmp1 Functions as a Nematicidal Virulence Factor. *Appl. Environ. Microbiol.* **2013**, *79*, 460–468. [CrossRef] [PubMed]
44. Peng, D.; Lin, J.; Huang, Q.; Zheng, W.; Liu, G.; Zheng, J.; Zhu, L.; Sun, M. A Novel Metalloproteinase Virulence Factor Is Involved in *Bacillus thuringiensis* Pathogenesis in Nematodes and Insects. *Environ. Microbiol.* **2016**, *18*, 846–862. [CrossRef] [PubMed]
45. Sick, A.J.; Schwab, G.E.; Payne, J.M. Genes Encoding Nematode-Active Toxins Cloned from Bacillus thuringiensis Isolate PS17. U.S. Patent 5,281,530, 1994.
46. Xu, C.; Chinte, U.; Chen, L.; Yao, Q.; Meng, Y.; Zhou, D.; Bi, L.J.; Rose, J.; Adang, M.J.; Wang, B.C.; et al. Crystal Structure of Cry51Aa1: A Potential Novel Insecticidal Aerolysin-Type β-Pore-Forming Toxin from *Bacillus thuringiensis*. *Biochem. Biophys. Res. Commun.* **2015**, *462*, 184–189. [CrossRef] [PubMed]
47. Peng, D.; Chai, L.; Wang, F.; Zhang, F.; Ruan, L.; Sun, M. Synergistic Activity between *Bacillus thuringiensis* Cry6Aa and Cry55Aa Toxins against *Meloidogyne incognita*. *Microb. Biotechnol.* **2011**, *4*, 794–798. [CrossRef]
48. Ni, H.; Zeng, S.; Qin, X.; Sun, X.; Zhang, S.; Zhao, X.; Yu, Z.; Li, L. Molecular Docking and Site-Directed Mutagenesis of a *Bacillus thuringiensis* Chitinase to Improve Chitinolytic, Synergistic Lepidopteran-Larvicidal and Nematicidal Activities. *Int. J. Biol. Sci.* **2015**, *11*, 304–3015. [CrossRef]
49. Tang, Y.; Zou, J.; Zhang, L.; Li, Z.; Ma, C.; Ma, N. Anti-Fungi Activities of *Bacillus thuringiensis* H3 Chitinase and Immobilized Chitinase Particles and Their Effects to Rice Seedling Defensive Enzymes. *J. Nanosci. Nanotechnol.* **2012**, *12*, 8081–8086. [CrossRef]
50. Tabashnik, B.E.; Brévault, T.; Carrière, Y. Insect Resistance to Bt Crops: Lessons from the First Billion Acres. *Nat. Biotechnol.* **2013**, *31*, 510–521. [CrossRef]
51. Petzold-Maxwell, J.L.; Alves, A.P.; Estes, R.E.; Gray, M.E.; Meinke, L.J.; Shields, E.J.; Thompson, S.D.; Tinsley, N.A.; Gassmann, A.J. Applying an Integrated Refuge to Manage Western Corn Rootworm (Coleoptera: Chrysomelidae): Effects on Survival, Fitness, and Selection Pressure. *J. Econ. Entomol.* **2013**, *106*, 2195–2207. [CrossRef]
52. Jeppson, L.R.; Keifer, H.H.; Baker, E.W. *Mites Injurious to Economic Plants*; Jeppson, L.R., Keifer, H.H., Baker, E.W., Eds.; California Univ. Press: Berkeley, CA, USA, 1975.
53. Alquisira-Ramírez, E.V.; Paredes-Gonzalez, J.R.; Hernández-Velázquez, V.M.; Ramírez-Trujillo, J.A.; Peña-Chora, G. In Vitro Susceptibility of *Varroa destructor* and *Apis mellifera* to Native Strains of *Bacillus thuringiensis*. *Apidologie* **2014**, *45*, 707–718. [CrossRef]
54. Dunstand-Guzmán, E.; Peña-Chora, G.; Hallal-Calleros, C.; Pérez-Martínez, M.; Hernández-Velazquez, V.M.; Morales-Montor, J.; Flores-Pérez, F.I. Acaricidal Effect and Histological Damage Induced by *Bacillus thuringiensis* Protein Extracts on the Mite *Psoroptes cuniculi*. *Parasites Vectors* **2015**, *8*, 285. [CrossRef]
55. Dunstand-Guzmán, E.; Hallal-Calleros, C.; Morales-Montor, J.; Hernández-Velázquez, V.M.; Zárate-Ramos, J.J.; Hoffman, K.L.; Peña-Chora, G.; Flores-Pérez, F.I. Therapeutic Use of *Bacillus thuringiensis* in the Treatment of Psoroptic Mange in Naturally Infested New Zealand Rabbits. *Vet. Parasitol.* **2017**, *238*, 24–29. [CrossRef] [PubMed]
56. Neethu, K.B.; Priji, P.; Unni, K.N.; Sajith, S.; Sreedevi, S.; Ramani, N.; Anitha, K.; Rosana, B.; Girish, M.B.; Benjamin, S. New *Bacillus thuringiensis* Strain Isolated from the Gut of Malabari Goat Is Effective against *Tetranychus macfarlanei*. *J. Appl. Entomol.* **2016**, *140*, 187–198. [CrossRef]
57. Krieg, A. Effectiveness of *Bacillus thuringiensis* Exotoxin on *Tetranychus telarius* (Acarina: Tetranychidae). *J. Invertebr. Pathol.* **1968**, *12*, 478. [CrossRef]
58. Hall, I.M.; Hinter, D.K.; Arakawa, K.Y. The Effect of the B-Exotoxin Fraction *Bacillus thuringiensis* on the Citrus Red Mite. *J. Invertebr. Pathol.* **1971**, *18*, 359–362. [CrossRef]
59. Hoy, M.A.; Ouyang, Y.-L. Toxicity of the B-Exotoxin of *Bacillus thuringiensis* to *Tetranychus pacificus* and *Metaseiulus occidentalis* (Acari: Tetranychidae and Phytoseiidae). *J. Econ. Entomol.* **1987**, *80*, 507–511. [CrossRef]
60. Royalty, R.N.; Hall, F.R.; Taylor, R.A.J. Effects of Thuringiensin on *Tetranychus urticae* (Acari: Tetranychidae) Mortality, Fecundity, and Feeding. *J. Econ. Entomol.* **1990**, *83*, 792–798. [CrossRef]
61. Royalty, R.N.; Hall, F.R.; Lucius, B.A. Effects of Thuringiensin on European Red Mite (Acarina: Tetranychidae) Mortality, Oviposition Rate and Feeding. *Pestic. Sci.* **1991**, *33*, 383–391. [CrossRef]
62. Böckenhoff, A.; Grundler, F.M.W. Studies on the Nutrient Uptake by the Beet Cyst Nematode *Heterodera schachtii* by in Situ Microinjection of Fluorescent Probes into the Feeding Structures in *Arabidopsis thaliana*. *Parasitology* **1994**, *109*, 249–255. [CrossRef]
63. Zhang, F.; Peng, D.; Ye, X.; Yu, Z.; Hu, Z.; Ruan, L.; Sun, M. In Vitro Uptake of 140 KDa *Bacillus thuringiensis* Nematicidal Crystal Proteins by the Second Stage Juvenile of *Meloidogyne hapla*. *PLoS ONE* **2012**, *7*, e38534. [CrossRef]
64. Zheng, D.; Zeng, Z.; Xue, B.; Deng, Y.; Sun, M.; Tang, Y.J.; Ruan, L. *Bacillus thuringiensis* Produces the Lipopeptide Thumolycin to Antagonize Microbes and Nematodes. *Microbiol. Res.* **2018**, *215*, 22–28. [CrossRef]
65. Baghaee-Ravari, S.; Mahdikani-Moghaddam, E. Efficacy of *Bacillus thuringiensis* Cry14 Toxin against Root Knot Nematode, *Meloidogyne javanica*. *Plant Prot. Sci.* **2015**, *51*, 46–51. [CrossRef]
66. Mohammed, S.H.; El Saedy, M.A.; Enan, M.R.; Ibrahim, N.E.; Ghareeb, A.; Moustafa, S.A. Biocontrol Efficiency of *Bacillus Thuringiensis* Toxins against Root-Knot Nematode, *Meloidogyne incognita*. *J. Cell Mol. Biol.* **2008**, *7*, 57–66.
67. Prasad, S.S.V.; Tilak, K.V.B.R.; Gollakota, R.G. Role of *Bacillus thuringiensis* Var. *thuringiensis* on the Larval Survivability and Egg Hatching of *Meloidogyne* spp., the Causative Agent of Root Knot Disease. *J. Invertebr. Pathol.* **1972**, *20*, 377–378. [CrossRef]

68. Devidas, P.; Rehberger, L.A. The Effect of Exotoxin (Thuringiensin) from *Bacillus thuringiensis* on *Meloidogyne incognita* and *Caenorhabditis elegans*. *Plant Soil* **1992**, *145*, 115–120. [CrossRef]
69. Ignoffo, C.M.; Dropkin, V.H. Deleterious Effects of the Thermostable Toxin of *Bacillus thuringiensis* on Species of Soil-Inhabiting, Myceliophagus, and Plant-Parasitic Nematodes. *J. Kansas Entomol. Soc.* **1977**, *50*, 394–398.
70. Liu, X.; Ruan, L.; Peng, D.; Li, L.; Sun, M.; Yu, Z. Thuringiensin: A Thermostable Secondary Metabolite from *Bacillus thuringiensis* with Insecticidal Activity against a Wide Range of Insects. *Toxins* **2014**, *6*, 2229–2238. [CrossRef]
71. Fira, D.; Dimkić, I.; Berić, T.; Lozo, J.; Stanković, S. Biological Control of Plant Pathogens by *Bacillus* Species. *J. Biotechnol.* **2018**, *285*, 44–55. [CrossRef]
72. Slama, H.B.; Cherif-Silini, H.; Bouket, A.C.; Qader, M.; Silini, A.; Yahiaoui, B.; Alenezi, F.N.; Luptakova, L.; Triki, M.A.; Vallat, A.; et al. Screening for *Fusarium* Antagonistic Bacteria from Contrasting Niches Designated the Endophyte *Bacillus halotolerans* as Plant Warden against *Fusarium*. *Front. Microbiol.* **2019**, *9*, 3236. [CrossRef]
73. He, C.N.; Ye, W.Q.; Zhu, Y.Y.; Zhou, W.W. Antifungal Activity of Volatile Organic Compounds Produced by *Bacillus methylotrophicus* and *Bacillus Thuringiensis* against Five Common Spoilage Fungi on Loquats. *Molecules* **2020**, *25*, 3360. [CrossRef]
74. Kim, P.I.; Bai, H.; Bai, D.; Chae, H.; Chung, S.; Kim, Y.; Park, R.; Chi, Y.T. Purification and Characterization of a Lipopeptide Produced by *Bacillus thuringiensis* CMB26. *J. Appl. Microbiol.* **2004**, *97*, 942–949. [CrossRef]
75. Roy, A.; Mahata, D.; Paul, D.; Korpole, S.; Franco, O.L.; Mandal, S.M. Purification, Biochemical Characterization and Self-Assembled Structure of a Fengycin-like Antifungal Peptide from *Bacillus thuringiensis* Strain SM1. *Front. Microbiol.* **2013**, *4*, 332. [CrossRef] [PubMed]
76. Béchet, M.; Caradec, T.; Hussein, W.; Abderrahmani, A.; Chollet, M.; Leclére, V.; Dubois, T.; Lereclus, D.; Pupin, M.; Jacques, P. Structure, Biosynthesis, and Properties of Kurstakins, Nonribosomal Lipopeptides from *Bacillus* spp. *Appl. Microbiol. Biotechnol.* **2012**, *95*, 593–600. [CrossRef] [PubMed]
77. Iriarte, J.; Bel, Y.; Ferrandis, M.D.; Andrew, R.; Murillo, J.; Ferré, J.; Caballero, P. Environmental Distribution and Diversity of *Bacillus thuringiensis* in Spain. *Syst. Appl. Microbiol.* **1998**, *21*, 97–106. [CrossRef]
78. Stewart, G.S.A.B.; Johnstone, K.; Hagelberg, E.; Ellar, D.J. Commitment of Bacterial Spores to Germinate. *Biochem. J.* **1981**, *198*, 101–106. [CrossRef]
79. Borodovsky, M.; McIninch, J. GENMARK: Parallel Gene Recognition for Both DNA Strands. *Comput. Chem.* **1993**, *17*, 123–133. [CrossRef]
80. Crickmore, N.; Zeigler, D.R.; Feitelson, J.; Schnepf, E.; Van Rie, J.; Lereclus, D.; Baum, J.; Dean, D.H. Revision of the Nomenclature for the *Bacillus thuringiensis* Pesticidal Crystal Proteins. *Microbiol. Mol. Biol. Rev.* **1998**, *62*, 807–813. [CrossRef]
81. Crickmore, N.; Berry, C.; Panneerselvam, S.; Mishra, R.; Connor, T.R.; Bonning, B.C. Bacterial Pesticidal Protein Resource Center. Available online: https://www.bpprc-db.org (accessed on 9 October 2022).
82. Malovichko, Y.V.; Nizhnikov, A.A.; Antonets, K.S. Repertoire of the *Bacillus thuringiensis* Virulence Factors Unrelated to Major Classes of Protein Toxins and Its Role in Specificity of Host-Pathogen Interactions. *Toxins* **2019**, *11*, 347. [CrossRef]
83. Marchler-Bauer, A.; Bo, Y.; Han, L.; He, J.; Lanczycki, C.J.; Lu, S.; Chitsaz, F.; Derbyshire, M.K.; Geer, R.C.; Gonzales, N.R.; et al. CDD/SPARCLE: Functional Classification of Proteins via Subfamily Domain Architectures. *Nucleic Acids Res.* **2017**, *45*, D200–D203. [CrossRef]
84. Madeira, F.; Pearce, M.; Tivey, A.R.N.; Basutkar, P.; Lee, J.; Edbali, O.; Madhusoodanan, N.; Kolesnikov, A.; Lopez, R. Search and Sequence Analysis Tools Services from EMBL-EBI in 2022. *Nucleic Acids Res.* **2022**, *50*, W276–W279. [CrossRef]
85. Bradford, M.M. A Rapid and Sensitive Method for the Quantification of Microgram Quantities of Protein Utilizing the Principle of Protein-Dye Binding. *Anal. Biochem.* **1976**, *72*, 248–254. [CrossRef]
86. Laemmli, U.K. Cleavage of Structural Proteins during the Assembly of the Head of Bacteriophage T4. *Nature* **1970**, *227*, 680–685. [CrossRef] [PubMed]
87. Hernandez, C.S.; Ferre, J.; Larget-Thiery, I. Update on the Detection of Beta-Exotoxin in *Bacillus thuringiensis* Strains by HPLC Analysis. *J. Appl. Microbiol.* **2001**, *90*, 643–647. [CrossRef] [PubMed]
88. Hussey, R.S.; Barker, K.R. A Comparison of Methods of Collecting Inocula of *Meloidogyne* spp., Including a New Technique. *Plant Dis. Rep.* **1973**, *57*, 1025–1028.
89. Domínguez-Arrizabalaga, M.; Villanueva, M.; Fernandez, A.B.; Caballero, P. A Strain of *Bacillus thuringiensis* Containing a Novel *Cry7Aa2* Gene That Is Toxic to *Leptinotarsa decemlineata* (Say) (Coleoptera: Chrysomelidae). *Insects* **2019**, *10*, 259. [CrossRef]
90. Griffitts, J.S.; Whitacre, J.L.; Stevens, D.E.; Aroian, R.V. Bt Toxin Resistance from Loss of a Putative Carbohydrate-Modifying Enzyme. *Science* **2001**, *293*, 860–864. [CrossRef]
91. Terefe, M.; Tefera, T.; Sakhuja, P.K. Effect of a Formulation of *Bacillus Firmus* on Root-Knot Nematode *Meloidogyne incognita* Infestation and the Growth of Tomato Plants in the Greenhouse and Nursery. *J. Invertebr. Pathol.* **2009**, *100*, 94–99. [CrossRef]
92. Sharma, R.D. *Bacillus thuringiensis*: A Biocontrol Agent of *Meloidogyne incognita* on Barley. *Nematol. Bras.* **1994**, *18*, 79–84.
93. Yu, Z.; Xiong, J.; Zhou, Q.; Luo, H.; Hu, S.; Xia, L.; Sun, M.; Li, L.; Yu, Z. The Diverse Nematicidal Properties and Biocontrol Efficacy of *Bacillus thuringiensis* Cry6A against the Root-Knot Nematode *Meloidogyne hapla*. *J. Invertebr. Pathol.* **2015**, *125*, 73–80. [CrossRef]
94. Lyousfi, N.; Lahlali, R.; Letrib, C.; Belabess, Z.; Ouaabou, R.; Ennahli, S.; Blenzar, A.; Barka, E.A. Improving the Biocontrol Potential of Bacterial Antagonists with Salicylic Acid against Brown Rot Disease and Impact on Nectarine Fruits Quality. *Agronomy* **2021**, *11*, 209. [CrossRef]

95. Sambrook, J.; Russell, D. *Molecular Cloning: A Laboratory Manual*, 3rd ed.; Cold Spring Harbor Laboratory Press: Cold Spring Harbor, NY, USA, 2001.
96. Finney, D.J. *Probit Analysis*, 3rd ed.; Cambridge University Press: Cambridge, UK, 1971.

MDPI
St. Alban-Anlage 66
4052 Basel
Switzerland
www.mdpi.com

Toxins Editorial Office
E-mail: toxins@mdpi.com
www.mdpi.com/journal/toxins

www.ingramcontent.com/pod-product-compliance
Lightning Source LLC
Chambersburg PA
CBHW061615210926
51850CB00009B/2327